gels

Special Issue Reprint

Hydrogel-Based Novel Biomaterials

Achievements and Prospects

Edited by
Ana Paula Serro, Ana Isabel Fernandes and Diana Silva

mdpi.com/journal/gels

MDPI

Hydrogel-Based Novel Biomaterials: Achievements and Prospects

Hydrogel-Based Novel Biomaterials: Achievements and Prospects

Editors

Ana Paula Serro
Ana Isabel Fernandes
Diana Silva

MDPI
Basel • Beijing • Wuhan • Barcelona • Belgrade • Novi Sad • Cluj • Manchester

Ana Paula Serro
Centro de Química
Estrutural (CQE)
Universidade de Lisboa
Lisboa
Portugal

Ana Isabel Fernandes
Egas Moniz Center for
Interdisciplinary Research
(CiiEM)
Egas Moniz School
of Health & Science
Monte de Caparica
Portugal

Diana Silva
Centro de Química
Estrutural (CQE)
Universidade de Lisboa
Lisboa
Portugal

Editorial Office
MDPI AG
Grosspeteranlage 5
4052 Basel, Switzerland

This is a reprint of the Special Issue, published open access by the journal *Gels* (ISSN 2310-2861), freely accessible at: www.mdpi.com/journal/gels/special_issues/novel_gel.

For citation purposes, cite each article independently as indicated on the article page online and using the guide below:

Lastname, A.A.; Lastname, B.B. Article Title. *Journal Name* **Year**, *Volume Number*, Page Range.

ISBN 978-3-7258-1646-0 (Hbk)
ISBN 978-3-7258-1645-3 (PDF)
https://doi.org/10.3390/books978-3-7258-1645-3

Contents

About the Editors vii

Ana Paula Serro, Diana Cristina Silva and Ana Isabel Fernandes
Hydrogel-Based Novel Biomaterials: Achievements and Prospects
Reprinted from: *Gels* **2024**, *10*, 436, doi:10.3390/gels10070436 1

David Grijalva Garces, Luise Josephine Appoldt, Jasmin Egner, Nico Leister and Jürgen Hubbuch
The Effect of Gelatin Source on the Synthesis of Gelatin-Methacryloyl and the Production of Hydrogel Microparticles
Reprinted from: *Gels* **2023**, *9*, 927, doi:10.3390/gels9120927 9

Alexandra Lupu, Luiza Madalina Gradinaru, Daniela Rusu and Maria Bercea
Self-Healing of Pluronic® F127 Hydrogels in the Presence of Various Polysaccharides
Reprinted from: *Gels* **2023**, *9*, 719, doi:10.3390/gels9090719 30

Tomás Pires, Andreia Sofia Oliveira, Ana Clara Marques, Madalena Salema-Oom, Célio G. Figueiredo-Pina and Diana Silva et al.
Effects of Non-Conventional Sterilisation Methods on PBO-Reinforced PVA Hydrogels for Cartilage Replacement
Reprinted from: *Gels* **2022**, *8*, 640, doi:10.3390/gels8100640 49

Pedro Francisco, Mariana Neves Amaral, Afonso Neves, Tânia Ferreira-Gonçalves, Ana S. Viana and José Catarino et al.
Pluronic® F127 Hydrogel Containing Silver Nanoparticles in Skin Burn Regeneration: An Experimental Approach from Fundamental to Translational Research
Reprinted from: *Gels* **2023**, *9*, 200, doi:10.3390/gels9030200 71

Thanyaporn Pinthong, Maytinee Yooyod, Jinjutha Daengmankhong, Nantaprapa Tuancharoensri, Sararat Mahasaranon and Jarupa Viyoch et al.
Development of Natural Active Agent-Containing Porous Hydrogel Sheets with High Water Content for Wound Dressings
Reprinted from: *Gels* **2023**, *9*, 459, doi:10.3390/gels9060459 91

Komal Ammar Bukhari, Imran Ahmad Khan, Shahid Ishaq, Muhammad Omer Iqbal, Ali M. Alqahtani and Taha Alqahtani et al.
Formulation and Evaluation of Diclofenac Potassium Gel in Sports Injuries with and without Phonophoresis
Reprinted from: *Gels* **2022**, *8*, 612, doi:10.3390/gels8100612 107

Bjad K. Almutairy, El-Sayed Khafagy and Amr Selim Abu Lila
Development of Carvedilol Nanoformulation-Loaded Poloxamer-Based In Situ Gel for the Management of Glaucoma
Reprinted from: *Gels* **2023**, *9*, 952, doi:10.3390/gels9120952 121

Mónica Guerra, Fábio F. F. Garrudo, Célia Faustino, Maria Emilia Rosa and Maria H. L. Ribeiro
Exploring Functionalized Magnetic Hydrogel Polyvinyl Alcohol and Chitosan Electrospun Nanofibers
Reprinted from: *Gels* **2023**, *9*, 968, doi:10.3390/gels9120968 136

Ana Camila Marques, Paulo Cardoso Costa, Sérgia Velho and Maria Helena Amaral
Injectable Poloxamer Hydrogels for Local Cancer Therapy
Reprinted from: *Gels* **2023**, *9*, 593, doi:10.3390/gels9070593 . . . . **156**

Mariana Ribeiro, Marco Simões, Carla Vitorino and Filipa Mascarenhas-Melo
Hydrogels in Cutaneous Wound Healing: Insights into Characterization, Properties, Formulation and Therapeutic Potential
Reprinted from: *Gels* **2024**, *10*, 188, doi:10.3390/gels10030188 . . . . **171**

About the Editors

Ana Paula Serro

Ana Paula Serro holds a degree in Chemical Engineering and a Ph.D. in Biomaterials, both from Instituto Superior Técnico (IST) - University of Lisbon, Portugal. Currently, she is an associate professor at the Chemical Engineering Department of IST and researcher at Centro de Química Estrutural (CQE). She is the coordinator of the Advanced Materials Group of CQE and head of the Biomaterials Research Group at IST. Her main research interests include the development and characterization of biomaterials, mainly for controlled drug release. She has conducted extensive research in hydrogels for therapeutic applications, including drug-loaded ophthalmic lenses (contact lenses and intraocular lenses), wound dressings, and cartilage-substitute materials, addressing topics related to sterilization, biotribology, and biomolecules adsorption, among others. She has been the principal researcher/institutional leader of several national and international projects (e.g., FCT, MERA.NET, and MARIE-CURIE) and has vast experience in student supervision (11 PhD and 45 MSc). She is the author of a wide number of publications in indexed journals and books and has been the guest editor of several special issues. The excellence of her research work has been recognized in several prizes, including a distinction by the University of Lisbon in 2023 and the nomination of one of her international projects as a "Success Story".

Ana Isabel Fernandes

Ana Isabel Fernandes is an associate professor of Pharmaceutics and head of the PharmSci Lab at Egas Moniz School of Health & Science, Portugal. She graduated in Pharmaceutical Sciences (University of Lisbon, PT) and holds a Ph.D. in Drug Delivery (University of London, UK). She has been involved in the study of polymers and polymeric systems with biomedical applications, namely in the delivery of therapeutic proteins and conventional drugs. Her current research is related to formulations in pediatrics, drug solubility enhancement by co-amorphization, and the 3D printing of pharmaceuticals, as well as nutraceuticals and the usage of lifestyle drugs. Over the years, she has been a scientific advisor, principal investigator, or collaborator in several externally financed projects (National Funding Agency—FCT); an editorial board member and guest editor of several journals; and has extensively reviewed and published in international refereed journals.

Diana Silva

Diana Cristina Silva is a junior researcher and invited assistant professor at Instituto Superior Técnico and Atlântica – Instituto Universitário, Portugal. She holds a degree in Bioengineering (Biomedical) (Instituto Politécnico de Portalegre, PT), as well as a Master's in Bioengineering and Nanosystems and a Ph.D. in Advanced Materials and Processing from Instituto Superior Técnico, PT. Diana's main research work is focussed on biomaterials, such as in drug release, material characterisation, surface modification, and interaction with biomolecules. She is also involved in the study of sterilisation procedures and related biological analysis (e.g., microbiological and sterility tests, cell viability, etc.). Diana has more than 20 articles published and is currently guest editing two special issues for *Gels*: (i) Design of Polymeric Hydrogels Biomaterials and (ii) Hydrogel-Based Novel Biomaterials: Achievements and Prospects. She has served has a reviewer for four different journals and published several peer-reviewed articles in international journals.

gels

Editorial

Hydrogel-Based Novel Biomaterials: Achievements and Prospects

Ana Paula Serro [1,*], Diana Cristina Silva [1,*] and Ana Isabel Fernandes [2,*]

1 Centro de Química Estrutural (CQE), Institute of Molecular Sciences, Instituto Superior Técnico, Universidade de Lisboa, 1049-001 Lisboa, Portugal
2 Egas Moniz Center for Interdisciplinary Research (CiiEM), Egas Moniz School of Health & Science, Campus Universitário, 2829-511 Caparica, Portugal
* Correspondence: anapaula.serro@tecnico.ulisboa.pt (A.P.S.); dianacristinasilva@tecnico.ulisboa.pt (D.C.S.); aifernandes@egasmoniz.edu.pt (A.I.F.)

check for updates

Citation: Serro, A.P.; Silva, D.C.; Fernandes, A.I. Hydrogel-Based Novel Biomaterials: Achievements and Prospects. *Gels* **2024**, *10*, 436. https://doi.org/10.3390/gels10070436

Received: 27 June 2024
Accepted: 28 June 2024
Published: 29 June 2024

In recent decades, hydrogels have garnered significant attention, thanks to their extensive biomedical and pharmaceutical applications [1,2]. These remarkable materials closely mimic biological tissues and exhibit unique behaviors due to their high-water content. The ability to customize their properties and enhance cell interactions by selecting specific monomers and synthesis techniques has propelled hydrogels to the forefront of biomaterial innovation.

Hydrogels' versatility has led to successful applications in areas as diverse as injectable particulate systems, contact lenses, cartilage substitutes, catheter linings, valves, suture threads, wound-healing dressings, skin grafts, and biosensors [3]. Furthermore, hydrogels are playing an increasingly vital role in tissue engineering, regenerative medicine, and targeted drug delivery systems, cementing their status as a cornerstone of modern biomedical science [4]. Figure 1 illustrates some of the areas where hydrogels are most applicable.

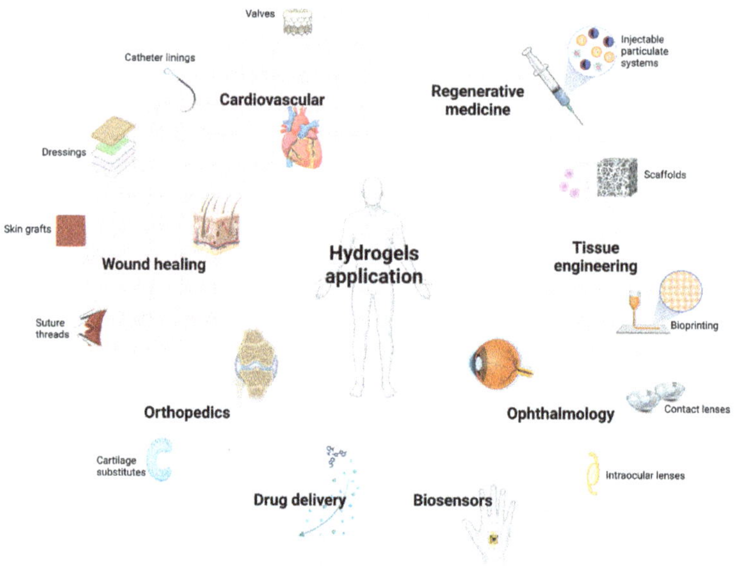

Figure 1. Main areas of application of hydrogels at the biomedical and pharmaceutical level.

Although significant efforts have been made to develop new hydrogels with improved properties and additional functionalities, several challenges remain and are currently the focus of intense research. These are being addressed through a multidisciplinary approach, combining knowledge of biology, chemistry, materials engineering and other disciplines, such as pharmaceutics. One major hurdle to overcome is the biocompatibility of hydrogels. Ensuring that hydrogels do not provoke adverse immune responses and can integrate seamlessly with biological tissues is crucial for their effective use in medical applications. Additionally, the mechanical properties of hydrogels need to be finely tuned to match the specific requirements of the different administration routes/tissues and clinical applications. For example, hydrogels used in cartilage replacement must be strong and elastic, while those used in wound dressings should be flexible and breathable. Depending on the case, other features may be of major importance, e.g., oxygen permeability, transparency or tribological behavior. Hydrogels with cross-linking/gelation capacity, which allows their formation in situ at the target site, may be advantageous in some situations. Appropriate degradation rates and bioactive surfaces that promote vascularization and support suitable tissue architecture may also be desirable. In all cases, the ability to prevent colonization by microorganisms, thereby reducing the risk of infection, is an added value. Their resistance to sterilization methods is also a transversal requirement, to ensure that they remain functional and safe after the sterilization procedures. Finaly, researchers are exploring new synthesis and functionalization methods, as well as novel materials to make hydrogel manufacturing more efficient and affordable [5]. The scalability and cost-effectiveness of hydrogel production are essential for their widespread clinical adoption.

This Special Issue presents a selection of the cutting-edge research that is being carried out by teams from diverse countries across Europe, Asia and America with the objective of developing new hydrogels that can meet the aforementioned requirements. It gathers eight original research articles covering topics that range from the production of hydrogels, addressing for example the influence of the nature of the raw materials and of the incorporation of additives, to the evaluation of the effects of unconventional sterilization methods. The characterization of the materials and their behavior is achieved through in vitro testing, experiments with animals and clinical trials. The presented works cover not only biotechnological applications, such as the immobilization of enzymes, scaffolds for cellular metabolism studies or tissue engineering, and biofabrication, but also therapeutic applications, e.g., ocular, local anti-cancer treatments and topical or transdermal drug delivery (Figure 2). The latter application is the most representative and covers areas such as wound or burn healing and treatment of sports injuries. The two reviews presented at the end of this collection provide an overview of the latest advances in the production and therapeutic potential of hydrogels in wound healing and cancer treatment. A summary of each of the works published in the Special Issue is given below.

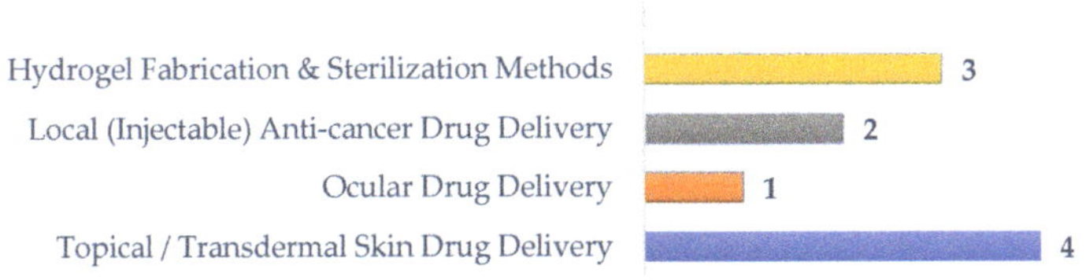

Figure 2. Topics covered in this Special Issue.

The first article evaluates the effect of the origin and variability of the raw materials on the synthesis of gelatin methacryloyl (GelMA) and production of hydrogels. Eight types of gelatin, from distinct sources (five porcine, one from fish, and two bovine) and with different bloom values, were made to react with methacrylic anhydride in the presence of urea to maintain room temperature and avoid GelMA gelation. The degree of functionalization (DoF) of the obtained product was influenced by the bloom strength and by the gelatin's donor species. Furthermore, the characteristics of the product were also affected by the batch-to-batch variability of the gelatins. In the second stage of the work, the authors produced hydrogel discs by UV photopolymerization using the synthesized GelMAs and investigated their viscoelastic behavior, concluding that it depends on the protein's DoF, bloom strength and origin. Additionally, a microfluidic device was used to generate droplets of GelMA solutions that were subsequently crosslinked to obtain hydrogel microparticles. The droplet size was controlled through the ratio continuous phase (GelMA solution) to the dispersed phase (sunflower seed oil). The swelling capacity of the obtained microparticles varied inversely with GelMA concentration. As a potential use of the microparticles, the authors advocate for cellular expansion and differentiation in stirred bioreactors, due to the high surface to volume ratio and to the fact that GelMA presents cell adhesion sites. Comparatively to underivatized gelatin, the proposed microparticles are easier to produce, as the crosslinking can be performed in a single stage by photopolymerization.

In another work, thermoresponsive hydrogels based on Pluronic® F127 containing different polysaccharides (xanthan gum, alginate, κ-carrageenan, gellan, or chitosan) were prepared for possible application in injectable biomaterials or bioinks. The rheological behavior of the materials was studied in different shear conditions and aqueous environments. Pluronic® F127 formed micellar networks with self-healing capacity, since after being submitted to high strain cycles, they were able to recover the initial structure. The viscoelastic behavior of the hydrogels containing xanthan gum was found to depend on the testing liquid. They exhibited shear-thinning behavior, yield stress, and enhanced self-healing capabilities. The experiments revealed that the best results were obtained for a xanthan gum concentration of 1%. However, the addition of salt led to aggregate formation and diminished the hydrogel performance. Hydrogels containing the same content of the remaining natural polymers were also tested. Although good results have been obtained with the hydrogels containing alginate, κ-carrageenan and gellan gum, their ability to recover the initial structure after application of high strains decreased. The addition of chitosan implied the use of acetic acid for its solubilization, which must have been responsible for the inferior viscoelastic performance and self-healing ability shown. The authors concluded that the hydrogel's structure and respective rheological properties can be tuned by choosing the appropriate polysaccharide.

Covering different aspects of the hydrogel production process, this Special Issue also includes an article on sterilization. This is a mandatory step to obtain biologically safe materials that can be implanted/used in direct contact with the body. The sterilization method and conditions must not only fulfill the sterility requirement but also ensure that the material maintains adequate properties for the intended application. Many hydrogels are thermo- and/or radiosensitive, precluding the use of standard industrial methods and warranting the quest for alternative sterilization methods. Pires et al. investigated the effect of three non-conventional methods, (a) microwave (MW); (b) high hydrostatic pressure (HHP), and (c) plasma (PM), on the properties of reinforced polyvinyl alcohol (PVA) hydrogels intended for articular cartilage replacement. The material was reinforced with poly(p-phenylene-2,6-benzobisoxazole) (PBO) nanofibers to assure the high mechanical performance needed in this demanding application. All the methods were effective at achieving the materials' sterility and did not affect the hydrogels' water content, nor their hydrophilicity. However, they induced some changes in crystallinity and/or crosslinking. MW was revealed to be the most suitable method for the studied hydrogel, since it further improved its mechanical properties, namely the hardness, stiffness and shear modulus. In addition, it reduced the friction coefficient observed against natural cartilage. The sterilized hydrogel kept its non-irritant and non-cytotoxic behavior. MW is deemed a suitable method for the sterilization of this type of hydrogel, due to the ease of use, low cost and short processing time (3 min).

Illustrating the application of hydrogels with bioactive agents for therapeutic purposes, Francisco et al. used Pluronic® F127 to produce a hydrogel that, due to the incorporation of silver nanoparticles (AgNPs), may be used in skin burn regeneration. In fact, AgNPs have demonstrated antibacterial activity and are promising in wound healing, a major public health issue lacking effective therapeutic strategies. In this work, different methods of nanoparticle production were tested and the one that yielded less aggregates and more homogeneous particles (dispersion index < 0.2) was selected. The AgNPs produced were characterized by atomic force microscopy (AFM), diffraction laser scattering (DLS) and zeta potential measurements. AgNPs were non-spherical, with an average size of 48.04 ± 14.87 nm and slightly negative surface charge (-0.79 ± 2.17); their solution was translucent yellow, with an absorption peak at 407 nm. Pluronic® F127, an amphiphilic poly(ethylene oxide)/poly(propylene oxide)/poly(ethylene oxide) triblock copolymer, was used to produce a thermoreversible hydrogel (i.e., upon administration, at body temperature, a stable gel matrix is formed), incorporating the AgNPs. Antibacterial activity was shown in vitro against known colonizers of infected skin burns (*Escherichia coli*, *Staphylococcus aureus* and *Pseudomonas aeruginosa*). In vitro skin permeation studies, using an artificial membrane to mimic the skin, demonstrated no AgNP permeation after 24 h, a good safety indicator. The semi-solid formulation was tested in vivo (mice) in a chemical skin burn model and its therapeutic effect and safety were compared to a commercial silver sulfadiazine cream. The skin regeneration performance of the topical hydrogel-loaded AgNPs at lower silver doses was comparable to the more concentrated commercial formulation, demonstrating the clinical potential of this approach.

Traditional hydrogels, despite their ability to absorb large amounts of water, typically exhibit delayed swelling after application and are not suitable for the treatment of moderate to heavily exuding wounds. Pinthong and co-workers developed a porous hydrogel incorporating Manuka honey (MH, 1 and 10% *w/w*) for such use. The hydrogel matrix was based on 2-acrylamido-2-methyl-1-propanesulphonic acid with the addition of pore-forming excipients—sodium bicarbonate (foaming agent), methacrylic acid (to promote CO_2 formation during gelation), and Poloxamer 407 (foam stabilizer)—to create a structure capable of rapidly absorbing significant amounts of water. The formation of a gas-blown porous hydrogel, with pores ranging from ~50 to 110 µm (scanning electron microscopy; SEM), the superior bulk swelling performance (~5000% vs. ~2000% increase in weight, by gravimetry) and initial surface absorption (10 µL vs. <1 µL, in <3000 ms, by contact angle measurement) compared to the non-porous counterpart were confirmed. Incorporation of MH resulted in smaller (~50 µm) and more homogeneous pores, increased porosity, more linear swelling behavior in the first 5 min, improved gel appearance and mechanical properties. The inherent properties of MH, which altered the system by increasing viscosity and lowering the pH, enhancing foam production during gelation, appear to be responsible for the improvements observed. Cell cytotoxicity was tested by the XTT (2,3-bis-(2-methoxy-4-nitro-5-sulfophenyl)-2H-tetrazolium-5-carboxanilide) assay, using normal human dermal fibroblasts, to evaluate cell viability during 24 h. Despite the acidic nature and hydrogen peroxide and flavonoid content of MH, cell viability was ~80-90% and the materials were regarded as non-toxic. In conclusion, the porous hydrogels fabricated offer a promising approach for advanced rapid-absorbing wound dressings.

To minimize discomfort, swelling, and inflammation following sports or accident injuries, non-steroidal anti-inflammatory drugs are typically administered. Permeation of the drug into the damaged tissues is paramount and dependent on formulation. The following work by Bukhari et al. evaluated the efficacy of diclofenac potassium (DK 2, 4 and 6%) gels, with and without phonophoresis application, in the management of such injuries. Marketed gels of the sodium salt of the drug (DS, 4%) and drug-free gels were used as controls. The patients ($n = 200$) were randomly allocated into five groups ($n = 20$ each) to which the different formulations were administered 3–4 times weekly, for 4 weeks, without phonophoresis; for another five similar groups, ultrasounds were applied in continuous mode (2:1), at a frequency of 0.8 MHz and an intensity of about 1.5 W/cm^2, to maximize the effect of the topical formulation. The patients were assessed (at baseline and sessions 4, 8, 12 and 16) by using the Numeric Pain Rating Scale (NPRS) and the Western Ontario McMaster Osteo-Arthritis (WOMAC) indexes for pain in disability and stiffness. Both scales showed significant dose-dependent pain relief in DK-treated groups, as compared to the group treated with DS gel. The additional application of phonophoresis resulted in faster and deeper pain relief due to the increased penetration of both DK and DS gels; the significant increases in benefits were dose-dependent and particularly stronger for DK gels. Moreover, phonophoresis was well tolerated by patients. The results demonstrated the superiority of the therapeutic scheme combining 6% DK gel and a physical enhancer of drug permeation, which is particularly suitable for the treatment of uncomplicated soft tissue injuries (e.g., plantar fasciitis, bursitis stress injuries, and tendinitis).

Glaucoma is a neurodegenerative disorder that may lead to vision loss or blindness because of damage to the optic nerve. One of the most significant risk factors for developing the disease is elevated intraocular pressure, which may be prevented using anti-hypertensive beta-adrenergic blockers such as carvedilol. However, due to its poor water solubility and extensive first-pass metabolism, carvedilol demonstrates low oral bioavailability, necessitating the use of alternative drug delivery routes (e.g., topical ocular) and formulations capable of remaining at the target site for long periods of time. To this end, Almutairy et al. fabricated a thermosensitive in situ gelling system for the ophthalmic delivery of carvedilol-loaded spanlastics (CRV-SPLs). SPLs are elastic colloidal carriers based on non-ionic surfactants and an edge activator, with enhanced corneal permeability of the entrapped drugs. The incorporation of such vesicles into an in situ forming hydrogel is anticipated to facilitate prolonged drug release, accurate dosage, straightforward administration, prolonged residence time and enhanced transcorneal permeation. Optimized SPLs with minimal diameter, high entrapment efficiency and drug permeability were prepared with Span 60:Brij 97 (80:20). Gels were formulated with thermosensitive polymers (poloxamer 407/poloxamer 188) and processing parameters (e.g., gelation temperature, muco-adhesion, rheological properties, and in vitro drug release pattern) were optimized. The gels produced were transparent (slightly less after loading with SPLs) with acceptable pH for ocular use (~neutrality) and high drug content (>97.3%). Ex vivo drug permeation studies using goat corneal membrane showed a marked drug permeation increase due to the nanoencapsulation in SPLs. Visual appearance, drug content, pH, and gelling capability were maintained over a period of 8 weeks, at 4 °C. In vivo pharmacokinetics, based on calculations of the amount of CRV quantified in the aqueous humor of rabbits over time, after a single administration, was analyzed using a non-compartment model. A rapid onset of action, and increased drug residence time and penetration through cornea were observed, resulting in higher AUC, $t_{1/2}$ and C_{max}. The optimized carvedilol-loaded in situ gel system was found to be efficacious in lowering intra ocular pressure, for up to 8 h after instillation, demonstrating its sustained action. Irritation, evaluated by the Draize rabbit eye test, was ruled out. In conclusion, the results demonstrated the potential of CRV-loaded spanlastic vesicles to enhance ocular bioavailability and serve as an alternative to conventional dosage forms in glaucoma treatment.

The high area/mass ratio and reactivity of polymeric nanofibers have been explored in controlled release and targeting of biomolecules in several diseases including cancer. Guerra and co-workers fabricated electrospun polyvinyl alcohol (PVA) and PVA/chitosan (CS) hydrogel nanofiber systems of lysozyme (Lys), crosslinked with boronic acids and functionalized with magnetic iron oxide nanoparticles (IONPs), for biological applications. The antibiotic, antitumor and immunomodulatory properties of Lys justify the choice of this enzyme. Magnetic nanovesicles have found usefulness in hyperthermia-based cancer therapy due to the sensitivity of some cancer cells to temperatures > 41 °C. Application of an alternating magnetic field to IONPs results in a temperature increase and irreversible damage and selective death of tumor cells (damage to normal cells is reversible). Electrospinning solutions were characterized in terms of specific conductivity and surface tension; morphological analysis of the electrospun nanofibers, as a function of processing parameters, was conducted by SEM. Most of the nanofibers were white, uniform, relatively thin and fragile and circular in shape (diameter ~ 3.3 cm) with an average mass of 4.12 mg. The joint use of PVA and CS seemed to improve the chemical and resistance properties of the nanofibers; crosslinking with boronic acids reduced its fragility. The effects of temperature and pH on the in vitro release and biological activity of the immobilized enzyme were determined by the *Micrococcus lysodeikticus* method. The nanofibers showed controlled release and Lys exhibited a high release rate at pH 6.74 and 45.5 °C. Since cancer cells exhibit lower pH and are more sensitive to temperature than normal cells, these features may prove useful in anticancer applications. The presence of IONPs does not seem to influence Lys release, but crosslinking is essential for controlled release over time. Enzymatic activity, ascertained by microbial lysis, presented a lag time and was not affected by the presence of magnetic particles. The cytotoxicity of the systems was evaluated in a widely used model of the intestinal mucosal barrier, an epithelial Caco-2 cell line, from colon carcinoma. After 10 days of contact, cell viability was evaluated by the MTT (3-[4,5-dimethylthiazol-2-yl]-2,5 diphenyl tetrazolium bromide) assay and these systems were found to be cytotoxic, especially those coated with IONPs. The biodegradability and bioresponsiveness of the nanofibers at high temperature highlight their potential for in situ application to reduce or inhibit the process of tumor metastasis.

Moving to the literature reviews, Marques et al. collected information on the recent advances over the past two decades, regarding the development of injectable poloxamer hydrogels for anti-cancer local treatments. Local cancer therapies offer several advantages relatively to systemic treatments, e.g., intravenous administration. In fact, intra/peri-tumoral injections ensure high levels of the therapeutic agents at the target site, while avoiding systemic toxicity, and their effect does not depend on the tumor vasculature. Injectable biodegradable hydrogels present the added advantage of eliminating the need for surgery and increase the retention of the drug in the tumor. Poloxamer-based gelling systems have raised special interest due to their biocompatibility, ease of preparation, thermoresponsiveness, and ability to incorporate various anticancer agents. The review describes the physicochemical and biological properties of injectable poloxamer hydrogels and summarizes their applications in local cancer therapy, namely in chemotherapy, phototherapy, immunotherapy, and gene therapy. Despite the promising preclinical results, such treatments are still at the proof-of-concept stage. To advance towards their use, poloxamers must be modified/combined with other polymers to reduce erosion and ensure precise drug delivery. Additionally, a more extensive characterization of the hydrogels is needed. Besides evaluating the rheological and biological behavior, insights into morphology and thermal properties should also be gathered. The determination of the therapeutic efficacy in animal models larger than those currently used (rodents) is also critical. Future prospects point out the increasing use of phototherapy and immunotherapy (in substitution of chemotherapy) and combination therapies as a new direction in cancer treatment. Injectable poloxamer hydrogels will gain prominence not only for drug delivery but also in fields such as tissue engineering and cartilage repair.

Finally, the concluding article offers a comprehensive overview and a critical analysis of several issues related to the use of hydrogels for skin regeneration and wound healing. After enumerating the advantages of hydrogels in wound care, the authors identify the most commonly used polymers in this field and provide a detailed description of their chemical structure, origin and extraction methods, as well as of their intrinsic properties and role in the wound healing process. Whenever possible, the authors provide explanations on the mechanisms of action of individual polymers and their effectiveness; the advantages and limitations of the different polymers are also discussed. Strategies for promoting wound healing, exploiting the intrinsic potential of hydrogels, are presented. These include the use of polymers that stimulate angiogenesis, or the incorporation of bioactive agents, such as drugs, antimicrobial substances or growth factors, which provide additional functionalities to the hydrogels. The usefulness of hydrogels as a 3D matrix for cell culture is also discussed, with particular focus on their potential to support the loading and recruitment of cells to the wound site, where they can proliferate and give rise to new tissue. Finally, future steps in the advancement of hydrogels for wound healing are critically analyzed with particular emphasis on the need for the material to be effective, safe, and environmentally friendly.

In conclusion, while substantial progress has been made in hydrogel development, overcoming the remaining challenges requires a continued interdisciplinary collaboration between specialists in different areas. The emergence of cutting-edge technologies will help to address issues related with biocompatibility, mechanical properties, enhanced functionality, and production efficiency, leading to the next generation of hydrogels with even greater potential to revolutionize healthcare. Ongoing research is pushing the boundaries of what hydrogels can achieve, exploring new directions [6] (Figure 3) such as **self-healing hydrogels** that repair themselves after damage [7], **bioactive hydrogels** that promote specific cellular responses [8] and **smart hydrogels**, which respond to environmental stimuli, such as pH, temperature, or light [9]. The latter have raised particular interest, due to their potential of enabling adaptive tissue engineering scaffolds and/or controlled and targeted release of drugs and other active agents, enhancing therapeutic outcomes. Current research trends also include novel **nanoarchitectured hydrogels**, incorporating nanoparticulates (as a second phase within the system) capable of accurately controlling drug delivery [10]. Recently, the potential of **three-dimensional printing of hydrogels** in producing living tissue structures or organs [11], or customized dosage forms [12], has been noted.

Figure 3. Trends in hydrogel research for biomedical applications.

Hydrogels will play a pivotal role in achieving more sophisticated medical treatments and devices, opening possibilities for personalized medicine, improved patient outcomes, and innovative solutions to complex medical challenges.

Author Contributions: A.P.S., D.C.S. and A.I.F. contributed equally to summarizing the papers and writing, illustrating, reviewing and editing the manuscript. All authors have read and agreed to the published version of the manuscript.

Funding: A.P.S. and D.C.S. acknowledge Fundação para a Ciência e Tecnologia (FCT) for providing funding through Milk4WoundCare (https://doi.org/10.54499/2022.03408.PTDC), SOL (https://doi.org/10.54499/PTDC/CTM-CTM/2353/2021), and through Centro de Química Estrutural Research Unit projects (https://doi.org/10.54499/UIDB/00100/2020 and the Institute of Molecular Sciences project (https://doi.org/10.54499/UIDP/00100/2020)). D.C.S. acknowledges FCT for her Junior Research contract (https://doi.org/10.54499/2022.08560.CEECIND/CP1713/CT0016). A.I.F. also thanks FCT/MCTES for the financial support provided to CiiEM (https://doi.org/10.54499/UIDB/04585/2020) through national funds.

Acknowledgments: The authors of this Special Issue who contributed high-quality research are gratefully acknowledged.

Conflicts of Interest: The authors declare no conflicts of interest.

List of Contributions

1. Grijalva Garces, D.; Appoldt, L.J.; Egner, J.; Leister, N.; Hubbuch, J. The effect of gelatin source on the synthesis of gelatin-methacryloyl and the production of hydrogel microparticles. *Gels* **2023**, *9*, 927. https://doi.org/10.3390/gels9120927.
2. Lupu, A.; Gradinaru, L.M.; Rusu, D.; Bercea, M. Self-healing of Pluronic® F127 hydrogels in the presence of various polysaccharides. *Gels* **2023**, *9*, 719. https://doi.org/10.3390/gels9090719.

3. Pires, T.; Oliveira, A.S.; Marques, A.C.; Salema-Oom, M.; Figueiredo-Pina, C.G.; Silva, D.; Serro, A.P. Effects of non-conventional sterilisation methods on PBO-reinforced PVA hydrogels for cartilage replacement. *Gels* **2022**, *8*, 640. https://doi.org/10.3390/gels8100640.
4. Francisco, P.; Neves Amaral, M.; Neves, A.; Ferreira-Gonçalves, T.; Viana, A.S.; Catarino, J.; Faísca, P.; Simões, S.; Perdigão, J.; Charmier, A.J.; et al. Pluronic® F127 hydrogel containing silver nanoparticles in skin burn regeneration: An experimental approach from fundamental to translational research. *Gels* **2023**, *9*, 200. https://doi.org/10.3390/gels9030200.
5. Pinthong, T.; Yooyod, M.; Daengmankhong, J.; Tuancharoensri, N.; Mahasaranon, S.; Viyoch, J.; Jongjitwimol, J.; Ross, S.; Ross, G.M. Development of natural active agent-containing porous hydrogel sheets with high water content for wound dressings. *Gels* **2023**, *9*, 459. https://doi.org/10.3390/gels9060459.
6. Bukhari, K.A.; Khan, I.A.; Ishaq, S.; Iqbal, M.O.; Alqahtani, A.M.; Alqahtani, T.; Menaa, F. Formulation and evaluation of diclofenac potassium gel in sports injuries with and without phonophoresis. *Gels* **2022**, *8*, 612. https://doi.org/10.3390/gels8100612.
7. Almutairy, B.K.; Khafagy, E.-S.; Abu Lila, A.S. Development of carvedilol nanoformulation-loaded poloxamer-based in situ gel for the management of glaucoma. *Gels* **2023**, *9*, 952. https://doi.org/10.3390/gels9120952.
8. Guerra, M.; Garrudo, F.F.F.; Faustino, C.; Rosa, M.E.; Ribeiro, M.H.L. Exploring functionalized magnetic hydrogel polyvinyl alcohol and chitosan electrospun nanofibers. *Gels* **2023**, *9*, 968. https://doi.org/10.3390/gels9120968.
9. Marques, A.C.; Costa, P.C.; Velho, S.; Amaral, M.H. Injectable poloxamer hydrogels for local cancer therapy. *Gels* **2023**, *9*, 593. https://doi.org/10.3390/gels9070593.
10. Ribeiro, M.; Simões, M.; Vitorino, C.; Mascarenhas-Melo, F. Hydrogels in cutaneous wound healing: Insights into characterization, properties, formulation and therapeutic potential. *Gels* **2024**, *10*, 188. https://doi.org/10.3390/gels10030188.

References

1. Selvaraj, S.; Dutta, V.; Gopalakrishnan, C.; Subbarayan, R.; Rana, G.; Radhakrishnan, A.; Elango, A.; Chauhan, A. Biomedical potential of hydrogels: A multifaceted approach to innovative medication delivery. *Emergent Mater.* **2024**, *7*, 721–763. [CrossRef]
2. Kesharwani, P.; Bisht, A.; Alexander, A.; Dave, V.; Sharma, S. Biomedical applications of hydrogels in drug delivery system: An update. *J. Drug Deliv. Sci. Technol.* **2021**, *66*, 102914. [CrossRef]
3. Kaith, B.S.; Singh, A.; Sharma, A.K.; Sud, D. Hydrogels: Synthesis, classification, properties and potential applications—A brief review. *J. Polym. Environ.* **2021**, *29*, 3827–3841. [CrossRef]
4. Hasirci, N.; Kilic, C.; Kömez, A.; Bahcecioglu, G.; Hasirci, V. Hydrogels in Regenerative Medicine. In *Gels Handbook*; World Scientific: Singapore, 2016; pp. 1–52.
5. Yang, J.M.; Olanrele, O.S.; Zhang, X.; Hsu, C.C. Fabrication of Hydrogel Materials for Biomedical Applications. *Adv. Exp. Med. Biol.* **2018**, *1077*, 197–224. [CrossRef] [PubMed]
6. Sánchez-Cid, P.; Jiménez-Rosado, M.; Romero, A.; Pérez-Puyana, V. Novel trends in hydrogel development for biomedical applications: A review. *Polymers* **2022**, *14*, 3023. [CrossRef] [PubMed]
7. Fan, L.; Ge, X.; Qian, Y.; Wei, M.; Zhang, Z.; Yuan, W.-E.; Ouyang, Y. Advances in synthesis and applications of self-healing hydrogels. *Front. Bioeng. Biotechnol.* **2020**, *8*, 654. [CrossRef] [PubMed]
8. Wu, L.; He, Y.; Mao, H.; Gu, Z. Bioactive hydrogels based on polysaccharides and peptides for soft tissue wound management. *J. Mater. Chem. B Mater. Biol. Med.* **2022**, *10*, 7148–7160. [CrossRef] [PubMed]
9. Bordbar-Khiabani, A.; Gasik, M. Smart Hydrogels for Advanced Drug Delivery Systems. *Int. J. Mol. Sci.* **2022**, *23*, 3665. [CrossRef] [PubMed]
10. Idumah, C.I.; Nwuzor, I.C.; Odera, S.R.; Timothy, U.J.; Ngenegbo, U.; Tanjung, F.A. Recent advances in polymeric hydrogel nanoarchitectures for drug delivery applications. *Int. J. Polym. Mater.* **2024**, *73*, 1–32. [CrossRef]
11. Kaliaraj, G.; Shanmugam, D.; Dasan, A.; Mosas, K. Hydrogels—A promising materials for 3D printing technology. *Gels* **2023**, *9*, 260. [CrossRef] [PubMed]
12. Rouaz-El Hajoui, K.; Herrada-Manchón, H.; Rodríguez-González, D.; Fernández, M.A.; Aguilar, E.; Suñé-Pou, M.; Nardi-Ricart, A.; Pérez-Lozano, P.; García-Montoya, E. Pellets and gummies: Seeking a 3D printed gastro-resistant omeprazole dosage for paediatric administration. *Int. J. Pharm.* **2023**, *643*, 123289. [CrossRef] [PubMed]

gels MDPI

Article

The Effect of Gelatin Source on the Synthesis of Gelatin-Methacryloyl and the Production of Hydrogel Microparticles

David Grijalva Garces [1,2], Luise Josephine Appoldt [2], Jasmin Egner [2], Nico Leister [3] and Jürgen Hubbuch [1,2,*]

1 Institute of Functional Interfaces, Karlsruhe Institute of Technology, 76344 Eggenstein-Leopoldshafen, Germany
2 Institute of Process Engineering in Life Sciences Section IV: Biomolecular Separation Engineering, Karlsruhe Institute of Technology, 76131 Karlsruhe, Germany
3 Institute of Process Engineering in Life Sciences Section I: Food Process Engineering, Karlsruhe Institute of Technology, 76131 Karlsruhe, Germany
* Correspondence: juergen.hubbuch@kit.edu

Abstract: Gelatin methacryloyl (GelMA) is widely used for the formulation of hydrogels in diverse biotechnological applications. After the derivatization of raw gelatin, the degree of functionalization (DoF) is an attribute of particular interest as the functional residues are necessary for crosslinking. Despite progress in the optimization of the process found in the literature, a comparison of the effect of raw gelatin on the functionalization is challenging as various approaches are employed. In this work, the modification of gelatin was performed at room temperature (RT), and eight different gelatin products were employed. The DoF proved to be affected by the bloom strength and by the species of gelatin at an equal reactant ratio. Furthermore, batch-to-batch variability of the same gelatin source had an effect on the produced GelMA. Moreover, the elasticity of GelMA hydrogels depended on the DoF of the protein as well as on bloom strength and source of the raw material. Additionally, GelMA solutions were used for the microfluidic production of droplets and subsequent crosslinking to hydrogel. This process was developed as a single pipeline at RT using protein concentrations up to 20 % (w/v). Droplet size was controlled by the ratio of the continuous to dispersed phase. The swelling behavior of hydrogel particles depended on the GelMA concentration.

Keywords: biomaterials; bloom value; gelatin; GelMA; hydrogel; microfluidics; microparticle

Citation: Grijalva Garces, D.; Appoldt, L.J.; Egner, J.; Leister, N.; Hubbuch, J. The Effect of Gelatin Source on the Synthesis of Gelatin-Methacryloyl and the Production of Hydrogel Microparticles. *Gels* **2023**, *9*, 927. https://doi.org/10.3390/gels9120927

Academic Editor: Diana Silva

Received: 26 October 2023
Revised: 7 November 2023
Accepted: 20 November 2023
Published: 24 November 2023

1. Introduction

Hydrogels are polymeric networks with a high water-binding and retaining capacity. Since the backbone of the hydrogels is crosslinked polymers, the structural stability of the hydrogel is preserved in aqueous phase [1]. These properties enable the transport of dissolved molecules within the physical structure which can be beneficial for a variety of biotechnological applications such as the immobilization of enzymes [2] and microorganisms [3,4] in bioreactors, as well as cell culture for studies of cellular metabolism [5]. For these diverse purposes, advanced manufacturing strategies are applied for the creation of defined physical structures such as microparticles in microfluidics [6] and tissue models in bioprinting [7].

A suitable biomaterial for the production of hydrogels is gelatin, which is extracted from collagen [8]. Furthermore, the molecular weight and molecular weight distribution of gelatin not only depend on the sources but also on the processing conditions such as treatment time, pH, and temperature. Gelatin extracted in acidic media, and media extracted using alkaline milieus, shows isoelectric points (IEP) at pH 8–9, and pH 4–5, respectively [9,10]. After processing, the protein backbone retains sites for cell adhesion as well as for enzymatic cleavage such as those present in collagen [8]. A challenging property of gelatin for certain applications is the transition of the gelatin solution to a gel below a physiological temperature. A way to handle the limited structural stability of hydrogels at elevated temperatures is the formation of covalent bonds between the proteins. For this purpose, gelatin is functionalized to gelatin methacryloyl (GelMA). The methacrylate and methacrylamide residues present in GelMA enable the creation of crosslinked networks via photopolymerization [11]. The first draft of the process was proposed by Van den Bulcke et al. [11]. The study included the addition of methacrylic anhydride (MAA) to the gelatin solution in phosphate-buffered saline (PBS) at pH 7.5 under stirring at 50 °C. Significant progress has been made by research groups to identify the effect of process parameters on the resulting degree of functionalization (DoF) of GelMA. Lee et al. [12] and Shirahama et al. [13] have presented a thorough characterization of the reaction using porcine gelatin. In these studies, the MAA-to-gelatin ratio was significantly reduced by using carbonate bicarbonate (CB) buffer at pH values above the IEP of porcine gelatin. This enhancement is due to the fact that free amino groups are not charged. Additionally, Shirahama et al. [13] studied the derivatization of gelatin in a temperature range from 35 to 50 °C with no difference in the produced DoF. Our previous study complemented the findings of both groups by producing porcine GelMA at room temperature (RT) while keeping the MAA-to-gelatin ratio at the same value [14]. Despite the improvement of the synthesis process concerning porcine GelMA, more work is required to compare the effect of raw material on the final product. To the best of our knowledge, a wide range of raw materials including a variation in species and bloom strength have only been reported once [15]. However, the used synthesis buffer was composed of 0.1 M CB buffer, lower than the optimum reported by Shirahama et al. [13]. Further studies have compared the use of porcine and bovine gelatin pairwise. However, making a comparison across studies is challenging. This is because the applied methods vary in terms of buffer composition and pH (PBS at pH 7.4 [16–18] or CB at pH 9 [15,19,20]), as well as buffering capacity (0.1 M [15,19] or 0.25 M CB [20]).

As GelMA contains cell adhesion sites, hydrogel microparticles can be used for cellular expansion and differentiation. Commonly used methods for the expansion of adherent cell types are based on the use of tissue culture (TC) flasks. This limits the production of large quantities of cells as the required physical space increases linearly with the number of required flasks. In contrast, a significant advantage is shown by the expansion of cells using microcarriers. Hydrogel microparticles offer a high growth surface-to-volume ratio and can be implemented into stirred bioreactors [21]. The application of GelMA when compared to underivatized gelatin has the advantage that crosslinking can be performed via photopolymerization in a single stage when producing hydrogel microparticles. In contrast, particle production with gelatin requires multiple stages [22]. The challenging property of GelMA solutions, however, is the sol–gel transition below 30 °C. This issue has been addressed in the literature by using relatively low concentrations of the protein, i.e., below 10 % (w/v) [23–25], or by heating the entire microfluidic systems [25,26]. In the first part of this work, we apply the previously presented method to produce GelMA at room temperature. To characterize the effect of the raw material on the produced GelMA, we use a wide range of gelatin products. Porcine gelatin of five different products was tested. The samples included two separate batches of the most commonly studied gelatin product, i.e., porcine gelatin, 300 g bloom strength. Additionally, fish gelatin as well as two bovine products with varying bloom values were incorporated into the study. Furthermore, the produced GelMA was used for the formulation of hydrogels. The elasticity as a function of the source of the raw material was characterized. As a second part of the study, fish and porcine GelMA were used for the microfluidic production of droplets and the subsequent crosslinking to hydrogel microparticles. The manufacturing of microparticles was performed on a single pipeline at room temperature. The resulting droplet size was controlled by variation in the feed ratio of continuous to disperse phase, as well as by variation in GelMA type and concentration. In addition, the swelling behavior of hydrogel microparticles in aqueous media was determined.

2. Results and Discussion

2.1. GelMA Synthesis and Characterization

As demonstrated previously, the dissolution of porcine gelatin of 300 g bloom strength in urea-containing buffer was possible solely under stirring at room temperature [14]. This method was applicable for dissolving porcine gelatin of various bloom strengths, and two different gelatin products from bovine tissue. All used gelatin products are listed in Table 1. The dissolution at room temperature was due to the fact that urea disrupts protein–protein hydrophobic interactions and causes gelatin to unfold to coils in solution [27–29]. Even though gelatin from cold-water fish does not form a physical gel above 5 to 10 °C due to the lower content of proline and hydroxyproline [9,30], the same synthesis buffer was used for the sake of comparability during the synthesis of GelMA. As the rheological behavior of protein solution affects the distribution of reactants during a stirred reaction, the viscosity of the gelatin solutions in the synthesis buffer was measured. Figure 1 provides the corresponding results.

Table 1. Overview of gelatin types for the synthesis of gelatin methacryloyl (GelMA). The products were purchased from Sigma-Aldrich; the corresponding product information is provided including source, Bloom strength is according to the manufacturer, as is the sample nomenclature used throughout this manuscript.

Product Number	Batch Number	Source	Bloom Strength	Nomenclature
G6144	SLCH4483	porcine	80–120 g	p80
G2625	SLCC4273	porcine	175 g	p175
G1890	SLCC7838	porcine	300 g	p300 I
G1890	SLBX2973	porcine	300 g	p300 II
39465	BCBW7164	porcine	ultrahigh	pUH
G7765	038K0681	fish	–	f
G6650	SLCM1231	bovine	50–120 g	b50
G9382	SLCF9893	bovine	225 g	b225

The viscosity of solutions containing porcine gelatin increased significantly from 11.80 ± 0.17 mPa s to 53.93 ± 0.25 mPa s with increasing bloom strength of the gelatin product. The latter value was shown by the solution using a porcine source labeled as gelatin with ultrahigh (UH) gel strength by the supplier. Two batches from the same product with a bloom strength of 300 g were acquired and used for the measurements of viscosity. The solution produced with p300 I and p300 II gelatin showed a viscosity of 46.63 ± 0.42 mPa s and 39.43 ± 0.25 mPa s, respectively. These two values were significantly different ($p < 0.05$). Fish gelatin solution showed a viscosity of 13.26 ± 0.06 mPa s. The viscosity of the solutions comprising bovine gelatin showed an increase in viscosity with increasing bloom strength of the product from 14.07 ± 0.06 mPa s to 28.60 ± 0.10 mPa s. Statistically significant differences between the viscosity values were found between all data sets ($p < 0.05$). The increase in viscosity of gelatin solutions with increasing bloom strength is in accordance with other studies [30,31]. This is because the bloom value correlates with the molecular weight (MW) of gelatin [9]. Therefore, an increasing molecular weight leads to increasing intramolecular friction and to a higher amount of entanglements of proteins in solution, and, thus, higher viscosity [32,33]. In the case of gelatin pUH, no bloom value is stated by the producer. However, it was assumed that the MW is higher than that of gelatin p300 I and p300 II due to the ultrahigh gel strength. This was confirmed by the higher viscosity of the solution. Similarly, no bloom value is provided for fish gelatin. This is because the determination of gel strength is performed following a standardized method at 21 °C. Therefore, no bloom value can be measured for this product. The viscosity of the fish gelatin solution was around the values of viscosity of porcine gelatin p80 and bovine gelatin b50. Thus, the MW of fish gelatin was around the same magnitude as that of bovine and porcine gelatin of lower bloom strengths, as has been observed in the literature [30].

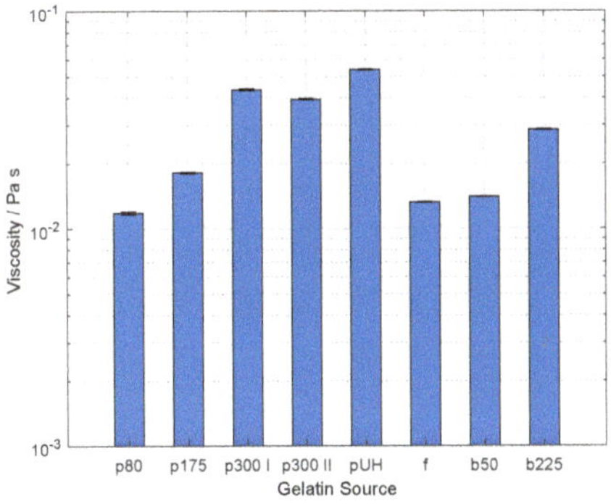

Figure 1. Viscosity of solution comprising gelatin at 10 % (w/v) in reaction buffer, i.e., 0.25 M carbonate bicarbonate (CB) buffer, and 0.25 M urea, measured at room temperature. Sample nomenclature is provided in Table 1. The viscosity increased with increasing bloom strength of both porcine and bovine gelatin. Additionally, the viscosity of gelatin solutions prepared with the same product but different batches, i.e., p300 I, and p300 II, showed a significant difference. Statistically significant differences between the viscosity values were found between all data sets ($p < 0.05$). Values are shown as mean and standard deviation. Each gelatin solution was tested three times from independently prepared samples.

Although it has been mentioned in the literature that the DoF of GelMA might vary when different types of gelatin are used [34], not many reports have been presented on this topic. In this study, GelMA was synthesized using different gelatin products with varying species of origin and various values of bloom strength. Table 1 provides relevant information on the tested products. The DoFs of the produced samples are shown in Figure 2. In the first part of the study regarding the synthesis of GelMA, the previous method using a urea-containing buffer to process gelatin at room temperature was simplified [14]. In contrast to the said study where MAA was continuously fed during the reaction, the complete amount of reactant was added at the beginning of the reaction in the present study. Additionally, the reaction time was shortened to 60 min. The GelMA sample p300 I was produced using the same gelatin product and batch. At a MAA-to-gelatin ratio from 100 µL g^{-1} (100 MA), the DoF exhibited a value of 0.899 ± 0.010. This value is not significantly different from the data shown previously with a value of 0.963 ± 0.027 [14]. Reaching a similar DoF in spite of the reduction in reaction time is comparable to the study by Shirahama et al. [13], as it was shown that the reaction is completed within 60 min when the complete volume of MAA is added at the starting point. As mentioned above, porcine gelatin with a bloom strength of 300 g has been widely studied for the production of GelMA [18,20,35]. Moreover, batch-to-batch variability is known to be a drawback of naturally derived polymers [36,37]. To test the effect of such inconsistencies, a second batch of the same product was used to synthesize GelMA p300 II. The DoF showed a value 0.832 ± 0.021 at 100 MA, significantly lower than that of GelMA p300 I ($p < 0.05$). Furthermore, the feasibility of using the developed method with porcine gelatin of varying bloom strength and the effects thereof were studied at 100 µL g^{-1} (100 MA). The DoF values of GelMA were 0.732 ± 0.014, and 0.810 ± 0.007 for the samples produced with gelatin of lower bloom strength, i.e., samples p80-100 MA, and p175-100 MA, respectively. These values differed significantly from each other and from the DoF of GelMA p300 I-100 MA ($p < 0.05$). Additionally, porcine gelatin with ultrahigh gel strength was modified to GelMA with a DoF of 0.910 ± 0.010. However, these data did not differ significantly from the data of the sample p300 I-100 MA.

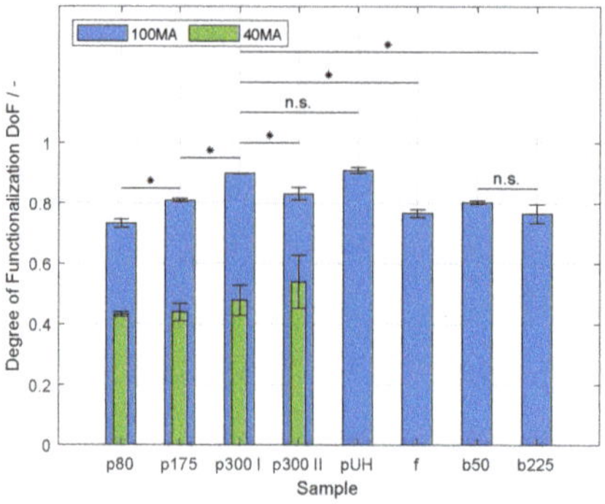

Figure 2. Degree of functionalization (DoF) of produced gelatin methacryloyl (GelMA). Sample nomenclature regarding the used raw materials is provided in Table 1. The DoF was determined by the trinitrobenzenesulfonic (TNBS) acid method [38]. At a methacrylic anhydride (MAA)-to-gelatin ratio of 100 µL g^{-1} (100 MA), asterisks denote a significant difference between synthesized samples ($p < 0.05$). No significant differences are denoted with the abbreviation n.s. ($p > 0.05$). Moreover, the DoF of porcine GelMA decreased significantly with decreasing MAA-to-gelatin ratio from 100 to 40 µL g^{-1} (40 MA) ($p < 0.05$). At a MAA-to-gelatin ratio of 40 µL g^{-1} (40 MA), no significant differences regarding the DoF were proven ($p > 0.05$). These differences are not shown for the purpose of clarity. Values are shown as mean and standard deviation. The functionalization of each gelatin type was carried out separately three times. The DoF of each batch was determined.

Gelatin p300 I and p300 II were derived from porcine skin with an acidic treatment (Type A) due to the high fat content of the tissue [10,30]. Individual differences within a species could lead to differences in MW and MW distribution. Additionally, slight differences in processing could also affect the properties of porcine gelatin as has been shown by Duconseille et al. [39]. The study showed that minor differences in the raw material and processing steps have a significant impact on the biochemical composition. In the reaction of gelatin to GelMA, the organic compound MAA is added to the aqueous gelatin solution. As both liquids are not miscible, thorough stirring is required to disperse the reactant to fine droplets. This issue has been addressed in the literature [13,35,40]. Hence, gelatin shows surface active properties leading to the adsorption of molecules to the created interface [41,42]. During stirring, MAA droplets are formed, which then collapse at different rates depending on the adsorption rate of gelatin to the interface and on the stabilization mechanism of the droplets. For instance, Shirahama et al. [13] mentioned that it was not feasible to evenly distribute MAA within a 1 % (w/v) gelatin solution as not enough protein was in solution to stabilize the MAA droplets. Furthermore, the adsorbed amount is dependent on the MW and MW distribution [43]. Additionally, the MW of the adsorbed protein affects the stabilization mechanism of the created droplets [41,43]. As the reaction took place in a buffered solution at pH 9, around the isoelectric point (IEP) of porcine gelatin, the stabilization mechanism is mostly steric. The magnitude of stabilization as well as the amount of adsorbed protein both increase with increasing MW. Consequently, the stabilization provided by gelatin of higher MW, i.e., higher bloom value, yields a higher interface and therefore a higher reaction rate leading to higher values of DoF. A study was presented by Aljaber et al. [15], where porcine gelatin of 300 g as well as 175 g bloom strength was used to produce GelMA, and showed a higher DoF for the material of higher bloom value, which is in accordance with the presented data in this study.

The possibility to transfer the developed approach to raw materials other than porcine gelatin was tested using fish gelatin from cold water as well as two bovine gelatin products with different gel strengths. The buffer system and the processing at room temperature proved to be applicable to fish and bovine gelatin. As mentioned above, the presence of urea in the buffer inhibits the formation of helical structures of the gelatin from bovine tissue, i.e., the transition from solution to a gel. At a MAA-to-gelatin ratio from 100 $\mu L\,g^{-1}$ (100 MA), the DoF of fish GelMA showed a value of 0.766 ± 0.013. Furthermore, the effect of various values of bloom strength was studied using bovine gelatin at 100 $\mu L\,g^{-1}$. The DoF values of GelMA were 0.804 ± 0.006, and 0.765 ± 0.013 for the samples produced with gelatin of lower bloom strength, i.e., samples b50-100 MA, and b225-100 MA, respectively. The DoF of both bovine GelMA samples as well as fish GelMA differed significantly from the DoF of GelMA p300 I-100 MA ($p < 0.05$). The effect of increasing bloom strength on the resulting DoF of bovine GelMA was not significant. As fish gelatin is extracted using acidic media [10], the IEP of the protein is similar to that of porcine gelatin. The determined DoF of GelMA f-100 MA is in the same range as the DoF of porcine GelMA with the lowest bloom strength, i.e., p80-100 MA. This comparable result is due to the fact that the MW of fish gelatin lies around the MW of gelatin p80, as was mentioned above regarding the results of the viscosity of both gelatin solutions. Hence, the stabilization of MAA droplets could take place at a similar magnitude. It has been shown by Lee et al. [12] that the reaction is most effective when the free amino groups are not charged; therefore, the pH during the reaction as well as the IEP of gelatin plays a significant role during the production of GelMA. In the literature, a higher DoF of bovine GelMA compared to that of porcine GelMA has been reported [16,18]. Both studies performed the reaction using phosphate-buffered saline, leading to the crucial difference in the surface charge of both proteins. The IEP of bovine gelatin lies around pH 4–5 due to the alkaline pre-treatment of bovine tissue where asparagine and glutamine are converted to aspartic acid and glutamic acid, respectively, [9,10]. Therefore, at pH 7 the reaction rate of bovine GelMA is much higher than the rate of porcine gelatin. Further studies producing GelMA using CB solutions have shown similar DoF values for porcine and bovine products. Lee et al. [19] prepared GelMA using porcine gelatin with 175 g bloom strength and bovine gelatin with 225 g. Both samples showed similar DoF values, which is in accordance with the presented study. Aljaber et al. [15] prepared GelMA using porcine gelatin with 300 g bloom strength, which had a higher DoF than the GelMA produced from bovine gelatin. Although the bloom strength of the bovine protein was not stated in that study, the results are in accordance with the results shown in this manuscript. As mentioned above, MAA and aqueous gelatin solutions are not miscible, and gelatin molecules are adsorbed to the created interface. The aqueous solution is buffered at pH 9; consequently, the bovine protein is negatively charged, making it less suitable for the stabilization of MAA droplets compared to the neutrally charged porcine gelatin [44]. This could lead to bigger droplets decreasing the amount of total interface for the reaction to GelMA, and, therefore decreasing the DoF of both bovine samples. The stabilization mechanism of MAA droplets could explain the missing difference regarding the DoF of GelMA b50 and b225. Additionally, proteins of lower MW show an electrostatic stabilizing effect because of the negatively charged surface, while the stabilizing mechanism of proteins with higher MW is rather steric [41]. As a result, both gelatin types could stabilize the created interface at similar magnitudes, thus showing similar DoF values.

The effect of the MAA-to-gelatin ratio was also studied using four gelatin raw materials, i.e., p80, p175, p300 I, and p300 II. The DoF of each GelMA sample at 40 $\mu L\,g^{-1}$ (40 MA) decreased significantly compared to each counterpart at 100 MA ($p < 0.05$). This result is in accordance with similar studies [12,13,20]. Holding the MAA-to-gelatin ratio constant at 40 MA, the DoF did not differ significantly by increasing bloom strength. As the volume of the reactant decreases, the created interface becomes smaller and the stabilization efficiency provided by the proteins is equally effective.

This study shows that the gelatin source as well as bloom strength and even batch-to-batch variations have a significant impact on the process. As the adsorption of the gelatin molecules at the interface to MAA is highly influenced by the MW and MW distribution, the setting of an optimal reactant ratio will depend on the used raw material. Our findings imply process parameters developed using a certain raw material cannot be simply transferred to the operation with a different one. In the case of GelMA, the process parameters to meet a certain DoF have to be adapted according to the gelatin material to be used. Further understanding of the reaction is required taking into account the properties of gelatin at the interface to the reactant. The stirring conditions should also be thoroughly studied, as the droplet size depends on the energy input to the process. As previously stated regarding the use of the protein in the field of tissue engineering, detailed information about the range of operating conditions to meet certain quality attributes is required. This is a requisite by regulatory authorities to reach clinical stages.

2.2. Hydrogel Characterization

GelMA solutions can be covalently crosslinked to hydrogels. This possibility is crucial when the intended application takes place at elevated temperatures. As shown in the literature, the elasticity of the produced hydrogels is influenced by the protein concentration and its DoF [11,14,45]. This study aimed to characterize the effect of the source as well as the effect of diverse values of bloom strength of the raw material on the resulting mechanical properties. For this purpose, hydrogels were prepared at 10 % (w/v) as described above, and the storage modulus was determined by oscillatory frequency sweeps on a rheometer. The associated values are presented in Figure 3.

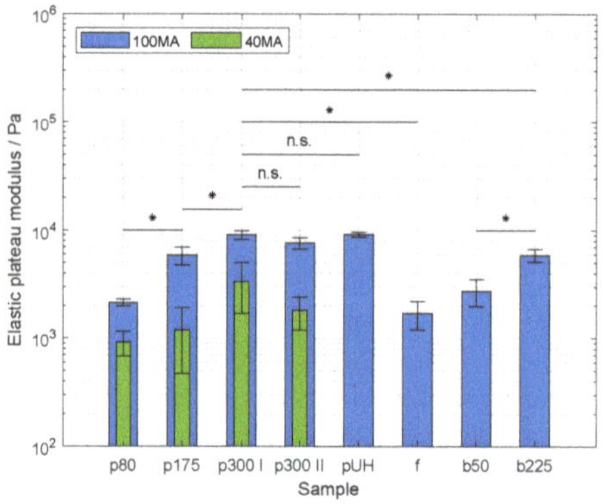

Figure 3. Elastic plateau modulus of gelatin methacryloyl (GelMA) hydrogels of various sources at 10 % (w/v). Sample nomenclature regarding the used raw materials is provided on Table 1. At a methacrylic anhydride (MAA)-to-gelatin ratio of 100 µL g^{-1} (100 MA), asterisks denote a significant difference between synthesized samples ($p < 0.05$). No significant differences are denoted with the abbreviation n.s. ($p > 0.05$). Furthermore, the elasticity of the hydrogels produced with porcine GelMA decreased significantly with decreasing MAA-to-gelatin ratio from 100 to 40 µL g^{-1} (40 MA) ($p < 0.05$). At a MAA-to-gelatin ratio of 40 µL g^{-1} (40 MA), no significant differences regarding the hydrogel elasticity were shown ($p > 0.05$). These differences are not shown for the purpose of clarity. Data are shown as mean and corresponding standard deviation. Each batch of GelMA was used for the formulation of hydrogels. Three samples of each batch were tested.

The elastic moduli of the hydrogels prepared with porcine GelMA 100 MA increased significantly with the bloom strength of the respective gelatin raw material ($p < 0.05$). The moduli were 2.15 ± 0.16 kPa, 5.85 ± 1.11 kPa, and 9.04 ± 0.85 kPa, for GelMA samples p80-100 MA, p175-100 MA, and p300 I-100 MA, respectively. The hydrogels produced with p300 II exhibited an elastic modulus of 7.60 ± 0.91 kPa, and the data did not differ significantly from the data produced with p300 I-100 MA, i.e., the same gelatin product used as raw material proceeding from a different batch. Similarly, the elastic moduli of GelMA hydrogels p300 I and p300 II did not differ significantly from that of hydrogels produced with GelMA pUH-100 MA. This effect is in accordance with the work of Aljaber et al. [15]. The study showed an increase in elastic as well as compressive moduli by increasing the bloom of the raw material from 175 g to 300 g. The increasing elasticity corresponds to higher crosslink density in the polymeric network. This resistance to the deformation is influenced by both covalent bonds and physical entanglements [33,46]. GelMA with higher bloom strength showed a higher DoF, and thus a higher amount of methacrylamide and methacrylate residues. Furthermore, as the bloom value of the protein increases, so too does the MW, which leads to an increment in the amount of physical entanglements, as well. Although significant differences in viscosity at the same gelatin concentration were measured meaning a difference in the MW and MW distribution, the missing difference by means of the elasticity of samples p300 I, p300 II, and pUH could arise from the crosslinking conditions in the present study. The irradiation dose was set to 2167 mJ cm^{-2}. This condition is much higher than the presented methods in similar studies [11,20,47]. As studied by O'Connell et al. [48], the reaction rate is proportional to the irradiance and photo-initiator concentration in free radical polymerization. As a result, the diffusivity of radicals and the accessibility of crosslinking sites are rapidly lowered by the increasing elasticity of the polymeric matrix, thus limiting the formation of covalent bonds.

Not only the elastic moduli of hydrogels made of porcine GelMA were determined, but also those of fish and bovine GelMA. The samples prepared with f-100 MA, and b50-100 MA showed elastic moduli values of 1.69 ± 0.50 kPa, and 2.75 ± 0.79 kPa, respectively. These data sets were not significantly different from each other. The elastic moduli of b225-100 MA hydrogels had a value of 5.88 ± 0.79 kPa, significantly higher than that of GelMA f-100 MA and b50-100 MA hydrogels ($p < 0.05$). The data measured from the b225-100 MA hydrogel were significantly lower than the data acquired from hydrogels p300 I, p300 II, pUH ($p < 0.05$). The lower elasticity of fish GelMA in comparison to porcine GelMA and bovine GelMA has been shown in the literature. While both Young et al. [18] and Aljaber et al. [15] state the use of porcine gelatin with 300 g bloom strength, only Young et al. mention the bloom value of bovine gelatin, i.e., 225 g. In both cases, hydrogels prepared with fish GelMA show the lowest elasticity. Due to the fact that the DoF of p80-100 MA, f-100 MA, and b50-100 MA were similar, the covalent crosslinks and chain entanglements contribute equally to the elasticity of the hydrogels. Moreover, the effect of increasing elasticity with increasing bloom strength of porcine gelatin is exhibited by the samples prepared with bovine GelMA, as well. As both bovine GelMA samples proved to have a similar DoF, the increasing elasticity of the hydrogel b225-100 MA is a consequence of the larger amount of physical entanglements due to the higher MW of the protein.

The effect of the DoF on the elasticity of hydrogels was characterized using porcine GelMA. The samples prepared with GelMA 40 MA were significantly less elastic than the counterparts produced with GelMA 100 MA ($p < 0.05$). The behavior is attributed to the fact of the lower amount of methacrylamide and methacrylate residues required for photo-crosslinking at an equal GelMA concentration. The effect of DoF on elasticity has been reported in similar studies [14,19,47,48]. Future research should include the characterization of the relationship between the properties of the GelMA backbone, i.e., MW and DoF, and hydrogel properties, i.e., elasticity. Additionally, the protein composition can also be taken into account as the protein sources vary in terms of species. Hence, variability regarding the amount of hydrophilic amino acids along the protein affects the mechanical properties of the hydrogel as well.

2.3. Microparticle Generation and Characterization

GelMA has proved to be a versatile material in a wide range of applications, e.g., three-dimensional cell culture in studies of disease and tissue engineering [49]. The formation of physical gels at room temperature imposes a challenge for the manufacturing of GelMA-containing products. Similar to gelatin solutions, GelMA solutions form physical gels due to inter- and intramolecular interactions leading to the formation of helical structures. Regarding the elasticity of physical gels, the effect is less pronounced in GelMA compared to raw gelatin; however, the transition temperature remains that of unmodified gelatin [11]. Because the gelation occurs below physiological temperature, the production of GelMA structures with techniques such as bioprinting [50], electrospinning [51], and microfluidics [52] relies on the addition of further polymers to the formulation in order to adapt the precursor solution to the particular method. Alternatively, the production equipment is heated above the gelation point [25,26]. The aim of this part of the study was the production of GelMA droplets using a microfluidic device at room temperature. For the production of droplets, fish GelMA was dissolved in ultrapure water as the protein solution did not form a physical gel at room temperature. In contrast, porcine GelMA was dissolved in 4 M urea solution to inhibit the gel formation as the process took place at room temperature, i.e., below the gel transition temperature. The processing of GelMA to hydrogel microparticles consists of two consecutive steps. Firstly, GelMA droplets are produced within an oil stream using a microfluidic device. Subsequently, the droplets in oil are covalently crosslinked to hydrogels under UV irradiation as fluid flows within light-transmitting tubing. A schematic draft of the process is provided in Figure 4. The tested samples and corresponding concentrations as well as the feed rates of both continuous and disperse phases are listed in Table 2.

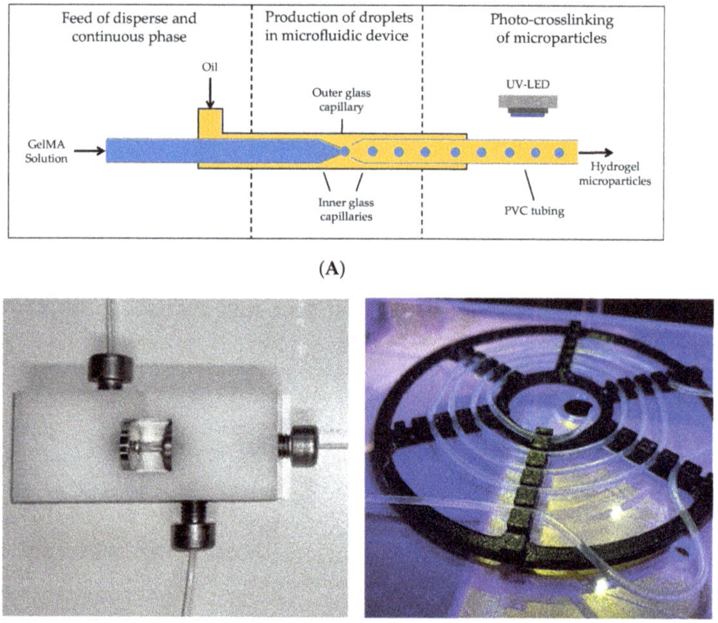

(A)

(B) (C)

Figure 4. Schematic of the microfluidic setup used in this study. (**A**) Disperse phase and continuous phase, i.e., Gelatin methacryloyl (GelMA) and sunflower seed oil, respectively, were fed into the microfluidic device using a syringe pump. The droplet production took place within an outer glass capillaries, where two inner capillaries were placed. The left and right inner capillaries had diameters of 170 µm and 340 µm, respectively. The dispersed droplets in oil were crosslinked to hydrogel microparticles under ultraviolet (UV) light-emitting diodes (LED). Figure adapted from Leister et al. [53]. The complete experimental setup was used at room temperature. (**B**) Image of the used microfluidic device. (**C**) Image of the light-transmitting tubing under UV irradiance for the crosslinking of hydrogel microparticles.

As the droplet formation process is influenced by the viscosity of both inner and outer phases, the viscosity of solutions prepared with fish GelMA and porcine GelMA, as well as the viscosity of sunflower seed oil, were measured. The data on the viscosity of GelMA solutions are shown in Figure 5. The viscosity of the oil exhibited a value of 60.5 ± 0.2 mPa s, which is shown as a line across the figure. A significant increase in viscosity was shown with an increasing concentration of both fish and porcine samples, i.e., samples f-100 MA, and p300 I-100 MA, respectively, ($p < 0.05$). The values of viscosity of the fish at the concentrations 10, 15 and 20 % (w/v) were 5.2 ± 0.1 mPa s, 9.9 ± 0.2 mPa s and 17.4 ± 0.6 mPa s, respectively. Solutions containing fish GelMA showed lower values of viscosity compared to those of porcine GelMA at the same protein concentration with values of 24.7 ± 2.0 mPa s, 61.1 ± 6.1 mPa s and 133.0 ± 19.2 mPa s. Furthermore, the viscosity of solutions of both GelMA types at 10 % (w/v) are significantly lower than the viscosity of the solution with the corresponding gelatin product shown in Figure 5.

Table 2. Composition of disperse phase as employed for the production of hydrogel microparticles as well as the feed rate and feed ratio of continuous phase to disperse phase. The feed rate of the continuous phase consisting of sunflower seed oil was set to 120 mL min^{-1}.

GelMA Sample	Concentration % (w/v)	Feed Rates mL min^{-1}	Feed Ratios x
f-100 MA	15	12, 24, 60	10, 5, 2
f-100 MA	20	12, 24, 60	10, 5, 2
p300 I-100 MA	10	12, 24, 60	10, 5, 2
p300 I-100 MA	15	12, 24, 60	10, 5, 2
p300 I-100 MA	20	8, 24, 60	15, 10, 5

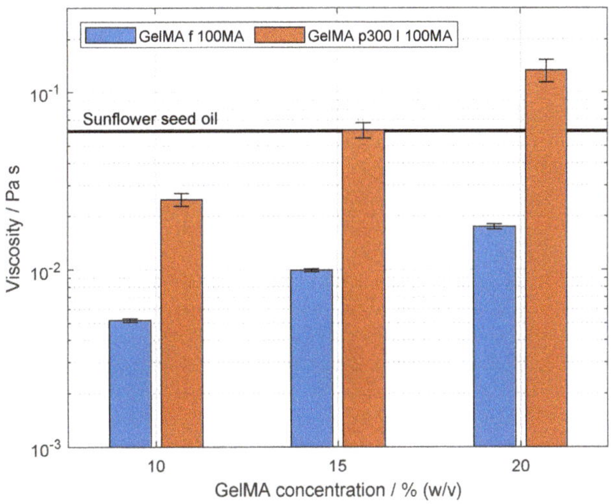

Figure 5. Viscosity of gelatin methacryloyl (GelMA) solutions and sunflower seed oil measured at room temperature. Sample nomenclature is provided in Table 2. The mean value and standard deviation of the viscosity of sunflower seed oil are shown as a black region. The viscosity of oil was measured three times from the same bulk. Fish GelMA was dissolved in ultrapure water, and porcine GelMA was dissolved in 4 M urea solution. The viscosity of both GelMA types was acquired at 10, 15 and 20 % (w/v). The viscosity of the solutions increased significantly with increasing concentration of both types of GelMA ($p < 0.05$). Significant differences were found between the viscosity of the solutions at a constant GelMA concentration ($p < 0.05$). The values are presented as mean and standard deviation. GelMA solutions were measured three times at each concentration. At a constant concentration, GelMA from an independently synthesized batch was used.

The effect of increasing porcine GelMA on the viscosity is in accordance with literature [45]. The viscosity of the solution is affected by the amount of bound water which increases with protein concentration. Additionally, the friction between protein chains and the number of physical entanglements of protein chains increases with the concentration [32,33]. The higher values of porcine GelMA solution compared to those of fish GelMA solution have not been reported in the literature, but it is expected since the viscosity of porcine gelatin solutions is higher than that of fish gelatin, as shown above. Additionally, the solution of fish GelMA did not contain urea since fish GelMA does not form a physical gel at room temperature. This fact could also account for the higher viscosity of the porcine GelMA solutions in this study as urea increases the viscosity as well [54]. The decreasing viscosity of the solution after the modification of gelatin to GelMA is in accordance with literature [16,45] and is attributed to the reduction in hydrophilic interaction between the GelMA backbone and the surrounding aqueous phase.

GelMA droplets were produced at the tip of glass capillaries within the microfluidic system. Exemplary images of the droplets at the break-up point are shown in Figure 6A,B. The droplets were formed in a co-flow configuration using the second capillary as a flow restriction to facilitate the droplet formation. Directly after droplet break-up, the particle size was determined using high-speed image acquisition and using automated image analysis to determine the droplet size. The effect of feed ratio as well as GelMA type and concentration were tested. The associated data are shown in Figure 6C. Droplets were generated with fish GelMA at 15 and 20 % (w/v). At both concentrations, increasing the feed ratio led to significantly lower droplet sizes. A similar effect was exhibited in the production of droplets with porcine GelMA solution at room temperature. For GelMA droplets at 10 and 15 % (w/v), the decreasing droplet size was significant. At 20 % (w/v) porcine GelMA, the same trend with respect to the lower concentrations was shown; however, the effect of increasing the feed ratio was not significant. In the literature, GelMA droplets for the production of microparticles have been studied at concentrations up to 10 % (w/v) [23,24,55]. Such low concentration has been used due to the thermal gelation of GelMA solutions at room temperature. Additionally, the heating of the microfluidic devices has been implemented in other studies to maintain the solutions as a liquid [25,26]. In the presented study, droplets of porcine GelMA solution with a protein concentration of 20 % (w/v) could be produced at room temperature. The processing without heating of the devices was feasible due to the presence of urea in the solution. Moreover, the effect of increasing feed rate leading to decreasing droplet size is in accordance with the literature [23,24,26,55]. This is due to the viscous drag of the oil phase in contrast to the decreasing inertial and interfacial force of the disperse phase [56]. Moreover, the study by Wang et al. [25] mentioned the higher droplet size for the solution comprising porcine GelMA compared to the solution containing fish GelMA. Furthermore, the study by Samanipour et al. [24] stated the increasing diameter of particles generated by increasing the GelMA concentration at the same feed rate. These effects were justified as due to an increase in viscosity—the former due to the higher viscosity of the protein of porcine origin and the latter due to the increment of the protein concentration. The droplet formation was in the dripping regime. In our study, this effect of the viscosity on particle size was partially exhibited, but it was not a trend overall. Additionally, the mechanism of droplet formation shifted with increasing viscosity of the protein solutions from dripping to jetting regime as shown in Figure 6A,B, respectively. In our study, the high GelMA concentration lowers the interface tension at a higher magnitude, and therefore the inner phase is more prone to forming a jet stream. Additionally, the viscosity of the oil was not considerably higher than the viscosities of the GelMA solutions. Especially for the sample at 20 % (w/v), where the viscosity of the inner phase exceeds the oil viscosity, the required feed ratios were even higher for the break-up of droplets. These conditions even lead to the widening of the jets, leading to higher droplet sizes [57].

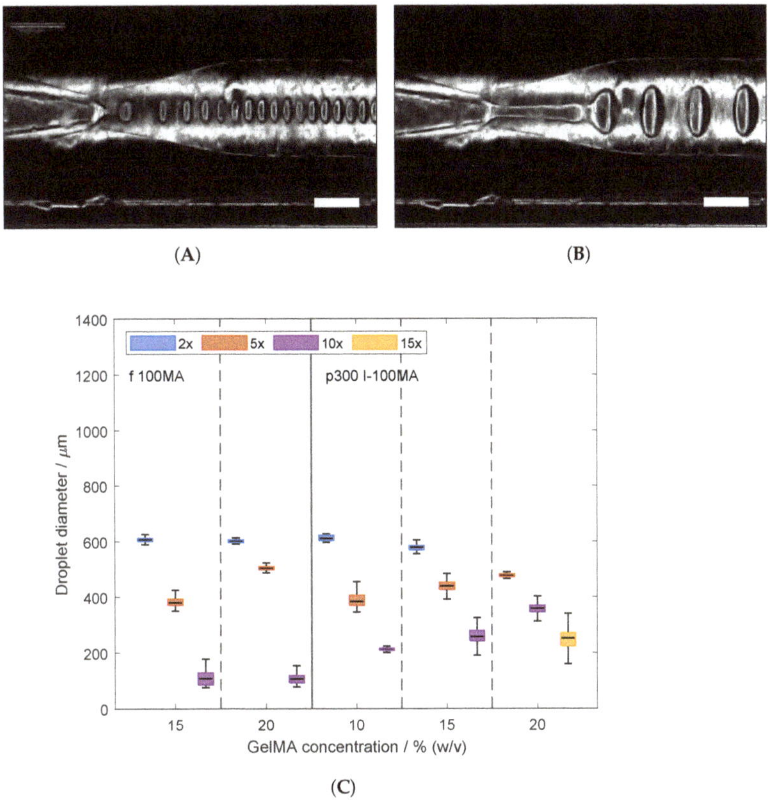

Figure 6. Production of gelatin methacryloyl (GelMA) droplets in a microfluidic device. Fish and porcine GelMA were used in the disperse phase, whereas sunflower seed oil was used in the continuous phase. Sample nomenclature is provided in Table 2. The feed rate of the continuous phase was set to 120 mL min^{-1}. (**A**,**B**) Microscopic images of the break-up points of GelMA droplets. Scale bar: 500 µm. (**A**) 15 (w/v) Fish GelMA with a feed rate of 12 mL min^{-1}, i.e., 10× feed ratio. (**B**) 20 (w/v) Fish GelMA with a feed rate of 24 mL min^{-1}, i.e., 5× feed ratio. (**C**) Droplet size of disperse phase composed of fish and porcine GelMA, i.e., samples f-100 MA and p300 I-100 MA, at different concentrations and different feed ratios. The droplet size was measured directly after formation using the "Droplet Morphometry and Velocimetry" (DMV) software [58]. The data were collected from at least 50 droplets of each GelMA sample. Values of the droplet size distribution are shown as a boxplot, where the middle line indicates the median and the edges of the boxes represent the 25 and 75 percentiles. Whiskers indicate maxima and minima within a 1.5-fold interquartile range. Moreover, the droplet size decreased with increasing feed ratio of continuous to disperse phase at each tested composition.

Particles generated at a feed ratio of 5× were collected for further analysis regarding the swelling behavior in DPBS. For this purpose, images were taken of the droplets in oil, and after equilibration in DPBS. These images are shown in Figure 7A,B, respectively. The particle diameters in both media were determined using an image processing and analysis workflow developed in Matlab®. Moreover, the volumetric swelling ratio was calculated according to Equation (2), and the associated results are shown in Figure 7C. The swelling ratio of fish GelMA particles decreased from 4.10 ± 1.00 to 1.35 ± 0.35 with increasing GelMA concentration. Similarly, the swelling behavior of particles composed of porcine GelMA decreased with increasing protein concentration. The volumetric swelling ratio of the 10 % (w/v), 15 % (w/v), and 20 % (w/v) hydrogel particles were 4.72 ± 0.77,

3.12 ± 0.05, and 2.81 ± 0.01. The effect of GelMA concentration on the swelling capacity of hydrogels has been shown in similar studies [59–61]. The swelling process is driven by the osmotic pressure difference between the aqueous phase within the polymeric network and the bulk phase. Counteracting the swelling process is the elasticity of the crosslinked network, which increases with increasing concentration of the protein [1,62]. As mentioned above, the increasing elasticity originates from the higher amount of both covalent bonds and physical entanglements [46]. The presented study shows the production of GelMA droplets and the subsequent crosslinking to hydrogel particles in a single step at room temperature. Fish GelMA and porcine GelMA were used for this purpose, including GelMA concentrations that have not been studied in the literature due to the complexity of the material and its thermal gelation at temperatures below physiological conditions. Further studies regarding droplet production and subsequent crosslinking should include a thorough characterization of the mechanisms of droplet formation including the calculation of dimensionless numbers such as the capillary and Weber number. The droplet formation is influenced by the composition of both phases, which depends on the intended application. Surfactants could be used for the stabilization of GelMA droplets, as the small molecules adsorb rapidly to newly created interfaces, and, hence, avoiding coalescence. For encapsulation of cells as well as biopharmaceuticals, fish GelMA at high concentrations could be used as it can be processed at room temperature without the use of urea as an additive. This is of significant importance as urea induces protein denaturation and cell disruption. Regarding cell delivery, research implies the biocompatibility of used surfactants; therefore, the determination of non-critical concentrations to avoid cytotoxic effects should be included. Furthermore, porcine GelMA can be implemented for the production of microcarriers. As previously reported, cells can attach to the GelMA hydrogels after the purification of GelMA. Hence, hydrogel microparticles can be used for the expansion of adherent cells. This approach increases the area-to-volume ratio of bioreactors compared to the commonly used TC flasks. Similarly, such microcarriers can be implemented for the selective differentiation of stem cells depending on the hydrogel formulation and its stiffness [22,63]. In the present study, the robustness of the production process is increased as the system is not sensitive to temperature fluctuations that could lead to the gelation of the GelMA-containing solutions. Particularly, hydrogel particles at concentrations of 15 and 20 % (w/v) were produced, higher than previously reported in the literature. Hence, stiffer hydrogels could be prepared, which is required for the differentiation and expansion of certain cell phenotypes. Additionally, as GelMA hydrogels can be enzymatically degraded, cells can be easily harvested and separated from the aqueous media.

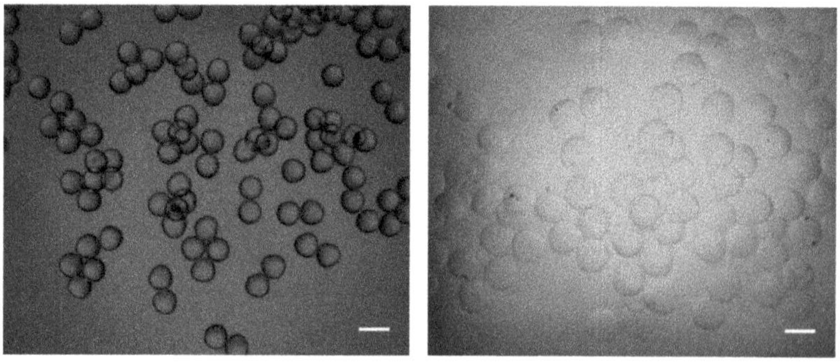

(A) **(B)**

Figure 7. *Cont.*

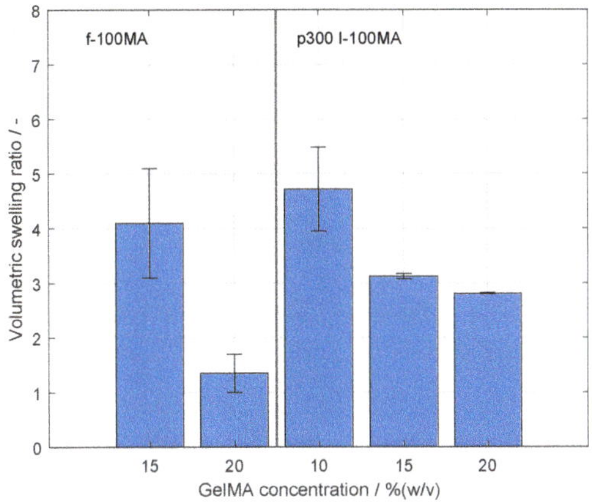

(C)

Figure 7. Volumetric swelling ratio of gelatin mathacryloyl (GelMA) microparticles. Fish and porcine GelMA particles were produced at a feed ratio of continuous to disperse phase of 5×. Sample nomenclature is provided in Table 2. (**A**,**B**) Microscopic images of GelMA microparticles. Scale bar: 1000 µm. (**A**) 20 (*w*/*v*) fish GelMA particles collected after photo-crosslinking in the sunflower seed oil phase. (**B**) 20 (*w*/*v*) fish GelMA after swelling to equilibrium in Dulbecco's phosphate-buffered saline (DPBS). (**C**) Volumetric swelling ratio (VSR) of GelMA particles composed of fish and porcine GelMA, i.e., samples f-100 MA and p300 I-100 MA. The swelling ratio was calculated according to Equation (2). Moreover, the swelling behavior decreased with increasing GelMA concentration of each type. Values are presented as mean and corresponding standard deviation. The particles size for the calculation of the swelling behavior was detected from at least 40 particles of each sample.

3. Conclusions

Gelatin methacryloyl is well established for the formulation of hydrogels, finding application in biotechnology, tissue engineering, and biofabrication. The studies on the manufacturing progress have been focused on the use of porcine gelatin as raw material. However, a comparison across the literature of the effects of raw materials on the final product is challenging, as various approaches are employed including differences in the composition of reaction buffer, pH, and buffering capacity. Additionally, the molecular weight of the protein has not been the focus of the reports. In the first part of this study, we produced GelMA at room temperature applying the previously reported method, where urea is used in the reaction buffer to inhibit the thermal gelation of the protein solution. This principle was successfully applied to the operation with a variety of raw materials other than the one used in our previous report. Moreover, insights were gained into the effects of batch-to-batch variability, as two different batches of the same porcine gelatin product were used, and the degree of functionalization of the two products differed. Furthermore, the bloom value, and hence the molecular weight of porcine gelatin, proved to have a significant impact on the degree of functionalization, which decreased with decreasing bloom strength. Additionally, fish gelatin and two bovine gelatin products with varying bloom values were modified to GelMA. The DoF of the products was lower than that of porcine GelMA with high bloom. Our findings underline the significant impact of the raw material on the processing of gelatin to GelMA. As the reactants are not miscible, stirring is required to disperse methacrylic anhydride droplets in the gelatin solution. The protein adsorption at the interface, where the reaction takes place, depends on the molecular weight, molecular weight distribution, as well as protein charge. Therefore, the optimization of process parameters is highly dependent on raw materials, and a developed process cannot simply be transferred to the operation with a different raw material. Further research should take into account the properties of gelatin at the interface to the reactant. Furthermore, the produced GelMA materials of varying species and bloom strength were used for the formulation of hydrogels, and the elasticity of the polymeric network was characterized. By variation in the degree of functionalization, GelMA hydrogels showed increasing elastic moduli. In addition, the molecular weight of the raw materials affected the elasticity. Decreasing the bloom strength of GelMA hydrogels led to less elastic behavior.

As a second part of the presented study, two GelMA types were used for the microfluidic generation of droplets and the subsequent crosslinking to hydrogel particles. Both processes were performed on a single pipeline at room temperature. Therefore, fish GelMA was dissolved in water and porcine GelMA was dissolved in urea solution to maintain a solution at room temperature. The droplets could be produced at higher GelMA concentrations than that found in the literature. Moreover, the droplet size decreased with an increasing feed ratio of the continuous to disperse phase at each tested concentration. The swelling behavior of crosslinked particles was characterized. Hydrogel particles exhibited a higher swelling degree with decreasing GelMA concentration. Future studies should include the formulation of GelMA using surfactants as well as the adaptation of the modular microfluidic device to stabilize the droplets by other means. Additionally, further understanding of the mechanisms of GelMA droplet generation is required, specifically how process parameters affect dripping and jetting regimes.

4. Materials and Methods

4.1. Synthesis and Characterization of Gelatin-Methacryloyl

4.1.1. Precursor Solution for the Synthesis of Gelatin-Methacryloyl

Gelatin products were purchased from Sigma-Aldrich (St. Louis, MI, USA), and the relevant product information is listed in Table 1. The buffer for the dissolution and synthesis of GelMA was prepared following the method by Grijalva Garces et al. [14]. Buffer components for the synthesis of GelMA were acquired from Merck (Darmstadt, Germany). The buffer composition was 0.25 M carbonate bicarbonate (CB) and 4 M urea. After the dissolution of the salts, the pH of the solution was adjusted to pH 9 using 1 M sodium hydroxide (NaOH) or 1 M hydrochloridic acid (HCl). Gelatin was dissolved in the synthesis buffer to a concentration of 10 % (w/v) at room temperature under stirring.

4.1.2. Rheological Characterization of Gelatin Solutions

The viscosity of the solutions at 10 % (w/v) gelatin in synthesis buffer was measured using a rotational rheometer (Physica MCR301, Anton Paar, Graz, Austria). For the characterization of the gelatin solutions, the configuration of the rheometer included a cone-plate geometry (diameter 60 mm, cone angle 0.5°), and a solvent trap in order to avoid sample drying during the measurements. The viscosity was determined within the shear rate range of 0.5 to 500 s^{-1}.

The viscosity values of the gelatin solutions provided below were determined as triplicates. Each value was measured from an independently prepared solution. Values are shown as mean and standard deviation.

4.1.3. Synthesis and Purification

The synthesis was performed at room temperature under stirring. The reaction was started by adding methacrylic anhydride (MAA, 94 %, Sigma-Aldrich) to the gelatin solutions. The MAA-to-gelatin ratio was 100 µL g^{-1} for all gelatin samples. Additionally, porcine gelatin was modified to GelMA with a ratio of 40 µL g^{-1}. Throughout this study, sample nomenclature includes 100 MA, and 40 MA depending on the used MAA-to-gelatin ratio. The reaction was carried out for 60 min. The process was terminated by two-fold dilution with ultrapure water and a subsequent pH adjustment to pH 7.4. The diluted reaction mixture was then dialyzed with a 3.5 kDa molecular weight cut-off tubing (Thermo Fisher Scientific, Waltham, MA, USA) in an ultrapure water reservoir. This purification took place for 4 days at 40 °C. GelMA solutions were frozen at −80 °C overnight and lyophilized. Solid GelMA samples were stored at room temperature until further use.

4.1.4. Determination of Degree of Functionalization

The degree of functionalization (DoF) of GelMA samples was determined based on the method by Habeeb [38]. Therefore, glycine (Sigma-Aldrich), gelatin materials, and GelMA samples were dissolved in ultrapure water. Glycine standards for the determination of a standard curve were prepared at 3, 5, 8, 10 and 20 µg mL^{-1}. Gelatin references and GelMA samples were dissolved in ultrapure water at 0.1, 0.3, 0.5 and 0.8 mg mL^{-1}. A 0.1 M CB buffer at pH 8.5 was used as a reaction buffer containing 0.01 % (*w*/*v*) trinitrobenzene-sulfonic (TNBS) acid (Sigma-Aldrich). A volume of 250 mL of the TNBS reagent solution was mixed with an equal volume of the gelation as well as GelMA samples. Incubation followed for 2 h at 40 °C. The reaction was terminated by addition of 250 µL of a 10 % (*w*/*v*) sodium dodecyl sulfate (Sigma-Aldrich) solution and 125 µL of a 1 M HCl solution. A microplate reader (infiniteM200, Tecan Group, Männedorf, Switzerland) for the measurement of the sample absorbance at 335 nm. The concentration of free amines in the samples was determined in comparison to a glycine calibration curve and normalized to the respective gelatin concentration. The DoF was calculated according to Equation (1). The difference between the number of free amino groups present in gelatin ($c_{NH_2,gelatin}$), i.e., before the functionalization, and the amount in the produced GelMA ($c_{NH_2,GelMA}$), i.e., after the reaction, was divided by the number of free amines in the raw gelatin.

$$\text{Degree of Functionalization } \% = \frac{c_{NH_2,gelatin} - c_{NH_2,GelMA}}{c_{NH_2,gelatin}} \quad (1)$$

The production and characterization of GelMA samples consisted of three experimental runs using each gelatin source. The measurement of absorbance in order to determine the DoF was performed for each independently synthesized batch. The values are shown as mean and standard deviation.

4.2. Hydrogel Characterization

4.2.1. Precursor Solution for the Production of Hydrogels

GelMA samples synthesized from different sources and MAA-to-gelatin ratios were used for hydrogel preparation. For this purpose, a solution containing 0.1 % (*w*/*v*) the photo-initiator lithium phenyl-2,4,6-trimethylbenzoylphosphinate (LAP, Sigma-Aldrich) was prepared in Dulbecco's phosphate-buffered saline (DPBS, without calcium and magnesium, 1×, pH 7.4, Thermo Fisher Scientific). Subsequently, the lyophilized material was dissolved to 10 % (*w*/*v*) in LAP containing DPBS at 40 °C. GelMA solutions were transferred to cylindrical polytetrafluoroethylene (PTFE) molds (diameter 10 mm, height 3 mm) by pipetting a volume of 235 µL. The samples were then crosslinked to hydrogels by exposure to an ultraviolet (UV) light-emitting diode (LED, 365 nm, OSRAM, Munich, Germany) with an irradiation intensity of 12 mW cm^{-2} for 3 min. GelMA hydrogels were equilibrated in DPBS until further analysis.

4.2.2. Mechanical Characterization

The viscoelastic properties of hydrogels were characterized using a rotational rheometer Physica MCR301. A plate-plate geometry (diameter 10 mm) and a solvent trap were part of the configuration of the rheometer. The cylindrical hydrogels were placed on the bottom plate and the top plate was positioned to a gap height of 2.5 mm. Storage and loss moduli were measured within the linear viscoelastic (LVE) regime covering the frequency range of 0.5 to 50 rad s^{-1}. A constant stress amplitude was set to 0.5 Pa. The data of the mechanical characterization were acquired from three experimental runs with three samples each. For each run, GelMA hydrogels from independently produced batches were prepared. The data are presented below as the mean and standard deviation of the elastic plateau modulus.

4.3. Microparticle Fabrication and Characterization

4.3.1. Precursor Solution for the Production of Microparticles

The manufacturing of GelMA hydrogel microparticles at room temperature was investigated. For these experiments, porcine GelMA, i.e., p300 I, and fish GelMA at a reactant ratio of 100 µL MAA per gram gelatin were used. A 4 M urea solution was prepared for the dissolution of porcine GelMA. Subsequently, the photoinitiator LAP was dissolved to 1 % (w/v). Lyophilized porcine GelMA was added to the mixture and dissolved under stirring. The precursor solution containing fish GelMA was prepared by dissolving the lyophilized material in ultrapure water containing 1 % (w/v) LAP. Samples from both sources were prepared at 10, 15 and 20 % (w/v). The precursor solutions were protected from light prior to their use in the disperse phase in the microfluidic device as described below. The tested samples and the used concentrations are summarized in Table 2. Sunflower seed oil (Sigma-Aldrich) was employed as continuous phase.

4.3.2. Rheological Characterization of Disperse and Continuous Phase

The shear-rate dependent viscosity of sunflower seed oil and the GelMA solutions was characterized as described in Section 4.1.2. For these measurements, the solvent trap served additionally as protection from light to avoid photo-crosslinking during the measurement.

The viscosity of the oil phase was measured as triplicates from the same bulk. The value is presented as the mean and associated standard deviation. Regarding the rheological characterization of the GelMA solutions, the viscosity was determined at each concentration as triplicates. Each value was acquired from the prepared solution from each independently synthesized batch.

4.3.3. Microfluidic Production of Droplets and Crosslinking to Microparticles

The setup for the production of GelMA droplets and the subsequent crosslinking process is shown schematically in Figure 4A. The disperse and continuous phases, i.e., GelMA solutions and sunflower seed oil, respectively, were filled in high-precision glass syringes (SETonic, Ilmenau, Germany). A Nemesys syringe pump was used to control the feed rates using the software QmixElements v20140605 (both CETONI, Korbussen, Germany). The rate of the oil phase was set to a constant value of 120 mL min^{-1} for all experiments. The tested GelMA concentrations and the corresponding feed rates, as well as the feed ratio, defined as the ratio of the feed rate of the continuous phase to that of the disperse phase, are listed in Table 2.

For the formation of GelMA droplets, a microfluidic device with glass capillaries was employed, as shown in Figure 4B. A detailed description of the equipment is provided by Leister et al. [53]. The setup consisted of one outer and two inner glass capillaries (World Precision Instruments, Friedberg, Germany). The inner capillaries (outer diameter 1 mm, inner diameter 0.58 mm) were modified by pulling with a micro-pipette puller (P-1000, Sutter Instruments, Novato, CA, USA). The tip diameter of the capillary for the disperse phase was 170 µm, while the tip diameter of the second capillary used as the outlet of dispersed droplets in oil was 340 µm. The inner capillaries as well as the outer capillary (length 15 mm, inner diameter 1.56 mm) were treated with 2-[methoxy(polyethyleneoxy)6-9-propyl]tris(dime thylamino)silane (Gelest Inc., Morrisville, PA, USA) in order to render the surface hydrophobic. The capillaries were attached in the polyoxymethylene (POM) module as published by Bandulasena et al. [64]. The distance between inner capillaries was set to 170 µm. The outlet of the microfluidic device was connected to a polyvinyl chloride (PVC) tubing (outer diameter 1.8 mm, inner diameter 1 mm, Deutsch & Neumann, Hennigsdorf, Germany). The tubing was arranged as a loop under four UV LEDs (OSRAM), where photo-crosslinking took place with a total irradiation intensity of 25.6 mW cm^{-2}. An image of the tubing placed under UV light is provided in Figure 4C. The produced hydrogel microparticles in oil were collected and stored at room temperature until further analysis.

4.3.4. Determination of Droplet Size

Image sequences of the formation of droplets at the break-up point were acquired using a monochrome camera (DMK 33U, The Imaging Source Europe, Bremen, Germany) equipped with a 1× lens (TMN 1.0/50, The Imaging Source Europe) using the software IC capture V2.5 (The Imaging Source Europe). The acquisition rate was set to 10 frames per second. The resulting droplet sizes were determined using "Droplet Morphometry and Velocimetry" (DMV) software [58] by analyzing at least 150 frames. The coordinates of the center of each droplet, as well as the respective droplet diameter in each frame, were exported. Detected objects with a center below or above the longitudinal axis of the capillaries at the direct proximity of the break-up point were considered outliers and removed from the distributions. The data distribution of at least 50 droplets per sample is shown below as box plots including median, upper, and lower quartile, as well as maxima and minima within a 1.5-fold interquartile range.

4.3.5. Determination of Hydrogel Swelling Behavior

For the determination of the swelling behavior of the hydrogel microparticles, the samples from both GelMA sources produced with a feed ratio of 5× were collected. DPBS was added to the particle/oil mixture and centrifuged at 500 rcf. The excessive oil and DPBS were removed keeping the particles in the bottom of the centrifuge tube. The microparticles were suspended in fresh DPBS and centrifuged. This wash series was performed four times. Microparticles were then equilibrated overnight in DPBS. Hydrogel microparticles and the corresponding oil phase and DPBS were placed on a microscopy slide for image acquisition. The imaging setup consisted of a monochrome camera (Genie Nano M2420 Mono, Teledyne Dalsa, Waterloo, ON, Canada) equipped with a 10× objective (Nikon, Tokyo, Japan). For the quantification of the particle size, an image processing and analysis workflow was developed using Matlab® R2023a (TheMathWorks Inc., Natick, MA, USA) with the library Image Processing Toolbox 11.7. The variation in pixel intensities compared to the nine-by-nine surrounding pixels was analyzed using the local entropy filter. The output images contain high-intensity values in the regions of high-intensity variation, i.e., the interface between microparticles and bulk media. Particle diameters were detected on said output processed images. The swelling behavior of the hydrogel microparticles was characterized by the ratio of media diameter after swelling in DPBS ($d_{p,DPBS}$) to the median diameter in oil ($d_{p,Oil}$), as shown in Equation (2). The data sets presented below correspond to the swelling ratio of at least 40 hydrogel particles of each sample.

$$\text{Volumetric Swelling Ratio }\% = \frac{d^3_{p,DPBS}}{d^3_{p,Oil}} \tag{2}$$

4.4. *Data Handling and Statistical analysis*

Image processing, data evaluation, data visualization, and statistical analysis of the data sets were performed with Matlab® R2023a (TheMathWorks Inc., Natick, MA, USA). One-way analysis of variance (ANOVA) was performed in order to determine significant differences. A p-value below 0.05 was considered as statistically significant.

Author Contributions: Conceptualization, D.G.G.; methodology, D.G.G. and N.L.; Investigation, D.G.G., L.J.A. and J.E.; formal analysis, D.G.G. and N.L.; data curation, D.G.G.; software, D.G.G.; writing—original draft preparation, D.G.G.; writing—review and editing, L.J.A., J.E., N.L. and J.H.; visualization, D.G.G.; supervision, J.H.; project administration, J.H.; funding acquisition, J.H. All authors have read and agreed to the published version of the manuscript.

Funding: This work was funded by the German Federal Ministry of Education and Research (BMBF) as project SOP-Bioprint under contract number 13XP5071B and supported by the KIT-Publication Fund of the Karlsruhe Institute of Technology.

Institutional Review Board Statement: Not applicable.

Informed Consent Statement: Not applicable.

Data Availability Statement: The raw data supporting the conclusions of this article as well as the written codes for Matlab® will be made available on request. Inquiries can be directed to the corresponding author.

Acknowledgments: We would like to thank Katharina Jonas for her valuable contribution in the form of preparation of the glass capillaries, and Goran Vladisavljević for providing the microfluidic module used in this study. Additionally, the authors are thankful for the thorough review of the manuscript by Svenja Strauß. We acknowledge support by the KIT-Publication Fund of the Karlsruhe Institute of Technology.

Conflicts of Interest: The authors declare that they have no known competing financial interests or personal relationships that could have appeared to influence the work reported in this paper.

Abbreviations

The following abbreviations are used in this manuscript:

CB	Carbonate-Bicarbonate
DoF	Degree of functionalization
DPBS	Dulbecco's phosphate-buffered saline
ECM	Extracellular matrix
GelMA	Gelatin-Methacryloyl
IEP	Isoelectric point
LED	Light-emitting diode
MAA	Methacrylic anhydride
MW	Molecular weight
RT	Room Temperature
TC	Tissue culture
TE	Tissue engineering
TNBS	2,4,6-Trinitrobenzenesulfonic acid solution
UV	Ultraviolet
VSR	Volumetric swelling ratio

References

1. Peppas, N.A.; Hilt, J.Z.; Khademhosseini, A.; Langer, R. Hydrogels in biology and medicine: From molecular principles to bionanotechnology. *Adv. Mater.* **2006**, *18*, 1345–1360. [CrossRef]
2. Kunkel, J.; Asuri, P. Function, Structure, and Stability of Enzymes Confined in Agarose Gels. *PLoS ONE* **2014**, *9*, e86785. [CrossRef]
3. Gutiérrez, M.C.; García-Carvajal, Z.Y.; Jobbágy, M.; Yuste, L.; Rojo, F.; Abrusci, C.; Catalina, F.; Del Monte, F.; Ferrer, M.L. Hydrogel scaffolds with immobilized bacteria for 3D cultures. *Chem. Mater.* **2007**, *19*, 1968–1973. [CrossRef]
4. Takei, T.; Ikeda, K.; Ijima, H.; Kawakami, K. Fabrication of poly(vinyl alcohol) hydrogel beads crosslinked using sodium sulfate for microorganism immobilization. *Process Biochem.* **2011**, *46*, 566–571. [CrossRef]
5. Tibbitt, M.W.; Anseth, K.S. Hydrogels as extracellular matrix mimics for 3D cell culture. *Biotechnol. Bioeng.* **2009**, *103*, 655–663. [CrossRef]
6. Daly, A.C.; Riley, L.; Segura, T.; Burdick, J.A. Hydrogel microparticles for biomedical applications. *Nat. Rev. Mater.* **2020**, *5*, 20–43. [CrossRef]
7. Murphy , S.V.; Atala, A. 3D bioprinting of tissues and organs. *Nat. Biotechnol.* **2014**, *32*, 773–785. [CrossRef] [PubMed]
8. Lee, K.Y.; Mooney, D.J. Hydrogels for Tissue Engineering. *Chem. Rev.* **2001**, *101*, 1869–1880. [CrossRef]
9. Schrieber, R.; Gareis, H. *Gelatine Handbook*; Wiley: Weinheim, Germany, 2007. [CrossRef]
10. Karim, A.A.; Bhat, R. Fish gelatin: Properties, challenges, and prospects as an alternative to mammalian gelatins. *Food Hydrocoll.* **2009**, *23*, 563–576. [CrossRef]
11. Van Den Bulcke, A.I.; Bogdanov, B.; De Rooze, N.; Schacht, E.H.; Cornelissen, M.; Berghmans, H. Structural and rheological properties of methacrylamide modified gelatin hydrogels. *Biomacromolecules* **2000**, *1*, 31–38. [CrossRef] [PubMed]
12. Lee, B.H.; Shirahama, H.; Cho, N.J.; Tan, L.P. Efficient and controllable synthesis of highly substituted gelatin methacrylamide for mechanically stiff hydrogels. *RSC Adv.* **2015**, *5*, 106094–106097. [CrossRef]
13. Shirahama, H.; Lee, B.H.; Tan, L.P.; Cho, N.J. Precise Tuning of Facile One-Pot Gelatin Methacryloyl (GelMA) Synthesis. *Sci. Rep.* **2016**, *6*, 31036. [CrossRef] [PubMed]
14. Grijalva Garces, D.; Radtke, C.P.; Hubbuch, J. A Novel Approach for the Manufacturing of Gelatin-Methacryloyl. *Polymers* **2022**, *14*, 5424. [CrossRef] [PubMed]

15. Aljaber, M.B.; Verisqa, F.; Keskin-Erdogan, Z.; Patel, K.D.; Chau, D.Y.; Knowles, J.C. Influence of Gelatin Source and Bloom Number on Gelatin Methacryloyl Hydrogels Mechanical and Biological Properties for Muscle Regeneration. *Biomolecules* **2023**, *13*, 811. [CrossRef]
16. Sewald, L.; Claaßen, C.; Götz, T.; Claaßen, M.H.; Truffault, V.; Tovar, G.E.; Southan, A.; Borchers, K. Beyond the Modification Degree: Impact of Raw Material on Physicochemical Properties of Gelatin Type A and Type B Methacryloyls. *Macromol. Biosci.* **2018**, *18*, 1–10. [CrossRef] [PubMed]
17. Pahoff, S.; Meinert, C.; Bas, O.; Nguyen, L.; Klein, T.J.; Hutmacher, D.W. Effect of gelatin source and photoinitiator type on chondrocyte redifferentiation in gelatin methacryloyl-based tissue-engineered cartilage constructs. *J. Mater. Chem. B* **2019**, *7*, 1761–1772. [CrossRef]
18. Young, A.T.; White, O.C.; Daniele, M.A. Rheological Properties of Coordinated Physical Gelation and Chemical Crosslinking in Gelatin Methacryloyl (GelMA) Hydrogels. *Macromol. Biosci.* **2020**, *20*, 2000183. [CrossRef]
19. Lee, B.H.; Lum, N.; Seow, L.Y.; Lim, P.Q.; Tan, L.P. Synthesis and characterization of types A and B gelatin methacryloyl for bioink applications. *Materials* **2016**, *9*, 797. [CrossRef]
20. Pepelanova, I.; Kruppa, K.; Scheper, T.; Lavrentieva, A. Gelatin-Methacryloyl (GelMA) Hydrogels with Defined Degree of Functionalization as a Versatile Toolkit for 3D Cell Culture and Extrusion Bioprinting. *Bioengineering* **2018**, *5*, 55. [CrossRef]
21. Tavassoli, H.; Alhosseini, S.N.; Tay, A.; Chan, P.P.; Weng Oh, S.K.; Warkiani, M.E. Large-scale production of stem cells utilizing microcarriers: A biomaterials engineering perspective from academic research to commercialized products. *Biomaterials* **2018**, *181*, 333–346. [CrossRef]
22. Ng, E.X.; Wang, M.; Neo, S.H.; Tee, C.A.; Chen, C.H.; Van Vliet, K.J. Dissolvable Gelatin-Based Microcarriers Generated through Droplet Microfluidics for Expansion and Culture of Mesenchymal Stromal Cells. *Biotechnol. J.* **2021**, *16*, 2000048. [CrossRef] [PubMed]
23. Jung, J.; Oh, J. Swelling characterization of photo-cross-linked gelatin methacrylate spherical microgels for bioencapsulation. *e-Polymers* **2014**, *14*, 161–168. [CrossRef]
24. Samanipour, R.; Wang, Z.; Ahmadi, A.; Kim, K. Experimental and computational study of microfluidic flow-focusing generation of gelatin methacrylate hydrogel droplets. *J. Appl. Polym. Sci.* **2016**, *133*, 43701. [CrossRef]
25. Wang, Z.; Tian, Z.; Menard, F.; Kim, K. Comparative study of gelatin methacrylate hydrogels from different sources for biofabrication applications. *Biofabrication* **2017**, *9*, 044101. [CrossRef] [PubMed]
26. Tang, T.; Liu, C.; Min, Z.; Cai, W.; Zhang, X.; Li, W.; Zhang, A. Microfluidic Fabrication of Gelatin Acrylamide Microgels through Visible Light Photopolymerization for Cell Encapsulation. *ACS Appl. Bio Mater.* **2023**, *6*, 2496–2504. [CrossRef] [PubMed]
27. Zou, Q.; Habermann-Rottinghaus, S.M.; Murphy, K.P. Urea effects on protein stability: Hydrogen bonding and the hydrophobic effect. *Proteins Struct. Funct. Genet.* **1998**, *31*, 107–115. [CrossRef]
28. Stumpe, M.C.; Grubmüller, H. Interaction of urea with amino acids: Implications for urea-induced protein denaturation. *J. Am. Chem. Soc.* **2007**, *129*, 16126–16131. [CrossRef] [PubMed]
29. Das, A.; Mukhopadhyay, C. Urea-mediated protein denaturation: A consensus view. *J. Phys. Chem. B* **2009**, *113*, 12816–12824. [CrossRef]
30. Leuenberger, B.H. Investigation of viscosity and gelation properties of different mammalian and fish gelatins. *Top. Catal.* **1991**, *5*, 353–361. [CrossRef]
31. Van Den Bosch, E.; Gielens, C. Gelatin degradation at elevated temperature. *Int. J. Biol. Macromol.* **2003**, *32*, 129–138. [CrossRef]
32. Münstedt, H.; Schwarzl, F.R. *Deformation and Flow of Polymeric Materials*; Springer: Berlin/Heidelberg, Germany, 2014. [CrossRef]
33. Osswald, T.; Rudolph, N. *Polymer Rheology*; Carl Hanser Verlag GmbH & Co. KG: München, Germany, 2014; pp. 101–141. [CrossRef]
34. Hoch, E.; Hirth, T.; Tovar, G.E.; Borchers, K. Chemical tailoring of gelatin to adjust its chemical and physical properties for functional bioprinting. *J. Mater. Chem. B* **2013**, *1*, 5675–5685. [CrossRef]
35. Loessner, D.; Meinert, C.; Kaemmerer, E.; Martine, L.C.; Yue, K.; Levett, P.A.; Klein, T.J.; Melchels, F.P.; Khademhosseini, A.; Hutmacher, D.W. Functionalization, preparation and use of cell-laden gelatin methacryloyl-based hydrogels as modular tissue culture platforms. *Nat. Protoc.* **2016**, *11*, 727–746. [CrossRef]
36. Wade, R.J.; Burdick, J.A. Engineering ECM signals into biomaterials. *Mater. Today* **2012**, *15*, 454–459. [CrossRef]
37. Ruedinger, F.; Lavrentieva, A.; Blume, C.; Pepelanova, I.; Scheper, T. Hydrogels for 3D mammalian cell culture: A starting guide for laboratory practice. *Appl. Microbiol. Biotechnol.* **2015**, *99*, 623–636. [CrossRef]
38. Habeeb, A.F. Determination of free amino groups in proteins by trinitrobenzenesulfonic acid. *Anal. Biochem.* **1966**, *14*, 328–336. [CrossRef]
39. Duconseille, A.; Andueza, D.; Picard, F.; Santé-Lhoutellier, V.; Astruc, T. Variability in pig skin gelatin properties related to production site: A near infrared and fluorescence spectroscopy study. *Food Hydrocoll.* **2017**, *63*, 108–119. [CrossRef]
40. Yue, K.; Trujillo-de Santiago, G.; Alvarez, M.M.; Tamayol, A.; Annabi, N.; Khademhosseini, A. Synthesis, properties, and biomedical applications of gelatin methacryloyl (GelMA) hydrogels. *Biomaterials* **2015**, *73*, 254–271. [CrossRef] [PubMed]
41. Olijve, J.; Mori, F.; Toda, Y. Influence of the molecular-weight distribution of gelatin on emulsion stability. *J. Colloid Interface Sci.* **2001**, *243*, 476–482. [CrossRef]
42. O'Sullivan, J.; Murray, B.; Flynn, C.; Norton, I. The effect of ultrasound treatment on the structural, physical and emulsifying properties of animal and vegetable proteins. *Food Hydrocoll.* **2016**, *53*, 141–154. [CrossRef]

43. Tadros, T.F. *Volume 1 Interfacial Phenomena and Colloid Stability, Basic Principles*; De Gruyter: Berlin, Germany, 2015. [CrossRef]
44. Tadros, T.F. *Volume 2 Interfacial Phenomena and Colloid Stability, Industrial Applications*; De Gruyter: Berlin, Germany, 2015. [CrossRef]
45. Hoch, E.; Schuh, C.; Hirth, T.; Tovar, G.E.; Borchers, K. Stiff gelatin hydrogels can be photo-chemically synthesized from low viscous gelatin solutions using molecularly functionalized gelatin with a high degree of methacrylation. *J. Mater. Sci. Mater. Med.* **2012**, *23*, 2607–2617. [CrossRef]
46. Anseth, K.S.; Bowman, C.N.; Brannon-Peppas, L. Mechanical properties of hydrogels and their experimental determination. *Biomaterials* **1996**, *17*, 1647–1657. [CrossRef]
47. Nichol, J.W.; Koshy, S.T.; Bae, H.; Hwang, C.M.; Yamanlar, S.; Khademhosseini, A. Cell-laden microengineered gelatin methacrylate hydrogels. *Biomaterials* **2010**, *31*, 5536–5544. [CrossRef]
48. O'Connell, C.D.; Zhang, B.; Onofrillo, C.; Duchi, S.; Blanchard, R.; Quigley, A.; Bourke, J.; Gambhir, S.; Kapsa, R.; Di Bella, C.; et al. Tailoring the mechanical properties of gelatin methacryloyl hydrogels through manipulation of the photocrosslinking conditions. *Soft Matter* **2018**, *14*, 2142–2151. [CrossRef] [PubMed]
49. Klotz, B.J.; Gawlitta, D.; Rosenberg, A.J.; Malda, J.; Melchels, F.P. Gelatin-Methacryloyl Hydrogels: Towards Biofabrication-Based Tissue Repair. *Trends Biotechnol.* **2016**, *34*, 394–407. [CrossRef] [PubMed]
50. Melchels, F.P.W.; Dhert, W.J.A.; Hutmacher, D.W.; Malda, J. Development and characterisation of a new bioink for additive tissue manufacturing. *J. Mater. Chem. B* **2014**, *2*, 2282. [CrossRef]
51. Yang, Y.; Xu, T.; Zhang, Q.; Piao, Y.; Bei, H.P.; Zhao, X. Biomimetic, Stiff, and Adhesive Periosteum with Osteogenic–Angiogenic Coupling Effect for Bone Regeneration. *Small* **2021**, *17*, 2006598. [CrossRef] [PubMed]
52. Colosi, C.; Shin, S.R.; Manoharan, V.; Massa, S.; Costantini, M.; Barbetta, A.; Dokmeci, M.R.; Dentini, M.; Khademhosseini, A. Microfluidic Bioprinting of Heterogeneous 3D Tissue Constructs Using Low-Viscosity Bioink. *Adv. Mater.* **2016**, *28*, 677–684. [CrossRef] [PubMed]
53. Leister, N.; Yan, C.; Karbstein, H.P. Oil Droplet Coalescence in W/O/W Double Emulsions Examined in Models from Micrometer- to Millimeter-Sized Droplets. *Colloids Interfaces* **2022**, *6*, 12. [CrossRef]
54. Halonen, S.; Kangas, T.; Haataja, M.; Lassi, U. Urea-Water-Solution Properties: Density, Viscosity, and Surface Tension in an Under-Saturated Solution. *Emiss. Control Sci. Technol.* **2017**, *3*, 161–170. [CrossRef]
55. Huang, J.; Fu, D.; Wu, X.; Li, Y.; Zheng, B.; Liu, Z.; Zhou, Y.; Gan, Y.; Miao, Y.; Hu, Z. One-step generation of core–shell biomimetic microspheres encapsulating double-layer cells using microfluidics for hair regeneration. *Biofabrication* **2023**, *15*, 025007. [CrossRef]
56. Guerrero, J.; Chang, Y.W.; Fragkopoulos, A.A.; Fernandez-Nieves, A. Capillary-Based Microfluidics—Coflow, Flow-Focusing, Electro-Coflow, Drops, Jets, and Instabilities. *Small* **2020**, *16*, 1904344. [CrossRef]
57. Utada, A.S.; Fernandez-Nieves, A.; Stone, H.A.; Weitz, D.A. Dripping to jetting transitions in coflowing liquid streams. *Phys. Rev. Lett.* **2007**, *99*, 094502. [CrossRef]
58. Basu, A.S. Droplet morphometry and velocimetry (DMV): A video processing software for time-resolved, label-free tracking of droplet parameters. *Lab Chip* **2013**, *13*, 1892–1901. [CrossRef] [PubMed]
59. Schuurman, W.; Levett, P.A.; Pot, M.W.; van Weeren, P.R.; Dhert, W.J.A.; Hutmacher, D.W.; Melchels, F.P.W.; Klein, T.J.; Malda, J. Gelatin-Methacrylamide Hydrogels as Potential Biomaterials for Fabrication of Tissue-Engineered Cartilage Constructs. *Macromol. Biosci.* **2013**, *13*, 551–561. [CrossRef] [PubMed]
60. Krishnamoorthy, S.; Noorani, B.; Xu, C. Effects of Encapsulated Cells on the Physical–Mechanical Properties and Microstructure of Gelatin Methacrylate Hydrogels. *Int. J. Mol. Sci.* **2019**, *20*, 5061. [CrossRef] [PubMed]
61. Shie, M.Y.; Lee, J.J.; Ho, C.C.; Yen, S.Y.; Ng, H.Y.; Chen, Y.W. Effects of gelatin methacrylate bio-ink concentration on mechano-physical properties and human dermal fibroblast behavior. *Polymers* **2020**, *12*, 1930. [CrossRef] [PubMed]
62. Rička, J.; Tanaka, T. Swelling of Ionic Gels: Quantitative Performance of the Donnan Theory. *Macromolecules* **1984**, *17*, 2916–2921. [CrossRef]
63. Jansen, K.A.; Donato, D.M.; Balcioglu, H.E.; Schmidt, T.; Danen, E.H.; Koenderink, G.H. A guide to mechanobiology: Where biology and physics meet. *Biochim. Biophys. Acta Mol. Cell Res.* **2015**, *1853*, 3043–3052. [CrossRef]
64. Bandulasena, M.V.; Vladisavljević, G.T.; Benyahia, B. Versatile reconfigurable glass capillary microfluidic devices with Lego® inspired blocks for drop generation and micromixing. *J. Colloid Interface Sci.* **2019**, *542*, 23–32. [CrossRef]

gels

Article

Self-Healing of Pluronic® F127 Hydrogels in the Presence of Various Polysaccharides

Alexandra Lupu *, Luiza Madalina Gradinaru, Daniela Rusu and Maria Bercea *

"Petru Poni" Institute of Macromolecular Chemistry, 41-A Grigore Ghica Voda Alley, 700487 Iasi, Romania; gradinaru.luiza@icmpp.ro (L.M.G.); rusu.daniela@icmpp.ro (D.R.)
* Correspondence: lupu.alexandra@icmpp.ro (A.L.); bercea@icmpp.ro (M.B.)

Abstract: Thermoresponsive Pluronic® F127 (PL) gels in water were investigated through rheological tests in different shear conditions. The gel strength was tuned with the addition of 1% polysaccharide solution. In the presence of xanthan gum (XG), the viscoelastic behavior of PL-based hydrogels was improved in aqueous environment, but the rheological behavior was less changed with the addition of XG in PBS solutions, whereas in the presence of 0.1 M NaCl, the viscoelastic parameters decreased. PL micellar networks exhibited a self-healing ability, recovering their initial structure after applying cycles of high strain. The rheological characteristics of the PL hydrogel changed with the addition of 1% polysaccharides (xanthan gum, alginate, κ-carrageenan, gellan, or chitosan). PL/polysaccharide systems form temperature-responsive hydrogels with shear thinning behavior, yield stress, and self-healing ability, being considered a versatile platform for injectable biomaterials or bioinks. Thus, in the presence of xanthan gum in aqueous medium, the gel strength was improved after applying a high strain (the values of elastic modulus increased). The other investigated natural polymers induced specific self-healing behaviors. Good performances were observed with the addition of gellan gum, alginate, and κ-carrageenan, but for high values of strain, the ability to recover the initial structure decreased. A modest self-healing behavior was observed in the presence of chitosan and xanthan gum dissolved in NaCl solution.

Keywords: Pluronic® F127; thermoresponsive hydrogels; xanthan gum; shear-thinning; yield stress; self-healing ability

Citation: Lupu, A.; Gradinaru, L.M.; Rusu, D.; Bercea, M. Self-Healing of Pluronic® F127 Hydrogels in the Presence of Various Polysaccharides. *Gels* **2023**, *9*, 719. https://doi.org/10.3390/gels9090719

Academic Editors: Ana Paula Serro, Ana Isabel Fernandes and Diana Silva

Received: 31 July 2023
Revised: 1 September 2023
Accepted: 2 September 2023
Published: 5 September 2023

1. Introduction

Hydrogels based on biomolecules are widely used materials in the biomedical field, due to their biocompatibility, low toxicity, high permeability to metabolites or nutrients, and water swelling ability. Among all polysaccharides, alginate, chitosan, hyaluronic acid, cellulose, dextran, and their derivatives have been intensively studied as hydrogel components during the last decades [1]. Polysaccharides have a low immunogenic profile, are easily handled, and have relatively low costs, aspects which make them preferred candidates for hydrogel fabrication [2]. One of the most important advantages of polysaccharide-based hydrogels is their self-healing ability, which implies that these materials have the capacity to self-repair and autonomously recover their original structure after damage [3]. The self-healing mechanisms require particular functionalities, in order to provide linkages between the injured parts of the gel. These mechanisms are based on dynamic interactions, which are either dynamic covalent crosslinking or physical interactions (supramolecular assembly) (Scheme 1), such as hydrophobic associations, ionic crosslinking, or hydrogen bonding networks [4]. Thus, self-healing hydrogels have attracted attention for tissue engineering applications, such as skin regeneration enhancement, due to their extracellular matrix (ECM)-like damage-healing properties [2].

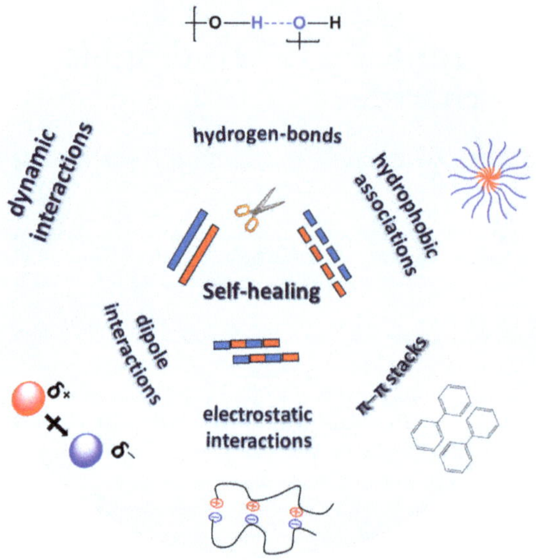

Scheme 1. Schematic presentation of various types of interactions responsible for the self-healing behavior of hydrogels.

In order to ensure their bio-applicability, hydrogels require stimuli-responsive properties, such as temperature-, light-, pH-, electric-, or magnetic field-responsive abilities. Polymers that exhibit thermoresponsive behavior present a low critical solution temperature (LCST), being able to reversibly change upon heating from a hydrophilic state (non-associative) to a hydrophobic state (associative) [5]. At low temperature (below LCST), aqueous solutions of temperature-responsive polymers are in a sol state (characterized by viscous flow), which allows them to be easily injected into a higher temperature environment, above the LCST (e.g., body temperature). Thus, due to the association of polymer chains, these gelators form a physical network (gel state with solid-like behavior) at the injection site [5].

Pluronics play an important role as polymeric excipients and they have been extensively used as stabilizers, solubility improvers, bioavailability enhancers, and thermosensitive drug carriers [6]. Pluronic® F127 (PL) is an FDA-approved, non-toxic and water-soluble triblock copolymer made of two hydrophilic poly(ethylene oxide) blocks and one hydrophobic poly(propylene oxide) block placed in the middle of the hydrophilic segments (PEO_{100}–PPO_{65}–PEO_{100}). PL is one of the most commonly used thermoresponsive polymers and exhibits a rapid sol–gel transition above the critical gelation temperature and concentration [7], when the hydrophilic PEO blocks stretch out, whereas the hydrophobic PPO sections start to associate and form a dense core, resulting a network structure [8]. Thus, due to their temperature-sensitive characteristics that allow effortless injectability, PL-based hydrogels are widely used in tissue engineering and drug delivery applications [9–11]. However, in contact with an excess of water, PL gels disintegrate within several days [12].

In this context, many efforts have been carried out to improve the gel performances, stability over time, viscoelastic and biological properties of PL gels [6,12–16]. Polysaccharide addition appears to be a convenient way to ensure the shape stability, gel stabilization against dissolution [12,17] and biocompatibility [18] of PL-based systems. The gels are stabilized by the interactions of long polysaccharide chains with PEO blocks, which can be combined with the hydrophobic interactions of hydrocarbon chains (>C16) with the core of the Pluronic micelles [12]. The PL micelles surrounded by the polysaccharide chains cannot be separated from the micelles network, and thus the composite gel is more stable in an excess of solvent (water or biological fluid) [12]. However, the stability of Pluronic gels slowed down in the presence of nonionic polysaccharide, and pullulan addition in concentrated systems caused a phase separation [18].

In recent years, PL-polysaccharides-based hydrogels have been discussed in many studies, especially from a biomedical and pharmaceutical point of view [18–20]. These materials are popular for drug-delivery systems, biosensor formulations, and tissue engineering applications. Cellulose, hyaluronic acid, chitosan, alginate, starch, and gellan gum are some of the most used polysaccharides in the preparation of hydrogels with injectable and self-healing properties. The functional groups of these polysaccharides, such as amino (–NH_2), carboxyl (–COOH), aldehyde (–CHO), and hydroxyl (–OH), are involved in hydrophobic interactions, which allow them to form physical networks [12,13,21–24].

The (cyto)toxicity of PL was deeply investigated, in order to evaluate the biocompatible properties and nontoxicity of the pure copolymer compared to other Pluronics (e.g., Pluronic P94) [25]. A series of studies dealt with the increase in the cytotoxicity of PL gels with chemical functionalization with different compounds [26,27]. For example, α-tocopherol (TOC) was used as a toxicity enhancer grafted on PL. As the study concluded, PL–SS–TOC showed low cytotoxicity, with an in vitro cell viability of about 85% after 48 h incubation. Thus, the results demonstrated the potential of PL–SS–TOC in the biomedical area [26]. Li et al. [27] reported the synthesis of a novel folated PL modified liposome (cur–FA–PL–Lps) by attaching folic acid (FA) to PL chains via a dehydration condensation reaction, followed by curcumin incorporation. Cytotoxicity studies of cur–FA–PL–Lps demonstrated a cell viability of about 90–100% after 24 h of incubation, which indicates that the complex was not toxic to cells.

The present paper first focused on the rheological behavior of PL hydrogels in the presence of xanthan gum (XG). Then, the self-healing behavior of PL/XG gels was compared with those of PL-based gels that contain other polysaccharides: alginate (Alg), κ-carrageenan (κ-Carr), gellan gum (GG), and chitosan (CS). The main interest was to obtain homogeneous and low-viscosity formulations at ambient temperature that undergo a sol–gel transition under physiological conditions, developing shear thinning and elastic networks with self-healing ability. How the addition of different polysaccharides changed the gel behavior after undergoing high deformations was investigated, as encountered in many applications. Pluronic® F127 was selected due to its wide-spread use in pharmaceutical formulations and its FDA approval for intravenous use in humans [25]. The Pluronics with higher PPO content exhibited cytotoxicity, due to their ability to incorporate in the cellular membrane, altering its viscosity, and also reducing the intracellular adenozin trifosfat concentration [25].

2. Results and Discussion

Table 1 presents the composition of the samples investigated in this paper. We used two PL samples in water, sample 1 of 20% (wt.) and sample 2 of 16.83% (wt.), to compare their behavior with PL/polysaccharide gels in similar conditions. Table 1 also includes the main rheological characteristics of the samples determined in different shear conditions, at constant temperature of 37 °C.

At low PL concentrations, micelles were isolated and the fluid was isotropic. For a sufficiently high concentration, by increasing the temperature, more spherical micelles were formed, increasing their density and determining their arrangement into a network [28,29].

PL sample of 20% concentration in water is in sol state at low temperature, with a predominantly viscous behavior. It undergoes a sol–gel transition around 21.6 °C and at higher temperatures a solid-like behavior prevails. This transition temperature slowly shifts to higher values as the copolymer concentration decreases; thus, for 16.83% PL, the transition was observed around 27 °C. The temperature-induced gelation of PL (Figure 1) was due to the intensification of hydrophobic interactions, and as the concentration of micelles increased, the degree of hydration of the ethylene oxide units decreased; thus, the hydrogels were rapidly formed by packing of micellar subunits [28,30] through an entropy-driven process [8,28] into a lattice arrangement (lamellar, hexagonal or cubic) [31–33]. This thermoreversible micellar mode of chains association (Scheme 2) makes PL attractive for various applications [34–36]. Several studies have proposed XG-PL-based hydrogels formulations with potential applications as ophthalmic drug-delivery systems [37,38], buccal drug-delivery systems [39], and for topical application and administration of different active principles [40–42]. Injectable Pluronic hydrogels were considered suitable systems to assist local cancer treatment (for chemotherapy, phototherapy, immunotherapy, or gene therapy). One limitation of their use is the high erosion rate of Pluronic hydrogels, which reduces the therapeutic efficacy [43].

The rheological behavior of XG-containing systems is influenced by the anions (from NaCl or PBS solutions) and also by the structure of polysaccharides in an aqueous environment. The effect of XG addition to 20% PL in water was investigated in detail for composite gels by adding aqueous solutions of polysaccharide (samples 6, 9–11), pH = 5, or salted solutions: 0.1 M NaCl solution (sample 7) and phosphate buffered saline (PBS) solution (sample 8), pH = 7.4. Samples with various XG concentrations added to PL gels were preliminary tested using rheology. An optimum concentration of 1% XG was found, with XG addition at this composition able to increase the network strength, as can be observed from the values of the rheological parameters shown in Table 1.

In aqueous solutions, XG macromolecules are able to form high viscous fluids, even at low concentrations, being used as a stabilizer or thickening agent in various applications, including in food, pharmaceutical, cosmetic, agricultural, textile, ceramic, and petroleum industry [44]. The behavior of the XG solutions was not significantly influenced by the temperature increase in the investigated range (up to 50 °C).

2.1. Gelation of Pluronic® F127 in the Presence of XG Chains

The effect of the addition of 1% xanthan gum (XG) in aqueous and salted (0.1 M NaCl and PBS) solutions on the gelation of PL was investigated as a function of temperature and at 37 °C using samples stored at 5 °C, in order to determine the transition temperature (Figure 1) and time (Figure 2). At low temperatures, the PL-based samples behaved as viscous fluids (liquid-like state), with an elastic modulus (G′) lower than the viscous modulus (G″), with the loss tangent having values higher than unity. The sol state of PL was characterized by low values of viscoelastic moduli (G′ and G″), which increased considerably in the presence of XG, when G′ became closer to G″ (tanδ decreased).

Table 1. Sample composition, gelation temperature (T_{gel}), and the values of rheological parameters determined in different shear conditions at 37 °C.

Sample Code	Polysaccharide Solution of 1% (wt.)	PL Concentrationin Water (% wt.)	T_{gel} (a)	G′ (b) (Pa)	G″ (b) (Pa)	tan δ (b)	σ_o (c) (Pa)	σ_o (d) (Pa)	γ_L (d) (%)
1	-	20	21.6	17,200	4200	0.2442	197.5	220	1.49
2	-	16.83	26.9	10,923	2780	0.2545	189	209	1.55
3	1% XG/water	-		20.9	7.03	0.3364	4.41	12.3	25.3
4	1% XG/0.1 M NaCl	-	-	32.52	8.47	0.2604	2.09	11.5	21.7
5	1% XG/PBS	-	-	56.4	13.1	0.2323	1.72	2.99	10.1
6	1% XG/water	20	25.4	13,100	763	0.0582	1080.4	1010	6.77
7	1% XG/0.1 M NaCl	20	23.6	10,800	1057	0.0979	273.1	338	4.58
8	1% XG/PBS	20	29.1	6643	1322	0.1990	104.9	133	1.38
9	0.5% XG/water	20	24.8	9710	832	0.0857	173	152	2.15
10	2% XG/water	20	28.0	9101	914	0.1004	159	180	2.17
11	4% XG/water	20	29.0	7864	793.4	0.1009	147	169	2.17
12	Alg/water	20	23.4	8990 (e)	2450 (e)	0.2725	306.4	422	2.74
13	κ-Carr/water	20	23.5	10,300 (e)	2990 (e)	0.2903	204.48	217	1.73
14	GG/water	20	23.1	8680 (e)	2310 (e)	0.2661	277	311	2.51
15	CS/1% acetic acid solution	20	24.3	8800 (e)	3300 (e)	0.3750	108	124	1.73

(a) temperature sweep test, heating rate of 1 °C/min; (b) frequency sweep test, γ = 1% and ω = 1 rad/s; (c) continuous shear flow (shear stress imposed); (d) amplitude sweep test, ω = 10 rad/s; (e) previous data [18].

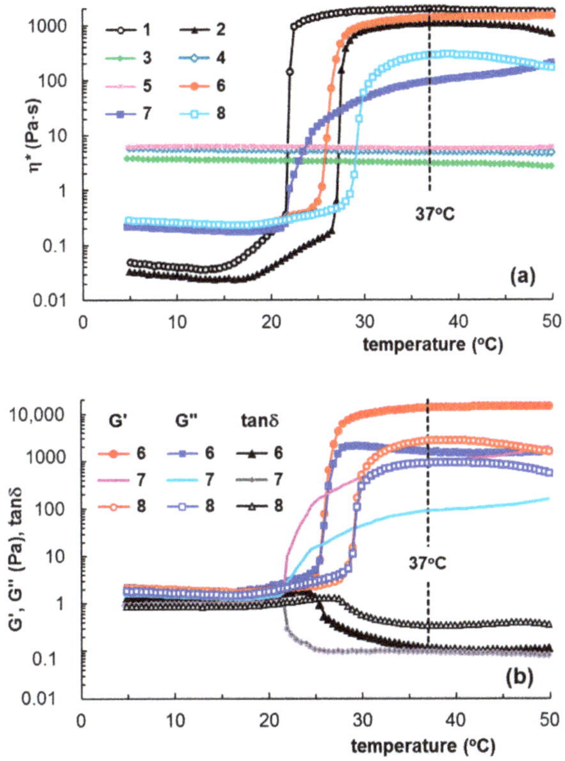

Figure 1. Temperature-induced gelation illustrated through the dependences of (**a**) complex viscosity (η^*) for samples 1–8; (**b**) viscoelastic parameters (G′, G′′ and tanδ) for PL/XG samples (heating rate of 1 °C/min, ω = 10 rad/s, γ = 1%). All samples were previously stored at 5 °C.

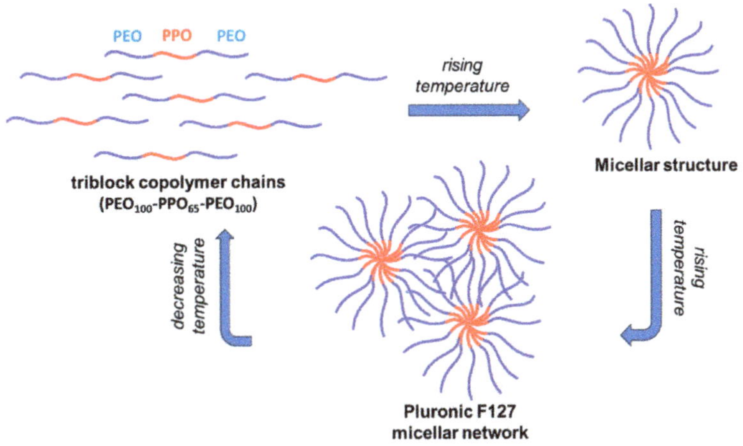

Scheme 2. Schematic presentation of temperature-induced gelation for Pluronic® F127.

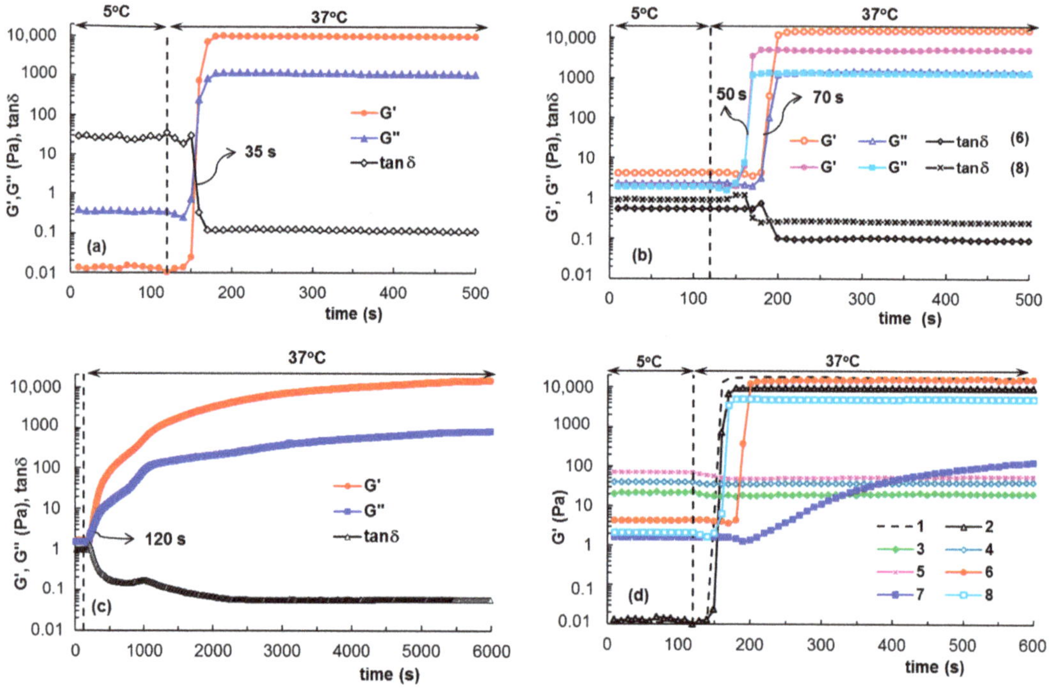

Figure 2. The evolution of viscoelastic parameters during gelation for (**a**) sample 2; (**b**) samples 6 and 8; (**c**) sample 7; (**d**) comparative behavior for samples 1–8. The samples were previously stored at 5 °C, introduced into the rheometer geometry, which was thermostated at 5 °C; firstly, they were monitored at 5 °C for 120 s; then the temperature was suddenly changed to 37 °C and the in situ gelation was followed over time.

By applying a controlled temperature rise (heating rate of 1 °C/min), the hydrophobic interactions inside the PL-containing samples increased and determined the micelles and polymicelles formation, which further generated a physical network [6,14,15]. Consequently, at the transition point, a jump of the viscoelastic parameters was registered, which allowed the identification of the transition temperature (Figure 1) or the transition time at constant temperature (Figure 2). After the transition point, G′ became higher than G″ and tanδ < 1, suggesting network formation (solid-like behavior). A visual illustration of the sol and gel states is presented in Figure 3.

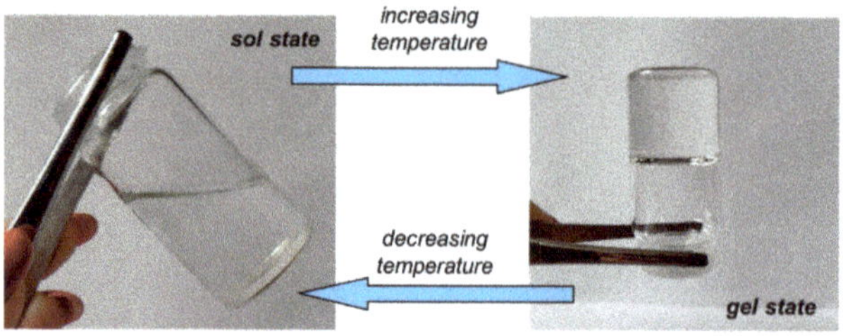

Figure 3. Illustration of the sol and gel states for PL-based hydrogels.

A well-defined transition point was observed for PL in the aqueous environment, when the sol–gel transition occurred over a narrow temperature domain (around 2 °C), which enlarged slightly in the presence of XG (around 4 °C). For the same polymer concentration, a further increase in temperature above the transition point determined an improvement in the gel strength in the presence of polysaccharide.

In aqueous solutions at low temperatures, the XG macromolecules formed random helix structures through hydrogen bonding [45]. Salt addition reduced the effective charge of XG macromolecule, determining a chain aggregation and the viscosity increases. During heating, the XG chains underwent a disruption of aggregates [46,47], slowing down the dynamics of PL gelation (sample 7, Figure 1a,b and Figure 2c).

Figure 4 shows SEM photographs obtained of the cross-section of samples 2, 3 and 5. The dry PL hydrogel (Figure 4a) exhibited a porous structure with pores of 1–5 μm. In the absence of PL, the XG sample presented large holes, higher than 20–30 μm (Figure 4b).

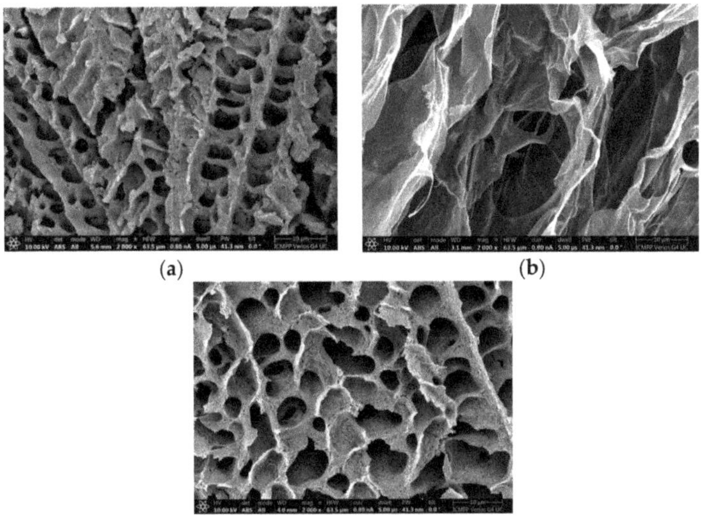

(a) (b) (c)

Figure 4. Scanning electron microscopy images of (**a**) sample 2; (**b**) sample 3, and (**c**) sample 5.

The pores become interconnected and their size increased to 5–10 μm through adding XG chains into the PL gel (Figure 4c). Similar structures were reported for PL and chitosan derivatives [48]. For the injectable hydrogels used as a scaffold for tissue engineering applications, the pores avoid the passage of nutrients to deliver them to the cells.

Besides temperature and concentration, environmental conditions influence sol–gel behavior. Light scattering studies evidenced that the nature and concentration of salts influences the critical micellization temperature of Pluronics, changing the hydrogen bonding in solution and the water structure [49,50]. In the presence of NaCl, corona-specific dehydration occurs, favoring the formation of micellar clusters [14].

Most pharmaceutical Pluronic-based gels contain phosphate buffers, and it was shown that sodium phosphates have a strong effect on copolymer micellization [49]. In the presence of anions from the buffers used as simulated body fluids, the water–water interactions are enhanced and the entropy increases during micellization. In a phosphate buffer solution, a less organized structure is formed by PL and the aggregates formed during micellization are smaller and more stable, as compared with those observed in pure water [51].

2.2. *Shear Flow Behavior at 37 °C*

Figure 5 shows the flow behavior in steady shear conditions for samples 1 to 8, which had previously been thermostated at 37 °C. The PL-based hydrogels exhibited a non-Newtonian behavior over the whole range of investigated shear rates (Figure 5a). In the presence of XG dissolved in water (sample 6), the hydrogel resistance to flow increased and higher values of shear viscosity were obtained. This may have been due to the competition between the hydrodynamic interactions exhibited by different macromolecules in shear conditions [52] or changes in the interactions of PL with water [14].

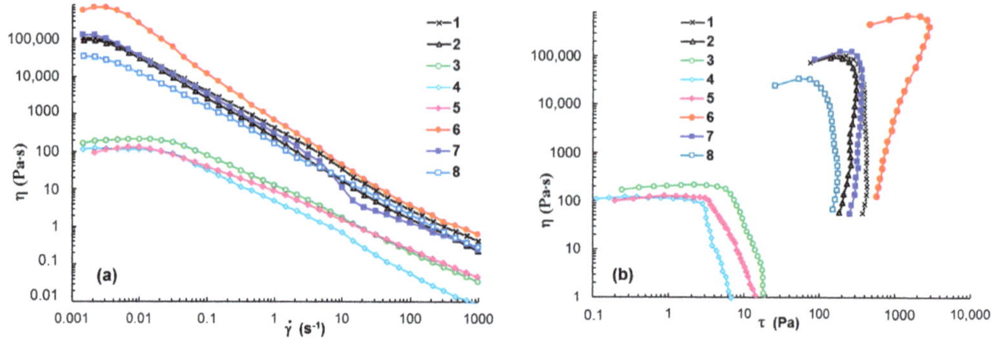

Figure 5. Shear viscosity as a function of (**a**) shear rate and (**b**) shear stress for samples 1–8 (Table 1) at 37 °C.

When XG was dissolved in 0.1 M NaCl (sample 7), a small increase in viscosity was observed for $\dot{\gamma} < 10\ s^{-1}$; above this limit, the gel showed flow instabilities and viscosity decreases, suggesting a loss of structural integrity under the action of the intense shearing forces.

The 1% XG in water or salted solutions exhibited Newtonian behavior at low shear rates ($\dot{\gamma} < 0.3\ s^{-1}$) and they behaved as non-Newtonian fluids with higher shear rate values, when the viscosity decreased with rising $\dot{\gamma}$.

The yield stress value (σ_o) is the minimum shear stress applied to a sample for starting the shear thinning flow when the rest of the structure is breaking down. Yield stress analysis is useful for injectable hydrogels [53,54] and bioinks [55,56] when the hydrogels are submitted to high shear forces. From the dependences of shear viscosity as a function of shear stress (Figure 5b), the yield stress values were determined. For the investigated PL-based hydrogels, the yield stress was of the order of a hundred Pa, in agreement with other data in the literature [29]. The high yield stress values of PL-based hydrogels were attributed to repulsive interactions between close-packed spherical micelles [57]. The addition of 1% XG in aqueous and NaCl solutions induced a higher σ_o value to the PL hydrogels. Similar effects were observed when adding aqueous solutions of Alg, κ-Carr and GG (Table 1). These systems had sufficient strength to resist tearing once they were stretched (σ_o varied from 104.9 Pa in PBS solution to 1080.4 Pa in aqueous medium), and thus they present shape fidelity in 3D printing [58]. The XG in PBS solutions and CS in 1% acetic acid decreased the yield stress value.

2.3. Delimitation of Linear and Non-Linear Viscoelastic Domains

Amplitude sweep tests were carried out to determine the upper strain amplitude values (γ_L) limiting the linear domain of viscoelasticity and the corresponding shear stress values (Table 1). In the linear range of viscoelasticity, G′ and G″ were independent of the applied stain, as shown in Figure 6 for samples 2, 6, 7, and 8 (ω = 10 rad/s). For the PL-based hydrogels, a linear range of viscoelasticity was reached from very low strain values (γ below 0.1%) up to a value of strain above 1%. It can be observed that the addition of 1% XG (Figure 6a) or other polysaccharides (Table 1) determined an increase of γ_L, i.e., the occurrence of an extended linear viscoelastic domain. Long XG chains in aqueous or salted solutions are in an entangled state and they are able to develop a greater resistance when increased strains are applied (Figure 6b), as compared with PL-based hydrogels with a micellar structure (Figure 6a).

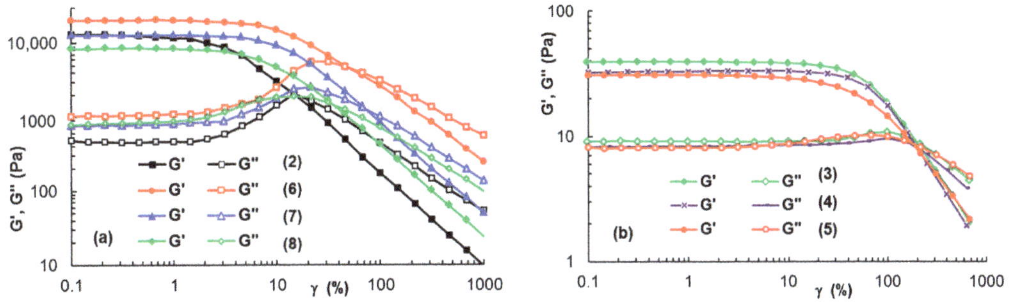

Figure 6. The viscoelastic moduli during amplitude sweep tests for (**a**) PL-containing gels; (**b**) XG solutions (10 rad/s, 37 °C).

The data obtained in the amplitude sweep tests also gave access to the yield stress, when the flow of the sample started [59]. Thus, σ_o can be considered as the upper value of shear stress, corresponding to the point that delimitates the linear and non-linear ranges of viscoelasticity. Usually, the value determined by this method overestimates the yield stress [54,56,60]. Thus, the σ_o values determined from these tests for XG solutions and PL-based hydrogels (Table 1) were higher compared with those determined in the continuous shear tests.

Samples withstanding higher deformations before yielding are considered more stretchable [56], thus the γ_L value can be correlated with the elasticity of materials. For a stress value below σ_o, the material only underwent reversible (elastic) strain, without viscous (permanent) deformation. Above σ_o, irreversible deformation of material occurs. From this point of view, XG addition to PL hydrogels seemed to be a suitable choice.

2.4. Self-Healing Behavior of Pluronic® F127 Hydrogels in the Presence of XG Chains

An important requirement of injectable hydrogels is self-healing ability; a fast recovering of the non-deformed state once they are injected in the damaged site, avoiding a sudden release of drugs or encapsulated cells after injection [24,61,62]. Similarly, it is necessary for 3D printed materials to possess high thixotropy, i.e., the viscosity (or other rheological parameters) quickly decreases when shear forces are applied, and the initial value is recovered very quickly when the shear forces are removed. From a structural point of view, the intermolecular interactions that ensure network formation recover rapidly to a high degree after printing [56]. When a hydrogel is submitted to high shear forces in a syringe needle, the network structure is strongly disturbed and the sample behaves as a fluid-like material. However, with the cessation of the external forces, the structural integrity must be recovered in order to maintain the functionality of biomaterials. Moreover, the shear forces applied to gels during their use affect the cell viability [54,63–65]. In such cases, soft materials with yield stress, shear thinning, and self-healing behavior are suitable materials.

Figure 7 gives the data obtained for PL hydrogel (sample 2) in the oscillatory three-step strains experiments (for ω = 10 rad/s), when the strain values were successively switched each 300 s from a low amplitude (γ = 1%) to a high amplitude value and again to a low γ value of 1%. The viscoelastic parameters (G', G'', and tanδ) were monitored as a function of time. High values of applied strain during the second step were chosen in the nonlinear viscoelastic regime (according to Figure 6): 50%, 100%, 300%, 500%, and 1000%, and in these conditions, a liquid-like behavior was registered (G' became lower that G'' and tanδ > 1). Sample 2 showed nearly complete recovery after applying high deformations, and the time required for structure recovery was very short (on the order of a few seconds).

Figure 8 presents the behavior of the entangled XG aqueous solution (sample 3). The strong intermolecular interactions determined a solid-like behavior at a low γ value (1%). They were perturbed by the high deformations during step 2 and partially recovered when the external forces were removed. A similar behavior was observed for samples 4 and 5, but the recovery during step 3 was smaller compared with that of sample 3.

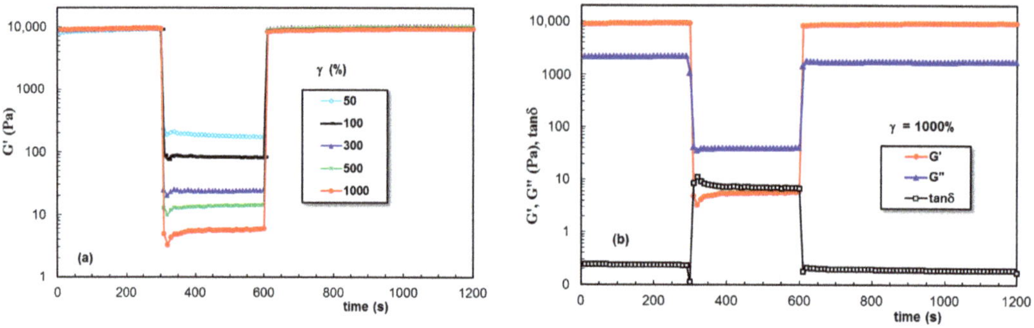

Figure 7. Illustration of thixotropic behavior for sample 2 at 37 °C: (**a**) Elastic modulus as a function of time during successive step strain measurements at low (1%), high (50%, 100%, 300%, 500%, 1000%), and again low (1%) values of γ; (**b**) G′, G″, and tanδ for γ = 1000%.

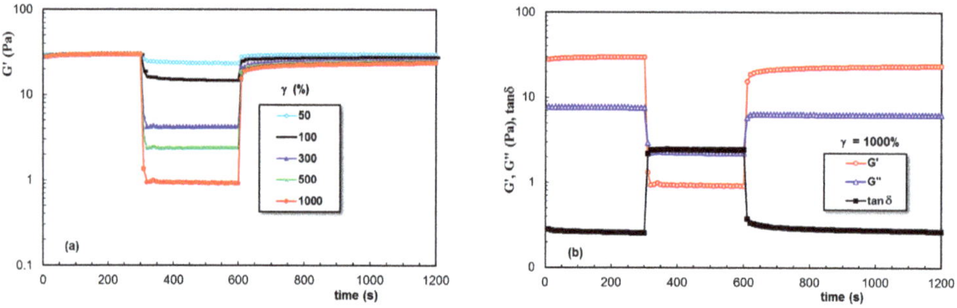

Figure 8. (**a**,**b**) As Figure 7, but for sample 3.

The data obtained for PL hydrogels in the presence of XG dissolved in aqueous, NaCl, and PBS solutions are given in Figures 9–11, respectively. A remarkable behavior was depicted for sample 6, for which the network strength increased after each deformation cycle. This feature was not found in the presence of NaCl, when the network became weaker after increased deformations were applied. However, a good stability and total recovery of rheological parameters was observed for sample 8.

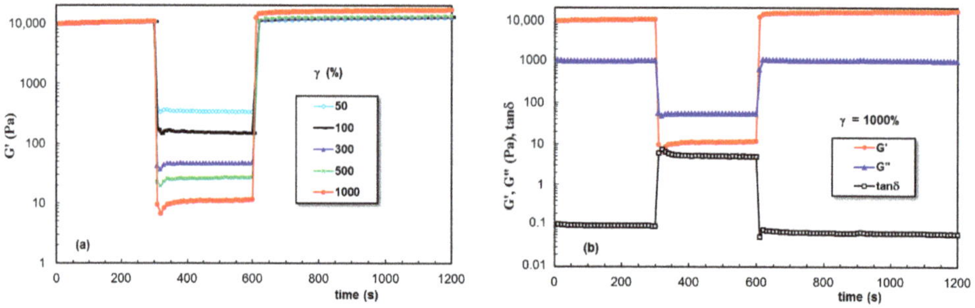

Figure 9. (**a**,**b**) As Figure 7, but for sample 6.

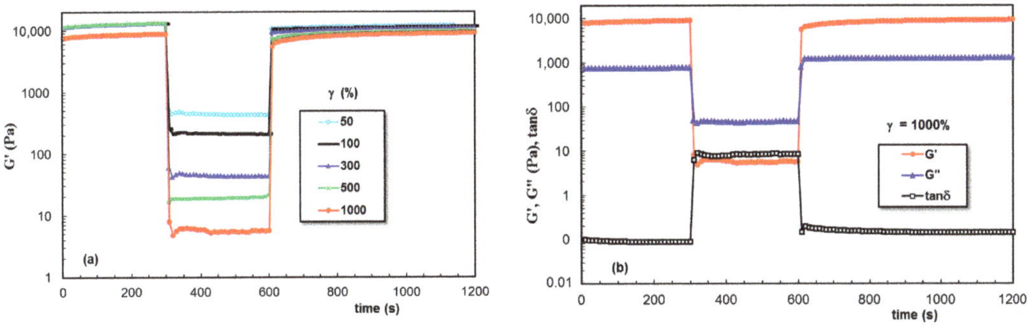

Figure 10. (**a**,**b**) As Figure 7, but for sample 7.

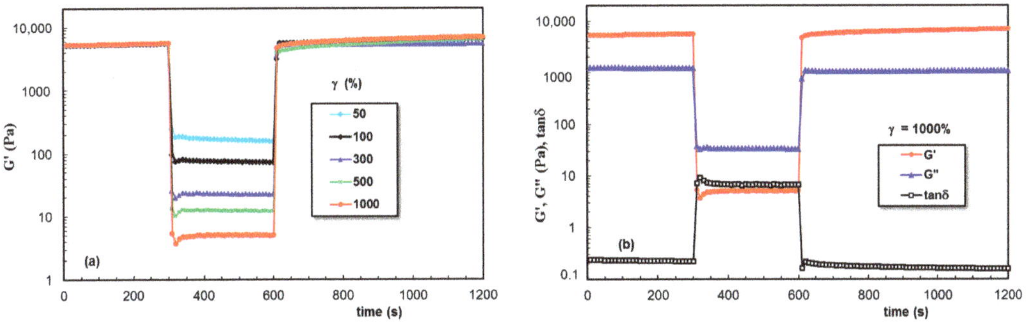

Figure 11. (**a**,**b**) As Figure 7, but for sample 8.

A possible explanation for these results could be the ability of XG chains to penetrate into the PL micellar network and to improve both the viscous and the elastic response, even in conditions of high applied strains. The XG concentration considerably influenced its viscosity and thermal history in aqueous solutions [66]. The addition of different amounts of XG to PL aqueous solutions (samples 6, 9–11) influenced the gelation temperature and rheological characteristics of the hydrogels (Table 1). It was observed that the addition of 1% XG induced improved rheological characteristics for PL/XG hydrogels compared to the neat PL sample.

2.5. Self-Healing Behavior of PL Hydrogels in the Presence of Other Polysaccharides

Due to the unusual thixotropic behavior observed for XG, the self-healing behavior was analyzed for PL gels containing other polysaccharides: Alg, κ-Carr, GG, and CS (Figure 12). For all these hydrogels (samples 12 to 15, Table 1), the high deformations applied during step 2 determined a more pronounced decrease in the viscoelastic parameters, suggesting that the network strength was smaller compared to the PL/XG hydrogels (Figures 9–11).

In order to compare the effect of each polysaccharide addition, we defined the self-healing efficiency as

$$SH\,(\%) = \frac{G'_{\gamma}}{G'_{rest}} \times 100 \quad (1)$$

where G'_{γ} is the elastic modulus registered after applying a given value of γ, and G'_{rest} is the initial value of G' obtained at low γ value, in the linear range of viscoelasticity.

Figure 13 presents the SH parameter for PL-based gels in the absence and in the presence of various polysaccharides, in conditions of various strain values applied during step 2. According to the thixotropic behavior investigated in similar conditions for all PL-containing samples, the pure PL hydrogels (in the absence of polysaccharides) presented a complete and rapid structure recovery. Besides XG addition, GG included into PL hydrogels determined an increase in PL network strength, but only for $\gamma \leq 300\%$; for γ values of 500% or 1000%, the G′ values decreased, but they were still above the initial value registered for γ = 1%. Alg or κ-Carr addition was efficient for low values of strain 50% or 100%, but the structure was disturbed at high values of strain (1000%). Similar effect was recently reported for PL samples in presence of carboxymethyl pullulan for concentrations below 1% [67]. PL gels in the presence of 1% CS or XG in NaCl solution presented modest self-healing ability, the structure was strongly perturbed by applying high strains that belonged to non-linear range of viscoelasticity, and the network strength decreased after each cycle of deformation.

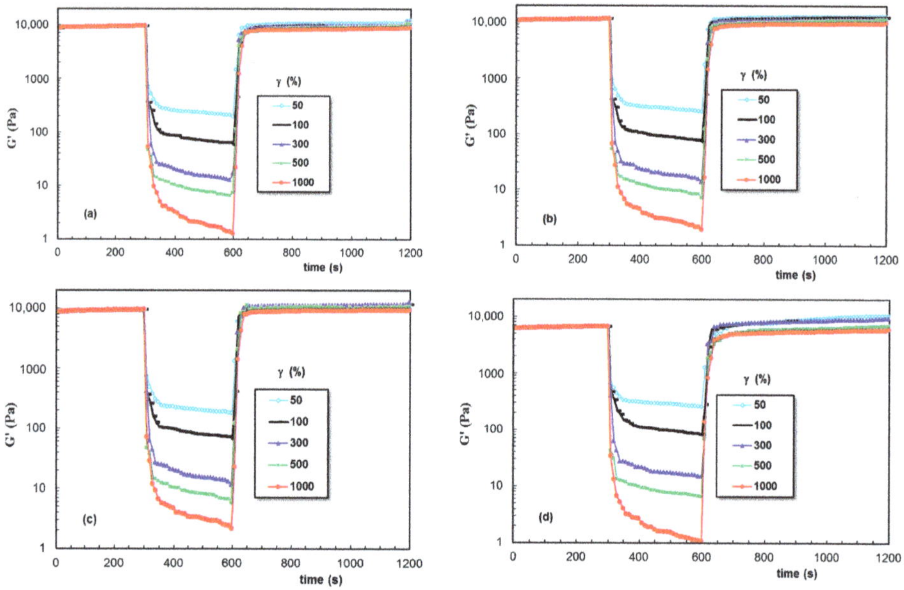

Figure 12. As Figure 7, but for samples (**a**) 12; (**b**) 13; (**c**) 14; (**d**) 15.

The addition of different XG amounts to PL samples revealed the existence of an optimum XG concentration (Figure 13b): thus, sample 6 (1% XG added to PL gels) presented a considerably improved gel strength (Table 1), which increased after applying high strains (the values of elastic modulus increased when increasing the applied strain). The macroscopic self-healing of this sample is illustrated in Figure 14. The PL/XG hydrogel was examined by taking pictures of two pieces of hydrogel, one of them colored with fluorescein (Figure 14a), at different times after putting them in contact (Figure 14b,c). This behavior was attributed to the re-formation of the hydrophobic interactions and hydrogen bonds through the spontaneous diffusion of PL micelles and slow rearrangements of long XG chains surrounded by PL micelles at the site of the applied strain.

The composite PL/XG gels were transparent and stable over time at room temperature. The addition of natural gums to the PL-based hydrogel influenced the gelation and the drug release profile of PL-based gels [38]. Thus, the main potential application of these hydrogels is their use as drug-delivery vehicles for ophthalmic [37,38], buccal [39,57], or transdermal drug delivery applications [40].

Another aspect concerns the influence of environmental conditions on the gel properties [49–51]. The solvent can change the mechanical performance of self-assembling hydrogels, as it was reported for triazole-linked lipid derivatives, suitable for topical formulations [68].

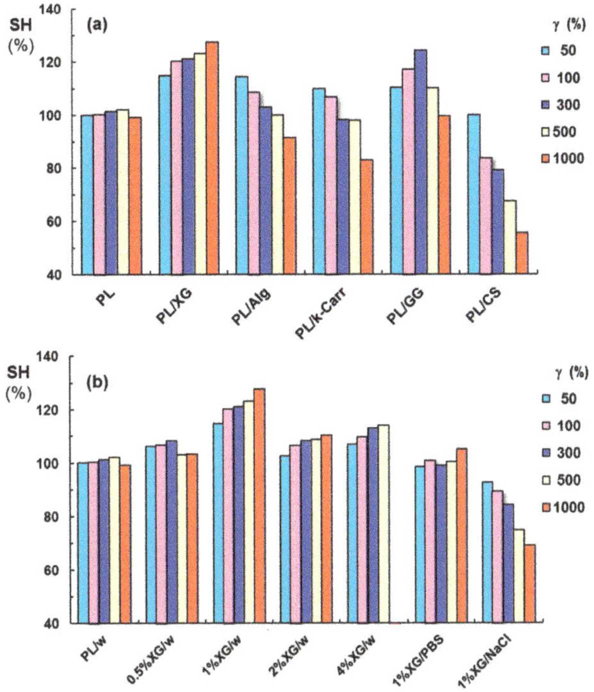

Figure 13. Illustration of the influence of polysaccharide addition on the self-healing efficiency of PL hydrogels after applying three step cycles of various levels of strain during step 2 (γ = 1% for the steps 1 and 3): (**a**) PL and PL/polysaccharides in water and (**b**) PL/XG hydrogels obtained by adding various XG concentrations in water (w) or 1% XG in salted solutions (0.1 M NaCl or PBS).

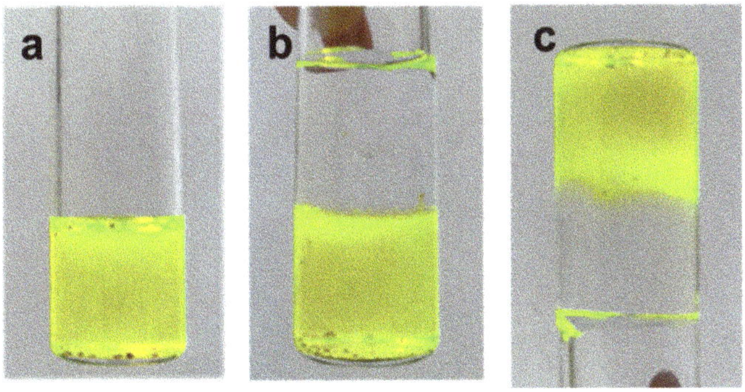

Figure 14. Macroscopic illustration of the self-healing behavior for sample 6: (**a**) sample 6 colored with fluorescein; (**b**) two pieces of hydrogel after they were put into contact; and (**c**) the aspect of the self-healed hydrogel after 12 h.

3. Conclusions

Hydrogels containing Pluronic® F127 in the presence various polysaccharides were prepared and investigated through rheological measurements. In situ gelation of PL hydrogels in the presence of XG was followed in detail. The PL/XG hydrogels were shear thinning, and presented yield stress and improved self-healing ability, important characteristics for injectable materials or bioinks. An optimum concentration of 1% XG was depicted in the rheological investigations. Salt addition determined the formation of aggregates and decreased the PL-based performances.

The excellent self-healing ability induced by XG chains was not observed when this biomacromolecule was replaced with other polysaccharides. The influence of alginate, κ-carrageenan, gellan, or chitosan addition was investigated in similar conditions. The data analysis suggested that the network structure and rheological properties of the PL micellar hydrogel can be tuned by selecting an appropriate polysaccharide structure. Thus, good results were obtained in the presence of gellan gum, alginate, and κ-carrageenan, but the ability to recover the initial structure decreased after applying high strains. A modest viscoelastic behavior and self-healing ability were observed when chitosan was added to PL gels and this may have been due to the presence of acetic acid in the system.

In conclusion, the present study was focused on obtaining homogeneous systems in sol state and on the investigation of the sol–gel transition and hydrogel properties, with an emphasis on viscoelastic characteristics and self-healing behavior. The current findings could provide a good starting point for discussion and further research.

4. Materials and Methods

4.1. Materials

All polymer samples, Pluronic® F127 (denoted PL), and polysaccharides: xanthan gum (XG), sodium alginate (Alg), kappa-carrageenan (κ-Carr), gellan gum (GG), and chitosan of low viscosity (CS) were purchased from Sigma-Aldrich Co. (Taufkirchen, Munich, Germany).

PL is a triblock copolymer with a central hydrophobic poly(propylene oxide) (PPO) block and two hydrophylic poly(ethylene oxide) (PEO) blocks, i.e., (PEO)x-b-(PPO)y-b-(PEO)x, with x = 100, y = 65. The XG sample had a high molecular weight: $M = 2 \times 10^6$ g/mol. Three polysaccharides presented closed values of molecular weight: GG: $M = 5 \times 10^5$ g/mol, κ-Carr: $M = 4.85 \times 10^5$ g/mol and Alg: $M = 4.70 \times 10^5$ g/mol. The CS sample had a M of 2×10^5 g/mol. The chemical structures of the polymers used in hydrogel preparation are given in Figure 15.

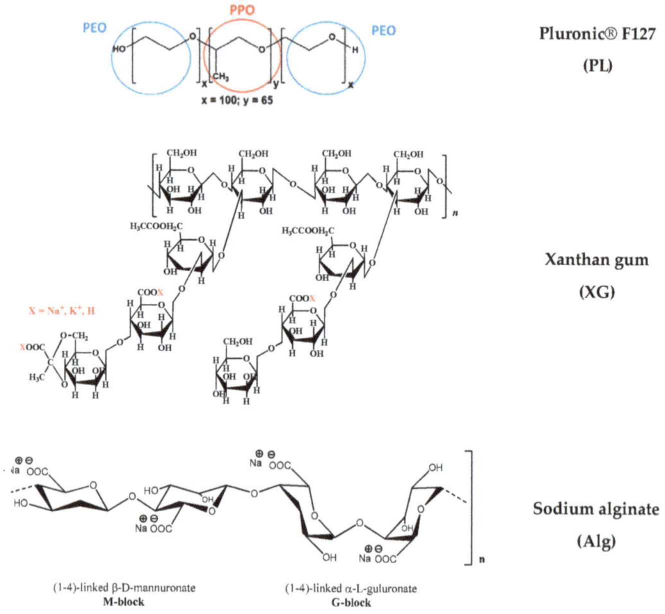

Figure 15. *Cont.*

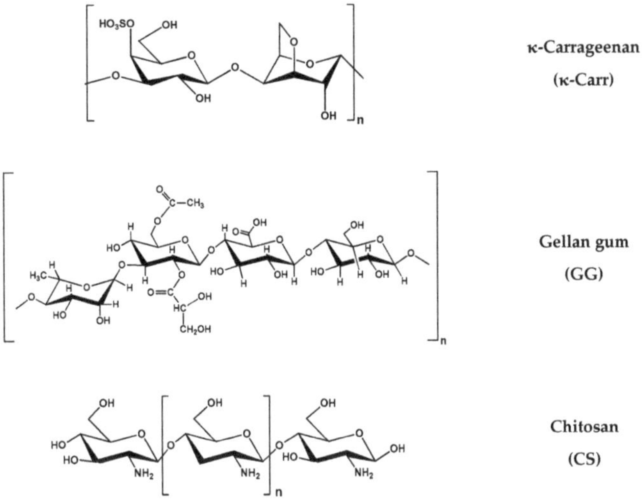

Figure 15. The chemical structure of the polymers.

4.2. Sample Preparation

PL and polysaccharides (excepting CS) were dissolved in Millipore water. Other XG samples were prepared in 0.1 M NaCl and PBS solution. CS was dissolved in 1% acetic acid solution. Stock solutions of PL (20%, wt. and 16.83%) and polysaccharides (1%) were prepared by mixing the polymer with the solvent. For XG, samples of various concentrations (0.5%, 1%, 2%, and 4%) were prepared in aqueous solutions.

PL solution was prepared at low temperature and then stored at 4 °C. Polysaccharide solutions were freshly prepared at room temperature using a weak magnetic stirring system and stored at 4 °C for 24 h, excepting the GG sample, which was kept at room temperature. The biomolecules were added to homogeneous 20% PL solutions and the total concentration of polymer (PL and polysaccharide) in all samples was 16.83% wt.

4.3. Rheological Investigation

Rheological measurements were carried out with a MCR 302 Anton-Paar rheometer (Graz, Austria) using a plane–plane geometry (the upper plate of 50 mm, gap of 500 μm) and Peltier device for temperature control.

The sol–gel transition was investigated using solutions stored at 5 °C and then introduced into the geometry of the rheometer, which had previously been thermostated at 5 °C. Temperature seep tests were carried out at a heating rate of 1 °C/min for an oscillation frequency (ω) of 10 rad/s and strain amplitude (γ) of 1%. The elastic (G') and viscous (G'') moduli were determined, and they give information about the stored and dissipated energy during one cycle of deformation, respectively. A useful parameter expressing the degree of viscoelasticity is the loss tangent (tanδ) determined as the G''/G' ratio.

During the gelation experiments, the temperature was first set at 5 °C for 120 s and then suddenly switched to 37 °C. The viscoelastic parameters G', G'', and tanδ were monitored as a function of time at a constant oscillation frequency (ω) of 10 rad/s and strain amplitude (γ) of 1%. The time-dependent self-assembly process was only observed with addition of chitosan/acetic acid solution and XG/0.1 M NaCl. For these systems, incubation at 37 °C for 24 h was required. For the other systems, the self-assembling was relatively fast, an incubation at 37 °C for 1 h was sufficient to reach equilibrium.

The gels thermostated at 37 °C were investigated in amplitude sweep tests carried out to determine the upper limit of strain, γ_L, for the linear viscoelastic regime and the yield stress value (σ_o). The shear viscosity was determined in stationary shear conditions for shear rates ($\dot{\gamma}$) varying from 0.01 s^{-1} to 1000 s^{-1}.

Self-healing tests were carried out in strain step oscillatory mode for ω = 10 rad/s and the step strains varied every 300 s from low (1%) to high values (50%, 100%, 300%, 500%, and 1000%), in the nonlinear range of viscoelasticity and again the low step of strain (1%).

4.4. Morphology of Hydrogels

The PL, XG, and composite PL/XG samples were frozen at −55 °C and dried using a freeze drier (BenchTop Pro with Omnitronics ™, SP Scientific) for 48 h. Dry samples were coated with a 6 nm platinum layer using a Leica EM ACE200 Sputter coater prior to microscopic examination to improve the electrical conductivity and prevent charge buildup. The morphology was analyzed using a Verios G4 UC Scanning Electron Microscope (Thermo Scientific, Brno, Czech Republic), operating at 10 kV in high-vacuum mode with a backscatter electron detector, ABS (Angular Backscattered Detector). The SEM images were observed at various magnifications.

Author Contributions: Conceptualization—M.B.; Methodology—A.L., L.M.G. and M.B.; Validation—L.M.G.; Formal analysis—A.L. and D.R.; Investigation—A.L., D.R. and M.B.; Writing—original draft—A.L., L.M.G. and M.B.; Writing—review and editing—A.L., D.R. and M.B.; Visualization—D.R. and A.L.; Supervision—L.M.G. and M.B. All authors have read and agreed to the published version of the manuscript.

Funding: This research received no external funding.

Institutional Review Board Statement: Not applicable.

Informed Consent Statement: Not applicable.

Data Availability Statement: Data are available on request.

Conflicts of Interest: The authors declare no conflict of interest.

References

1. Chen, J.; Nichols, B.; Norris, A.; Frazier, C.E.; Edgar, K.J. All-polysaccharide, self-healing injectable hydrogels based on chitosan and oxidized hydroxypropyl polysaccharides. *Biomacromolecules* **2020**, *21*, 4261–4272. [CrossRef]
2. Xuan, H.; Wu, S.; Fei, S.; Li, B.; Yang, Y.; Yuan, H. Injectable nanofiber-polysaccharide self-healing hydrogels for wound healing. *Mater. Sci. Eng. C* **2021**, *128*, 112264. [CrossRef] [PubMed]
3. Zheng, B.D.; Ye, J.; Yang, Y.C.; Huang, Y.Y.; Xiao, M.T. Self-healing polysaccharide-based injectable hydrogels with antibacterial activity for wound healing. *Carbohydr. Polym.* **2022**, *275*, 118770. [CrossRef] [PubMed]
4. Maiz-Fernández, S.; Pérez-Álvarez, L.; Ruiz-Rubio, L.; Vilas-Vilela, J.L.; Lanceros-Mendez, S. Polysaccharide-based in situ self-healing hydrogels for tissue engineering applications. *Polymers* **2020**, *12*, 2261. [CrossRef] [PubMed]
5. Safakas, K.; Saravanou, S.F.; Iatridi, Z.; Tsitsilianis, C. Thermo-responsive injectable hydrogels formed by self-assembly of alginate-based heterograft copolymers. *Gels* **2023**, *9*, 236. [CrossRef]
6. Grela, K.P.; Bagińska, I.; Burak, J.; Marciniak, D.M.; Karolewicz, B. Natural gums as viscosity-enhancers in Pluronic® F-127 thermogelling solutions. *Pharmazie* **2019**, *74*, 334. [CrossRef]
7. Wanka, G.; Hoffmann, H.; Ulbricht, W. Phase diagrams and aggregation behavior of poly(oxyethylene)–poly(oxypropylene)–poly(oxyethylene) triblock copolymers in aqueous solutions. *Macromolecules* **1994**, *27*, 4145–4159. [CrossRef]
8. Alexandridis, P.; Holzwarth, J.F.; Hatton, T.A. Micellization of poly(ethylene oxide)-poly(propylene oxide)-poly(ethylene oxide) triblock copolymers in aqueous solutions: Thermodynamics of copolymer association. *Macromolecules* **1994**, *27*, 2414–2425. [CrossRef]
9. Kushan, E.; Senses, E. Thermoresponsive and injectable composite hydrogels of cellulose nanocrystals and Pluronic F127. *ACS Appl. Bio. Mater.* **2021**, *4*, 3507. [CrossRef]
10. Hu, Q.; Xie, X.; Liao, K.; Huang, J.; Yang, Q.; Zhou, Y.; Liu, Y.; Deng, K. An injectable thermosensitive Pluronic F127/hyaluronic acid hydrogel loaded with human umbilical cord mesenchymal stem cells and asiaticoside microspheres for uterine scar repair. *Int. J. Biol. Macromol.* **2022**, *219*, 96–108. [CrossRef]
11. Zhang, K.; Xue, K.; Loh, X.J. Thermo-responsive hydrogels: From recent progress to biomedical applications. *Gels* **2021**, *7*, 77. [CrossRef] [PubMed]
12. Kjoniksen, A.L.; Calejo, M.T.; Zhu, K.Z.; Nystrom, B.; Sande, S.A. Stabilization of Pluronic gels in the presence of different polysaccharides. *J. Appl. Polym. Sci.* **2014**, *131*, 40465. [CrossRef]

13. Lee, M.; Shin, G.H.; Park, H.J. Solid lipid nanoparticles loaded thermoresponsive Pluronic–xanthan gum hydrogel as a transdermal delivery system. *J. Appl. Polym. Sci.* **2017**, *135*, 46004. [CrossRef]
14. Dey, J.; Kumar, S.; Nath, S.; Ganguly, R.; Aswal, V.K.; Ismail, K. Additive induced core and corona specific dehydration and ensuing growth and interaction of Pluronic F127 micelles. *J. Colloid Interface Sci.* **2014**, *415*, 95–102. [CrossRef]
15. Desai, P.R.; Jain, N.J.; Sharma, R.K.; Bahadur, P. Effect of additives on the micellization of PEO/PPO/PEO block copolymer F127 in aqueous solution. *Colloids Surf. A Physicochem. Eng. Asp.* **2001**, *178*, 57–69. [CrossRef]
16. Bercea, M.; Darie, R.N.; Nita, L.E.; Morariu, S. Temperature responsive gels based on Pluronic F127 and poly(vinyl alcohol). *Ind. Eng. Chem. Res.* **2011**, *50*, 4199–4206. [CrossRef]
17. Shriky, B.; Vigato, A.A.; Sepulveda, A.F.; Pompermayer Machado, I.; de Araujo, D.R. Poloxamer-based nanogels as delivery systems: How structural requirements can drive their biological performance? *Biophys. Rev.* **2023**. [CrossRef]
18. Lupu, A.; Rosca, I.; Gradinaru, V.R.; Bercea, M. Temperature induced gelation and antimicrobial properties of Pluronic F127 based systems. *Polymers* **2023**, *15*, 355. [CrossRef]
19. Ramya, K.A.; Kodavaty, J.; Dorishetty, P.; Setti, M.; Deshpandea, A.P. Characterizing the yielding processes in Pluronic-hyaluronic acid thermoreversible gelling systems using oscillatory rheology. *J. Rheol.* **2019**, *63*, 215–228. [CrossRef]
20. Jaquilin, R.; Oluwafemi, O.S.; Thomas, S.; Oyedeji, A.O. Recent advances in drug delivery nanocarriers incorporated in temperature-sensitive Pluronic F-127—A critical review. *J. Drug Deliv. Sci. Technol.* **2022**, *72*, 103390. [CrossRef]
21. Gioffredi, E.; Boffito, M.; Calzone, S.; Giannitelli, S.M.; Rainer, A.; Trombetta, M.; Mozetic, P.; Chiono, V. Pluronic F127 hydrogel characterization and biofabrication in cellularized constructs for tissue engineering applications. *Procedia CIRP* **2016**, *49*, 125–132. [CrossRef]
22. Ou, Y.; Tian, M. Advances in multifunctional chitosan-based self-healing hydrogels for biomedical applications. *J. Mater. Chem. B* **2021**, *9*, 7955–7971. [CrossRef] [PubMed]
23. Yang, Y.; Xu, L.; Wang, J.; Meng, O.; Zhong, S.; Gao, Y.; Cui, X. Recent advances in polysaccharide-based self-healing hydrogels for biomedical applications. *Carbohydr. Polym.* **2022**, *283*, 119161. [CrossRef] [PubMed]
24. Rusu, A.; Nita, L.E.; Bercea, M.; Tudorachi, N.; Diaconu, A.; Pamfil, D.; Rusu, D.; Ivan, F.E.; Chiriac, A. Interpenetrated polymer network with modified chitosan in composition and self-healing properties. *Int. J. Biol. Macromol.* **2019**, *132*, 374–384. [CrossRef] [PubMed]
25. Arranja, A.; Schroder, A.P.; Schmutz, M.; Waton, G.; Schosseler, F.; Mendes, E. Cytotoxicity and internalization of Pluronic micelles stabilized by core cross-linking. *J. Control. Release* **2014**, *196*, 87–95. [CrossRef] [PubMed]
26. Liu, Y.L.; Fu, S.; Lin, L.F.; Cao, Y.H.; Xie, X.; Yu, H.; Chen, M.W.; Li, H. Redox-sensitive Pluronic F127-tocopherol micelles: Synthesis, characterization, and cytotoxicity evaluation. *Int. J. Nanomed.* **2017**, *12*, 2635–2644. [CrossRef]
27. Li, Z.; Xiong, X.; Peng, S.; Chen, X.; Liu, W.; Liu, C. Novel folated Pluronic F127 modified liposomes for delivery of curcumin: Preparation, release and cytotoxicity. *J. Microencapsul.* **2020**, *37*, 220–229. [CrossRef]
28. Alexandridis, P.; Alan Hatton, T. Poly(ethylene oxide)-poly(propylene oxide)-poly(ethylene oxide) block copolymer surfactants in aqueous solutions and at interfaces: Thermodynamics, structure, dynamics, and modeling. *Colloids Surf. A Physicochem. Eng. Asp.* **1995**, *96*, 1–46. [CrossRef]
29. Jalaal, M.; Cottrell, G.; Balmforth, N.; Stoeber, B. On the rheology of Pluronic F127 aqueous solutions. *J. Rheol.* **2017**, *61*, 139–146. [CrossRef]
30. Shaikhullina, M.; Khaliullina, A.; Gimatdinov, R.; Butakov, A.; Chernov, V.; Filippov, A. NMR relaxation and self-diffusion in aqueous micellar gels of Pluronic F-127. *J. Mol. Liq.* **2020**, *306*, 112898. [CrossRef]
31. Linemann, R.; Läuger, J.; Schmidt, G.; Kratzat, K.; Richtering, W. Linear and nonlinear rheology of micellar solutions in the isotropic, cubic and hexagonal phase probed by rheo-small-angle light scattering. *Rheol. Acta* **1995**, *34*, 440–449. [CrossRef]
32. Chaibundit, C.; Ricardo, N.M.P.S.; de Costa, F.M.L.L.; Yeates, S.G.; Booth, C. Micellization and gelation of mixed copolymers P123 and F127 in aqueous solution. *Langmuir* **2007**, *23*, 9229–9236. [CrossRef] [PubMed]
33. Ivanova, R.; Lindman, B.; Alexandridis, P. Evolution in structural polymorphism of Pluronic F127 poly(ethylene oxide)-poly(propylene oxide) block copolymer in ternary systems with water and pharmaceutically acceptable organic solvents: from "glycols" to "oils. *Langmuir* **2000**, *16*, 9058–9069. [CrossRef]
34. Bodratti, A.M.; Alexandridis, P. Formulation of poloxamers for drug delivery. *J. Funct. Biomater.* **2018**, *9*, 11. [CrossRef]
35. Chen, Z.; Han, Z.; Cai, P.; Mo, X.; Zhang, Y.; Wu, J.; Sun, B. Application of gel suspension printing system in 3D bio-printing. *Mater. Lett.* **2023**, *341*, 134235. [CrossRef]
36. Nita, L.E.; Chiriac, A.P.; Bercea, M.; Nistor, M.T. Static and dynamic investigations of poly(aspartic acid) and Pluronic F127 complex prepared by self-assembling in aqueous solution. *Appl. Surf. Sci.* **2015**, *359*, 486–495. [CrossRef]
37. Shastri, D.H.; Prajapati, S.T.; Patel, L.D. Design and development of thermoreversible ophthalmic in situ hydrogel of Moxifloxacin HCl. *Curr. Drug Deliv.* **2010**, *7*, 238–243. [CrossRef]
38. Bhowmik, M.; Kumari, P.; Sarkar, G.; Bain, M.K.; Bhowmick, B.; Mollick, M.M.R.; Mondal, D.; Maity, D.; Rana, D.; Bhattacharjee, D.; et al. Effect of xanthan gum and guar gum on in situ gelling ophthalmic drug delivery system based on poloxamer-407. *Int. J. Biol. Macromol.* **2013**, *62*, 117–123. [CrossRef]
39. Zeng, N.; Seguin, J.; Destruel, P.L.; Dumortier, G.; Maury, M.; Dhotel, H.; Bessodes, M.; Scherman, D.; Mignet, N.; Boudy, V. Cyanine derivative as a suitable marker for thermosensitive in situ gelling delivery systems: In vitro and in vivo validation of a sustained buccal drug delivery. *Int. J. Pharm.* **2017**, *534*, 128–135. [CrossRef]

40. Djekic, L.; Martinovic, M.; Dobricic, V.; Calija, B.; Medarevic, D.; Primorac, M. Comparison of the Effect of Bioadhesive polymers on stability and drug release kinetics of biocompatible hydrogels for topical application of ibuprofen. *J. Pharm. Sci.* **2018**, *108*, 1326–1333. [CrossRef]
41. Sicurella, M.; Pula, W.; Musial, K.; Cieslik-Boczula, K.; Sguizzato, M.; Bondi, A.; Drechsler, M.; Montesi, L.; Esposito, E.; Marconi, P. Ethosomal gel for topical administration of dimethyl fumarate in the treatment of HSV-1 infections. *Int. J. Mol. Sci.* **2023**, *24*, 4133. [CrossRef] [PubMed]
42. Sguizzato, M.; Pepe, A.; Baldisserotto, A.; Barbari, R.; Montesi, L.; Drechsler, M.; Mariani, P.; Cortesi, R. Niosomes for Topical application of antioxidant molecules: Design and in vitro behavior. *Gels.* **2023**, *9*, 107. [CrossRef]
43. Marques, A.C.; Costa, P.C.; Velho, S.; Amaral, M.H. Injectable poloxamer hydrogels for local cancer therapy. *Gels* **2023**, *9*, 593. [CrossRef] [PubMed]
44. Nsengiyumva, E.M.; Alexandridis, P. Xanthan gum in aqueous solutions: Fundamentals and applications. *Int. J. Biol. Macromol.* **2022**, *216*, 583–604. [CrossRef] [PubMed]
45. Morris, E.R. Ordered conformation of xanthan in solutions and "weak gels": Single helix, double helix–or both? *Food Hydrocoll.* **2019**, *86*, 18–25. [CrossRef]
46. Craing, D.M.Q.; Kee, A.; Tamburic, S.; Barnes, D. An investigation into the temperature dependence of the rheological synergy between xanthan gum and locust bean gum mixtures. *J. Biomat. Sci. Polym. Ed.* **1997**, *8*, 377–389. [CrossRef]
47. Bercea, M.; Morariu, S. Real-time monitoring the order-disorder conformational transition of xanthan gum. *J. Mol. Liq.* **2020**, *309*, 113168. [CrossRef]
48. Yap, L.-S.; Yang, M.-C. Evaluation of hydrogel composing of Pluronic F127 and carboxymethyl hexanoyl chitosan as injectable scaffold for tissue engineering applications. *Colloids Surf. B* **2016**, *146*, 204–211. [CrossRef]
49. Pandit, N.; Trygstad, T.; Croy, S.; Bohorquez, M.; Koch, C. Effect of salts on the micellization, clouding, and solubilization behavior of Pluronic F127 solutions. *J. Colloid Interface Sci.* **2000**, *222*, 213–220. [CrossRef]
50. Anderson, B.C.; Cox, S.M.; Ambardekar, A.V.; Mallapragada, S.K. The effect of salts on the micellization temperature of aqueous poly(ethylene oxide)-b-poly(propylene oxide)-b-poly(ethylene oxide) solutions and the dissolution rate and water diffusion coefficient in their corresponding gels. *J. Pharm. Sci.* **2002**, *91*, 180–188. [CrossRef]
51. Perinelli, D.R.; Cespi, M.; Pucciarelli, S.; Casettari, L.; Palmieri, G.P.; Bonacucina, G. Effect of phosphate buffer on the micellization process of Poloxamer 407: Microcalorimetry, acoustic spectroscopy and dynamic light scattering (DLS) studies. *Colloids Surf. A Physicochem. Eng. Asp.* **2013**, *436*, 123–129. [CrossRef]
52. Brunchi, C.E.; Bercea, M.; Morariu, S. Hydrodynamic properties of polymer mixtures in solution. *J. Chem. Eng. Data* **2010**, *55*, 4399–4405. [CrossRef]
53. Jons, C.K.; Grosskopf, A.K.; Baillet, J.; Yan, J.; Klich, J.H.; Saouaf, O.M.; Appe, E.A. Yield-stress and creep control depot formation and persistence of injectable hydrogels following subcutaneous administration. *Adv. Funct. Mat.* **2022**, *32*, 2203402. [CrossRef]
54. Bertsch, P.; Diba, M.; Mooney, D.J.; Leeuwenburgh, S.C.G. Self-healing injectable hydrogels for tissue regeneration. *Chem. Rev.* **2023**, *123*, 834–873. [CrossRef]
55. Lee, J.; Oh, S.J.; An, S.H.; Kim, W.D.; Kim, S.H. Machine learning-based design strategy for 3D printable bioink: Elastic modulus and yield stress determine printability. *Biofabrication* **2020**, *12*, 035018. [CrossRef]
56. Bercea, M. Rheology as a tool for fine-tuning the properties of printable bioinspired gels. *Molecules* **2023**, *28*, 2766. [CrossRef]
57. Yu, Y.B.; Cheng, Y.; Tong, J.Y.; Zhang, L.; Wei, Y.; Tian, M. Recent advances in thermo-sensitive hydrogels for drug delivery. *J. Mater. Chem. B* **2021**, *9*, 2979. [PubMed]
58. Schwab, A.; Levato, R.; D'Este, M.; Piluso, S.; Eglin, D.; Malda, J. Printability and shape fidelity of bioinks in 3D bioprinting. *Chem. Rev.* **2020**, *120*, 11028–11055. [CrossRef] [PubMed]
59. Dinkgreve, M.; Paredes, J.; Denn, M.M.; Bonn, D. On different ways of measuring "the" yield stress. *J. Non-Newton. Fluid Mech.* **2016**, *238*, 233–241. [CrossRef]
60. Gradinaru, L.M.; Bercea, M.; Lupu, A.; Gradinaru, V.R. Development of polyurethane/peptide-based carriers with self-healing properties. *Polymers* **2023**, *15*, 1697. [CrossRef]
61. Lupu, A.; Gradinaru, L.M.; Gradinaru, V.R.; Bercea, M. Diversity of bioinspired hydrogels: From structure to applications. *Gels* **2023**, *9*, 376. [CrossRef]
62. Rad, E.R.; Vahabi, H.; Formela, K.; Saeb, M.R.; Thomas, S. Injectable poloxamer/graphene oxide hydrogels with well-controlled mechanical and rheological properties. *Polym. Adv. Technol.* **2019**, *30*, 2250–2260. [CrossRef]
63. Mappa, T.A.; Liu, C.-M.; Tseng, C.-C.; Ruslin, M.; Cheng, J.-H.; Lan, W.-C.; Huang, B.-H.; Cho, Y.-C.; Hsieh, C.-C.; Kuo, H.-H.; et al. An innovative biofunctional composite hydrogel with enhanced printability, rheological properties, and structural integrity for cell scaffold applications. *Polymers* **2023**, *15*, 3223. [CrossRef] [PubMed]
64. Shi, J.; Wu, B.; Li, S.; Song, J.; Song, B.; Lu, W.F. Shear stress analysis and its effects on cell viability and cell proliferation in drop-on-demand bioprinting. *Biomed. Phys. Eng. Express* **2018**, *4*, 045028. [CrossRef]
65. Ishida-Ishihara, S.; Takada, R.; Furusawa, K.; Ishihara, S.; Haga, H. Improvement of the cell viability of hepatocytes cultured in three-dimensional collagen gels using pump-free perfusion driven by water level difference. *Sci. Rep.* **2022**, *12*, 20269. [CrossRef] [PubMed]
66. Nsengiyumva, E.M.; Heitz, M.P.; Alexandridis, P. Thermal hysteresis phenomena in aqueous xanthan gum solutions. *Food Hydrocoll.* **2023**, *144*, 108973. [CrossRef]

67. Popescu, I.; Constantin, M.; Bercea, M.; Cosman, B.; Suflet, D.M.; Fundueanu, G. Poloxamer/carboxymethyl pullulan aqueous systems - Miscibility and thermogelation studies using viscometry, rheology and dynamic light scattering. *Polymers* **2023**, *15*, 1909. [CrossRef]
68. Sharma, K.; Joseph, J.P.; Sahu, A.; Yadav, N.; Tyagi, M.; Singh, A.; Pal, A.; Kartha, K.P.R. Supramolecular gels from sugar-linked triazole amphiphiles for drug entrapment and release for topical application. *RSC Adv.* **2019**, *9*, 19819–19827. [CrossRef]

gels MDPI

Article

Effects of Non-Conventional Sterilisation Methods on PBO-Reinforced PVA Hydrogels for Cartilage Replacement

Tomás Pires [1], Andreia Sofia Oliveira [1,2,3], Ana Clara Marques [4], Madalena Salema-Oom [3], Célio G. Figueiredo-Pina [3,5,6], Diana Silva [1,*] and Ana Paula Serro [1,3]

1 Centro de Química Estrutural (CQE), Institute of Molecular Sciences, Departamento de Engenharia Química, Instituto Superior Técnico, Universidade de Lisboa, Av. Rovisco Pais, 1049-001 Lisbon, Portugal
2 Instituto de Engenharia Mecânica (IDMEC), Instituto Superior Técnico, Universidade de Lisboa, Av. Rovisco Pais 1, 1049-001 Lisbon, Portugal
3 Centro de Investigação Interdisciplinar Egas Moniz (CiiEM), Instituto Universitário Egas Moniz, Quinta da Granja, Monte da Caparica, 2829-511 Caparica, Portugal
4 CERENA, DEQ, Instituto Superior Técnico, Universidade de Lisboa, Avenida Rovisco Pais, 1049-001 Lisboa, Portugal
5 CDP2T, Escola Superior de Tecnologia de Setúbal, Instituto Politécnico de Setúbal, 2910-761 Setúbal, Portugal
6 CeFEMA, Instituto Superior Técnico, Universidade de Lisboa, 1049-001 Lisbon, Portugal
* Correspondence: dianacristinasilva@tecnico.ulisboa.pt

Abstract: Articular cartilage (AC) degradation is a recurrent pathology that affects millions of people worldwide. Polyvinyl alcohol (PVA) hydrogels have been widely explored for AC replacement. However, their mechanical performance is generally inadequate, and these materials need to be reinforced. Moreover, to be used in a clinical setting, such materials must undergo effective sterilisation. In this work, a PVA hydrogel reinforced with poly(p-phenylene-2,6-benzobisoxazole) (PBO) nanofibres was submitted to three non-conventional sterilisation methods: microwave (MW), high hydrostatic pressure (HHP), and plasma (PM), in order to evaluate their impact on the properties of the material. Sterilisation was achieved in all cases. Properties such as water content and hydrophilicity were not affected. FTIR analysis indicated some changes in crystallinity and/or crosslinking in all cases. MW was revealed to be the most suitable method, since, unlike to PM and HHP, it led to a general improvement of the materials' properties: increasing the hardness, stiffness (both in tensile and compression), and shear modulus, and also leading to a decrease in the coefficient of friction against porcine cartilage. Furthermore, the samples remained non-irritant and non-cytotoxic. Moreover, this method allows terminal sterilisation in a short time (3 min) and using accessible equipment.

Keywords: articular cartilage substitutes; PVA hydrogels; PBO nanofibres; sterilisation; microwave; high hydrostatic pressure; plasma

Citation: Pires, T.; Oliveira, A.S.; Marques, A.C.; Salema-Oom, M.; Figueiredo-Pina, C.G.; Silva, D.; Serro, A.P. Effects of Non-Conventional Sterilisation Methods on PBO-Reinforced PVA Hydrogels for Cartilage Replacement. *Gels* **2022**, *8*, 640. https://doi.org/10.3390/gels8100640

Academic Editor: Esmaiel Jabbari

Received: 9 September 2022
Accepted: 5 October 2022
Published: 9 October 2022

1. Introduction

Osteoarthritis is a degenerative joint disease that affects more than 500 million people worldwide and is recognized as the main cause of permanent disability. Rheumatoid arthritis, injuries, or overuse can also lead to joint damage. The hip and knee are the most commonly affected joints, due to the large loads sustained [1]. Articular cartilage (AC) is a highly hydrated connective tissue that covers the ends of bones in diarthrodial joints. It consists of a porous elastic solid matrix, containing interstitial fluid, and exhibits high load-bearing and low-friction capabilities. Such features arise from its biphasic composite structure, in conjunction with the lubrication characteristics of the synovial fluid (SF) [2,3]. The capability of cartilage to repair and regenerate is limited, due to its aneural and avascular nature [4–6]. As such, the progression of cartilage degradation is generally inevitable.

Presently, the available treatment options for injured AC range from physiotherapy and/or the use of anti-inflammatories or viscosupplementation, when the damage is mild, to total joint arthroplasty, in the more severe cases [7,8]. Since the latter is a quite invasive procedure, new treatment options have been pursued, to manage the problem in earlier stages [9]. Cartilage replacement, by filling the defector in the form of a hemiarthroplasty counter face, has been considered, as well as the possibility of using an interposition device to alleviate pain and/or correct joint deformity [10]. Regardless, replicating the unique biphasic and heterogeneous structure of AC is a challenging task. Among the studied materials, hydrogels have emerged as a promising alternative, owing to their high-water contents, porous structure, and viscoelastic mechanical properties that resemble those of cartilage tissue [11,12].

Polyvinyl alcohol (PVA) has received special attention due to its excellent biocompatibility and non-toxicity. However, simple PVA hydrogels usually present insufficient mechanical properties to resist the high loads experienced by diarthrodial joints [13]. The preparation procedure may critically affect the performance of these materials. Cast-drying and freeze-thawing are the most commonly used methods to obtain these hydrogels. In a previous work [14], the authors found that PVA hydrogels obtained by cast-drying better mimic the natural cartilage tissue, in terms of morphology, water content, and tribomechanical properties. However, an improvement of these properties was still required. Different strategies have been used for this purpose, e.g., post-production physical treatments, such as thermal annealing [10,15], or the incorporation of reinforcement agents into the hydrogel matrix, such as nanofibres or nanoparticles [16–19]. The addition of nano/microfibres to a PVA hydrogel polymer network represents a promising method for cartilage substitution, since this can mimic the collagen fibres present in the extracellular matrix of natural cartilage.

In a recent work, Oliveira et al. [20] synthesized a PVA hydrogel reinforced with poly(p-phenylene-2,6-benzobisoxazole (PBO, commercial name Zylon®), which demonstrated excellent mechanical properties for cartilage substitution, when the damage is still circumscribed. In particular, it presented a large water content, high mechanical stiffness, low friction, and excellent biocompatibility. PBO fibres show an extremely high tensile strength (~5.8 GPa), approximately 1.6-times higher than Kevlar®. Its tensile elastic modulus depends on the type of fibre, ranging from 180 or 270 GPa for "as spun" (AS) or "high modulus" (HM) fibres, respectively. PBO's compressive strength is significantly lower than its tensile strength, falling between 0.469 and 0.561 GPa. PBO also shows high thermal stability (up to 650 °C) [21]. The PBO fibres can be tailored into nanofibres through acid-catalysed hydrolysis. The high aspect ratio of the nanofibres can lead to multiple fibre interconnections, promoting gelation. Furthermore, this process creates functional sites in the nanofibres (i.e., carboxyl, amino and phenolic hydroxyl groups), which can be combined with PVA molecules through hydrogen bonding [20,22].

Sterilisation is a mandatory step for biomaterials intended for AC replacement, in order to minimize the incidence of medical device-related infections. Terminal sterilisation methods can induce changes to several material properties (e.g., aspect, size, colour, chemical structure, physical integrity, and biocompatibility). Many hydrogels are sensitive to conventional sterilisation methods (e.g., steam and heat or gamma-radiation) [23]. The presence of water in these materials makes their sterilisation even more challenging, as the water molecules may have harmful effects, such as the breaking of chemical bonds [24]. Thus, non-conventional sterilisation methods have been explored, to overcome possible detrimental damage to biomaterials during processing.

Microwave irradiation (MW) is an accessible method that relies on the exposition of the materials to electromagnetic radiation in the microwave frequency range (300 MHz and 300 GHz). This induces rotation of the polar molecules (as water) and ionic polarization, leading to a rapid increase in temperature [25], which results in the irreversible denaturation of enzymes and structural proteins essential for microorganisms' viability and reproduction [26]. Another method recently investigated for the sterilisation of hydrogels is high hydrostatic pressure (HHP) [27,28]. In this method, the packaged materials are immersed in a water bath and submitted to high pressures (e.g., 100–1000 MPa) at relatively low temperatures (50–70 °C). The high applied pressures induces changes in the microorganisms' cell membranes that lead to the leakage of the intracellular contents and also denature key enzymes [29]. Low-temperature glow discharge plasma treatment has also been investigated for its potential to sterilise polymeric materials. Gas plasma (PM) is created through the application of strong electromagnetic fields in an inert gas atmosphere, forming free radicals that cause etching of the cells' surfaces, leading to the release of the inner contents from living cells [30,31].

This work aimed to study the effect of these three non-conventional sterilisation methods (i.e., MW, HHP, and PM) on a PBO nanofibre reinforced PVA hydrogel intended to be used as cartilage replacement material. The efficacy of sterilisation was accessed, and the sterilised hydrogels were characterised in terms of their structural and mechanical properties, wettability, swelling capacity, and rheological behaviour, to evaluate possible effects of the different sterilisation procedures. Finally, the biocompatibility of the hydrogels was ascertained via irritability (Hen's egg test–chorioallantoic membrane, HET–CAM test) and cytotoxicity assays.

2. Results and Discussion

PVA has been, to date, the most investigated synthetic hydrogel for cartilage replacement. However, PVA alone presents mechanical and tribological properties below those of natural cartilage [32]. Therefore, the reinforcement of PVA with fibres, and/or copolymerization with other compounds, has been attempted, to improve its properties [33]. In this work, PVA/PBO hydrogels were produced through the cast-drying method, which leads to the formation of hydrogen bonds between PVA chains [34] and of those of PBO [20]. The PBO fibres were solubilized in a TFA (trifluoroacetic acid)/MSA (methanesulfonic acid) mixture, to carry out acid-catalysed hydrolysis, tailoring them into nanofibres.

Sterilisation is a mandatory step, to ensure the biological safety of cartilage replacement materials. However, it can change their properties and impair their performance in extreme cases. Conventional sterilisation methods often rely on exposure to heat and highly energetic radiation (e.g., steam heat, gamma irradiation), but PVA hydrogels are generally sensitive to such agents [35]. In this sense, this work aimed to evaluate the effect of non-conventional sterilisation methods on PVA/PBO hydrogels, to identify a possible alternative that, while being effective, does not harm them. The materials were exposed to microwaves, HHP, and Argon (Ar) plasma, and the efficacy of sterilisation was verified. In all cases, they were processed in the hydrated state, except for plasma, where samples were first dried. Such a procedure was adopted, as the presence of water, even in small amounts, makes the argon plasma unstable and dangerously increases the temperature (which may reach temperatures of several hundreds of °C) [36], potentially damaging the hydrogels. Therefore, plasma-treated samples require a posterior hydration, which must be performed in aseptic conditions.

2.1. Microbiological Load and Sterility Assessment

Determination of the microbiological load of the non-sterile PVA/PBO samples showed a low bioburden value, in the order of 10^2 CFU/sample for bacteria, and of 7.4 ± 2.8 CFU/sample for fungi. This was expected, taking into account that the method of synthesis of the PVA/PBO hydrogels involved highly acid solutions. Nevertheless, the potential for infection with such a bioburden is still of high risk.

The sterility tests did not allow visualizing any signs of contamination in the culture media for all sterilisation processes after the 14 days of incubation (Figure 1), demonstrating that all the studied methods ensured biologically safe materials. Exposure to plasma for 3 or 5 min (MW-3 and MW-5, respectively), led to sterile materials in both cases, whereby only MW-3 samples were kept in the study.

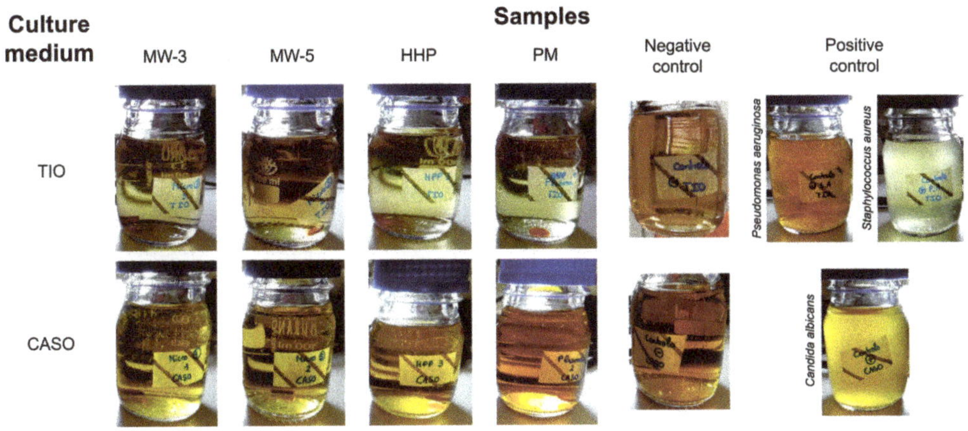

Figure 1. Culture media for bacteria and fungi after 14 days of incubation for MW-3, MW-5, HHP, and PM PVA/PBO hydrogel samples. The negative and positive controls are also shown.

2.2. Morphology

Morphological analysis of the hydrogels' surfaces by SEM (Figure 2) showed some irregularities on non-sterile samples, which had almost disappeared after HHP treatment. MW led to an even smoother surface. This effect may have resulted from the temperature to which the samples were submitted during the sterilisation procedures, and in the case of HHP, also from the high pressure: 70 °C, 600 MPa for 10 min for HHP; and 100 °C, atmospheric pressure for almost 3 min for MW. In contrast, the plasma induced some heterogeneity, leading to the appearance of protuberances and an increase in the surface's roughness. Moreover, etching-derived holes could also be observed in these samples. Etching is a common side effect of the plasma procedure, being primarily caused by the enhanced energies of the ions and radicals that strike the surfaces [30,37].

2.3. Chemical Structure

Fourier-transform infrared spectroscopy (FTIR) analysis was carried out on the non-sterile and sterile hydrogels, to attain information regarding possible chemical modifications induced by the sterilisation procedures. Figure 3 shows the FTIR spectra of the different hydrogel types.

The hydrogels' spectra demonstrated several similarities to the PVA spectrum observed in a previous study [20]. This was expected, since PVA had a much higher proportion in the hydrogel composition. Additionally, no new peaks arose from the sterilisation procedures.

The large band observed at ca. 3280 cm^{-1} is linked to the stretching of O–H bonds. The vibrational bands observed between 3000 and 2840 cm^{-1} refer to alkyl group C–H bonds stretching, and that at ca. 1085 cm^{-1} to C-O-C, i.e., C-O in secondary alcohols, such as those typical of PVA structures [17,38]. The region between 1700 and 1500 cm^{-1} exhibits peaks typically attributed to the stretching of C-C bonds and the covalent double bonds C=N and C=C from PBO, and the C=O present in PVA [20,39]. The most significant difference was the higher intensity of the peaks within the region 1700–1500 cm^{-1} for MW-3 samples, compared to non-sterilised ones. This is in agreement with what was observed in other studies regarding the effect of microwaves on PVA and PVA/graphene nanocomposites [40]. These authors found an increase in the C=O and C=C peak intensity after 3 min irradiation at 200 W and hypothesized that microwaves led to side chain scission, enhancing the subsequent bonds between the polymer chains, i.e., the formation of a more cross-linked structure. At lower wavenumbers, but above 1300 cm^{-1}, several peaks were identified, corresponding to C-N stretching of PBO and CH_2 bending of PVA, while the peak at 704 cm^{-1} was ascribed to the C-H bending of PBO [39].

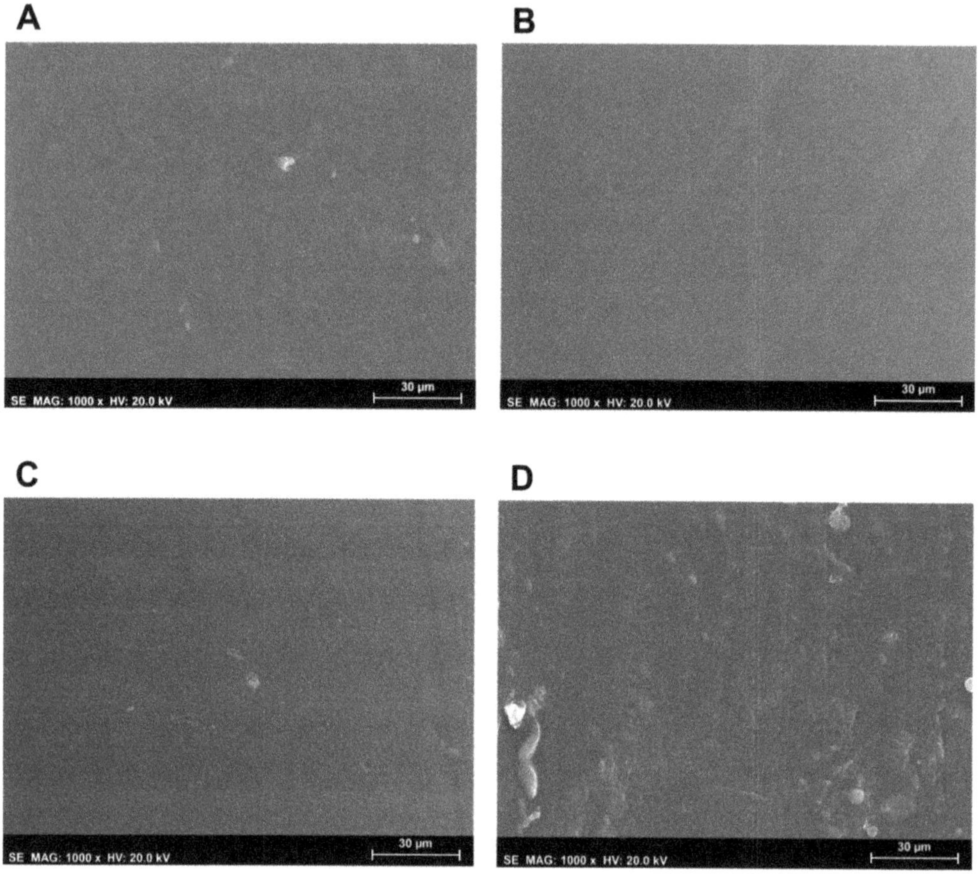

Figure 2. SEM micrographs (1000× magnification, scale bar 30 µm) of the PVA/PBO hydrogels: non-sterile (**A**), MW-3 (**B**), HHP (**C**), and PM (**D**) samples.

According to the literature [41], FTIR allows inferring the degree of crystallinity of hydrogels through the analysis of the peak at 1141 cm^{-1}. The intensity of this peak is influenced by the crystalline portion of the polymeric chains, which is related to the symmetric C–C stretching mode or stretching of the C–O of a portion of the chain, where an intramolecular hydrogen bond is formed between two neighbouring OH groups that are on the same side of the plane as the carbon chain. Following a procedure similar to that adopted in [41], the ratio of band intensities (height) of the 1141 cm^{-1} (C-O crystallinity) and 1085 cm^{-1} band (C-O-C bond stretching in PVA structure, secondary alcohol), was calculated for each sample (Table 1). It should be noted that this latter band did not significantly vary with the treatments. This analysis revealed that samples MW-3 and HHP were the ones exhibiting a higher crystallinity, followed by the PM sample. In fact, the highest peak ratio value observed for HHP was likely be due to the extremely high pressures occurring during the HHP process, which might decrease the distance between the polymeric chains, inducing the rupture of some intra/inter-molecular bridges and/or the formation of new bonds, namely hydrogen bonds [42,43], and favouring order at long distance, i.e., the formation of crystalline regions.

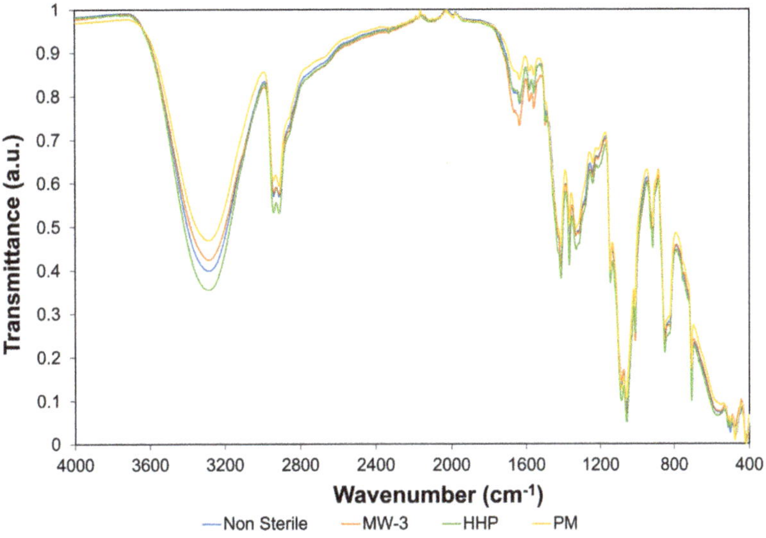

Figure 3. FTIR-ATR spectra of the non-sterile and sterile PVA/PBO hydrogels, in the region of 4000–400 cm^{-1}.

Table 1. Peak intensity ratios of relevant peaks from Figure 3.

Sample	Peak Ratio I_{1141} cm^{-1}/I_{1085} cm^{-1}	Peak Ratio I_{3280} cm^{-1}/I_{1085} cm^{-1}	Peak Ratio I_{1056} cm^{-1}/I_{1085} cm^{-1}
NS	0.635	1.33	1.14
MW-3	0.673	1.35	1.20
HHP	0.675	1.37	1.08
PM	0.662	1.13	1.00

Regarding the O-H stretching vibration peak (ca. 3280 cm^{-1}), although no significant change in its wavenumber was observed, a decrease in terms of intensity was found for the plasma samples (PM), as Table 1 reveals through the peak ratio of the OH bonding at 3280 cm^{-1} and that of the C-O bond at 1085 cm^{-1} (I_{3280} cm^{-1}/I_{1085} cm^{-1}). The effect of plasma on the materials strongly depended on the conditions used (e.g., power, time, gas flux). The literature is not consensual on the influence of argon plasma on PVA-based materials: while some authors [44] found significant changes in the FTIR spectra, such as the appearance of new peaks associated with amide, carboxylic acid, and OH/NH functionalities, others [45] reported minor changes.

Concerning the samples MW-3 and HHP, a slight increase in the ratio I_{3280} cm^{-1}/I_{1085} cm^{-1} was observed relative to the non-sterile samples. The higher degree of crystallinity of these samples (referred above) is probably associated with the higher number of OH groups, important for establishing H bonds.

Finally, Table 1 shows a decrease for the PM samples in the relative intensity of the peak at 1056 cm^{-1}, ascribed to C-O bond stretching in primary alcohols, when compared to that of C-O bond stretching in the secondary alcohols typical of the PVA macromolecular chains (I_{1056} cm^{-1}/I_{1085} cm^{-1}). This is in line with the lower amount of OH groups found for the PM samples, i.e., a lower peak ratio I_{3280} cm^{-1}/I_{1085} cm^{-1}. In turn, the MW-3 samples exhibited a higher value for this peak ratio, suggesting more primary alcohol moieties than the non-sterilised samples, which might eventually indicate some chain breaking.

2.4. Water Content and Wettability

The hydration process kinetics were studied over time (Figure 4). No significant differences were observed between the sterile and non-sterile materials ($p = 0.8793$). A sharp rise in *WC* was observed in the first 5 h, with equilibrium being reached after 24 h (values in the range 73–74.5%). Such values fall within the range observed for AC (65–80%) [46].

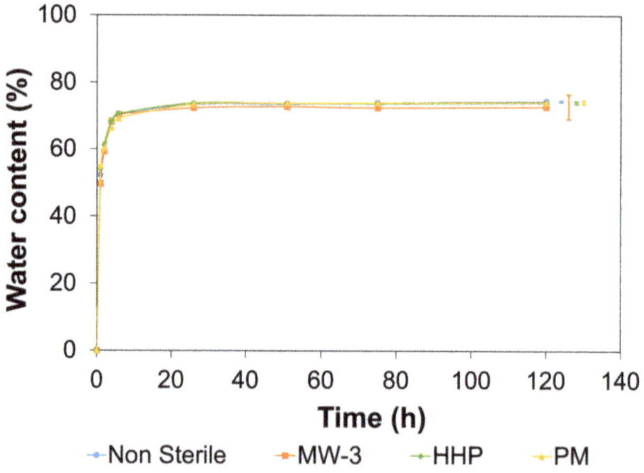

Figure 4. Water content over time for non-sterile and sterile PVA/PBO hydrogels. The error bars represent the ± mean standard deviations ($n = 3$).

The wettability of polymers is often defined as a crucial parameter for cell attachment [47]. The optimal contact angle values seem to depend on the type of cells, but research suggests that moderate values are more desirable than having hydrogels with a very high or low wettability [48]. The untreated PVA/PBO hydrogels led to an average contact angle value of 40 ± 3° (in the range of those found in literature [15,49]), showing an hydrophilic behaviour. The sterilisation methods did not cause meaningful modifications of the samples' wettability (MW-3 39 ± 3°, HHP 39 ± 4°, PM 40 ± 4°).

2.5. Mechanical Behaviour

The mechanical properties of cartilage replacement materials are critical, since in the body they will experience a variety of mechanical forces, namely compression, tension, and shear [50]. In order to minimize strain mismatch at the implant–tissue interface, the materials must present properties as similar as possible to the natural tissue.

2.5.1. Compressive Tests

Figure 5A shows typical compressive stress-strain curves obtained for the PVA/PBO hydrogels.

Since hydrogels are biphasic materials, in the initial stages of the compression experiments, the applied load was essentially supported by the liquid contained in the hydrogels. Raising the load leads to the exudation of the fluid from the matrix, being the load transferred to the solid phase. This is reflected by an increase in the slope of the curves.

Figure 5B depicts the compressive tangent modulus between 5 and 35% strains. Although all samples showed an increase of this parameter over the entire strain range tested, the rate of increase was different for the various tested sterilisation treatments. This leads to distinct compressive behaviours, depending on the compressive strain. Bellow 15%, the HHP and MW-3 samples exhibited a higher tangent compressive modulus, which might have been due to the higher crystallinity level of these samples, as detected by FTIR analysis. Above this strain, the MW-3 samples stood out, with superior values, which may have resulted from the higher content of intramolecular covalent double bonds.

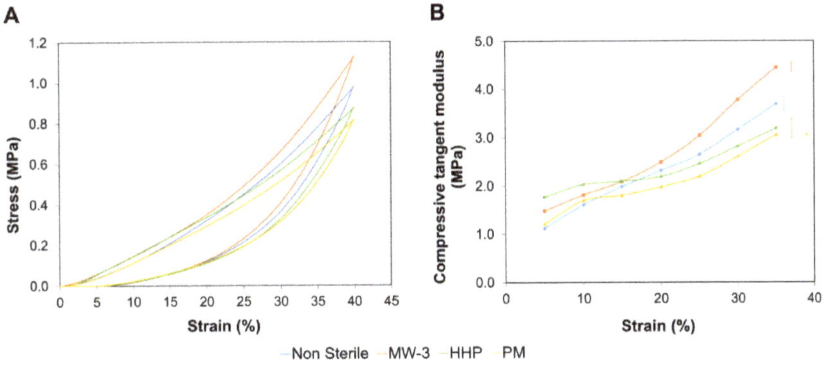

Figure 5. Typical compressive stress-strain curves (**A**) and compressive tangent modulus (**B**) for non-sterile and sterile PVA/PBO hydrogels. The error bars represent the mean ±standard deviations ($n = 3$).

2.5.2. Tensile Tests

Representative tensile stress-strain curves of all tested samples are shown in Figure 6, as well as their tensile tangent modulus. All the sterilisation treatments led to an increase in the tangent tensile modulus, due to the higher crystallinity level, as concluded from the FTIR spectra. The highest value was observed for MW-3 samples, possibly due to the greater content of covalent double bonds C=N and C=C from PBO, and the C=O present in PVA. It should be emphasized that, contrarily to the compressive tangent modulus, the tensile modulus decreased with strain. Indeed, both compressive and tensile stresses promoted, in the initial stage, the release of water from the hydrogels. However, as the stress was increased, compression led to the interpenetration of the polymer chains, resulting in the materials' hardening, while stretching may cause the rupture of hydrogen bonds (that ensure the intermolecular physical crosslinking), allowing the slipping of the chains over each other and resulting in a lower material stiffness [15].

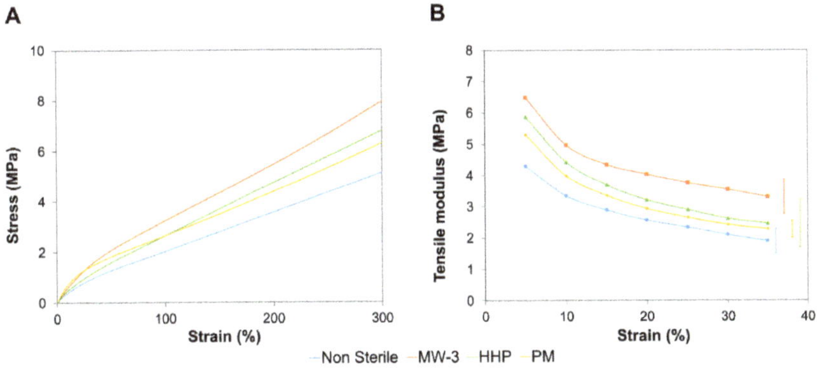

Figure 6. Typical tensile stress-strain curves (**A**), and tangent tensile modulus (**B**) for non-sterilised and sterilised PVA/PBO hydrogels. The error bars represent the mean ±standard deviations ($n = 3$).

2.5.3. Rheological Tests

Oscillatory shear mode testing was conducted, by carrying out a frequency sweep analysis in the range of 0.1–50 Hz within the LVER at a fixed shear strain of $\gamma = 0.1\%$

(determined through an amplitude sweep assay). Figure 7A shows that for all hydrogels, G' was about one order of magnitude higher than G'', which indicates that, independently of having been sterilised or not, they mainly exhibited a solid elastic behaviour, within the frequency range studied. Measurements of G' and G'' at a fixed strain (γ = 0.1%) and frequency (ω = 1 Hz) were carried out as a function of time for 1 h (Figure 7B). Again $G' > G''$ for all hydrogels. Microwave treatment resulted in an increase of both parameters, but this was more accentuated for G', showing that after this treatment the material could store a higher amount of energy. These samples showed a larger loss factor (Figure 7C), and therefore a higher damping capacity, which may have resulted from the larger amount of primary alcohol moieties, suggesting the presence of smaller segments embedded within the long polymeric PVA chains.

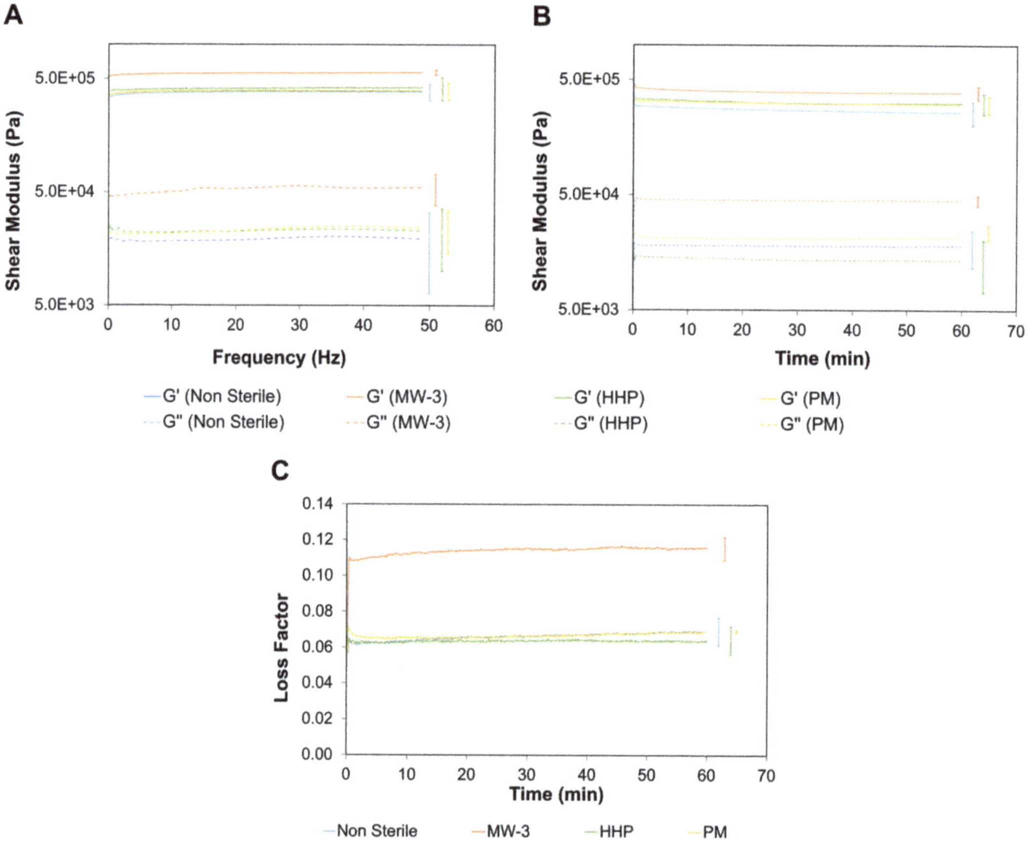

Figure 7. Storage (G′) and loss (G″) moduli for non-sterile and sterile PVA/PBO hydrogels (**A**) as a function of the frequency (0.1–50 Hz) and (**B**) of time (1 h) with a frequency of 1 Hz, both at a fixed strain rate γ = 0.1%. Loss factor variation with time (**C**). The error bars are the ± standard deviation (n = 3).

It should be underlined that both the non-sterile and sterile PVA/PBO hydrogels showed a shear modulus value ($G^* = (G'^2 + G''^2)^{1/2}$) within the reported range for AC. In fact, although the reported values for the G^* of AC vary between studies, they are typically in a range of 0.2–2.5 MPa [51]. Studies of lamb cartilage plugs on a similar device as the one used in this study (1 Hz frequency and effective shear strain amplitude of 0.023%) reported a storage moduli in the range of 0.4–0.6 MPa [52].

2.5.4. Hardness Tests

The Shore A hardness value of the non-sterile hydrogel was 45.0 ± 1.3 (Figure 8). HHP treatment did not led to a statistically significant variation of this property (p = 0.1102), while plasma treatment slightly affected it (p = 0.0191). However, MW-3 led to an increase of ≈18 (p < 0.001). This is in agreement with the superior mechanical behaviour observed for these samples and may have been a result of the higher crystallinity and, possibly, higher degree of cross-linking.

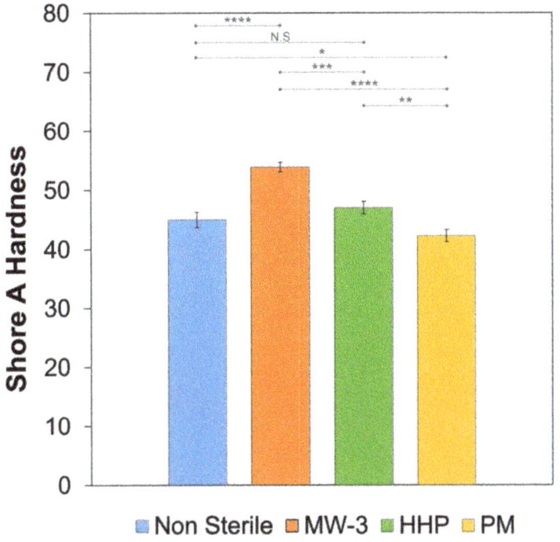

Figure 8. Shore A hardness values for the non-sterile and sterile PVA/PBO hydrogels. The error bars are the ± standard deviation (n = 3). Statistical analysis was performed using Student-t test, with significance set at * $p < 0.05$, ** $p < 0.01$, *** $p < 0.005$, **** $p < 0.001$. N.S. = not significant.

2.6. *Tribological Behaviour*

The coefficient of friction of the non-sterile hydrogel was measured against porcine cartilage samples, using a normal load of 10, 20, and 30 N. Porcine cartilage is significantly similar to human cartilage, and therefore is often used as a model to study the effects of friction between cartilage surfaces and synthetic materials. DD water and SSF were used as lubricants, the latter to better simulate the pseudo-synovial fluid. The average values obtained after stabilization of the coefficient of friction are given in Figure 9A. They increased with the normal load (p = 0.0056 and p = 0.0035 for DD water and SSF, respectively). This was expected, due to the higher deformation of the hydrogel, which should increase the contact area between the sliding surfaces and, therefore, the adhesion forces. In addition, lower coefficient of friction values were observed with SSF lubricant. The presence of hyaluronic acid and albumin significantly increased the viscosity of the fluid (17.5 ± 0.7 mPa.s compared to ≈1 mPa.s for water at room temperature), enhancing the hydrodynamic component of the lubrication. This contributed to a mixed lubrication regime over the boundary the one that should occur with DD water [53]. Moreover, the adsorption of the biomolecules on the surfaces gives rise to a protective layer that will contribute to decreasing the friction between the surfaces.

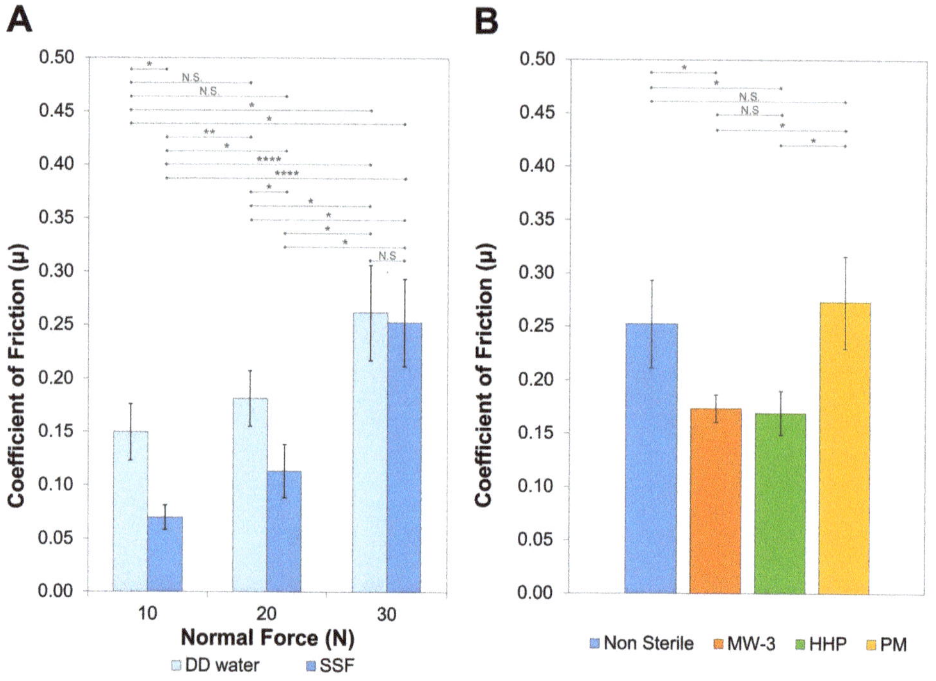

Figure 9. Coefficient of friction values for the non-sterile PVA/PBO hydrogels at different applied loads, using DD water and SSF as lubricants (**A**), and of the non-sterile and sterile hydrogels with 30 N of normal load and SSF as lubricant (**B**). Porcine cartilage was used as the counter body. The error bars represent the mean ± standard deviations (n = 3). Statistical analysis was performed using a Student-t test, with significance set at * $p < 0.05$, ** $p < 0.01$, **** $p < 0.001$. N.S. = not significant.

The effect of the sterilisation procedures was only studied for the 30 N load (average contact pressure ≈1 MPa) and using SSF as lubricant (Figure 9B). MW-3 and HHP samples led to coefficients of friction about 30% lower than the non-sterile hydrogels (p = 0.0406 and p = 0.0412, respectively), while PM samples showed a slight increase (≈8%), but this was not statistically significant. The coefficient of friction was affected by multiple factors, for example, the surfaces' adhesion, deformation, and roughness. In addition, in the presence of biological molecules, this also depends on the characteristics of the adsorbed layer (rigid or viscoelastic), which is determined by the surface chemical structure, polarity, and roughness. Although the interpretation of the friction results was not straightforward, they fell within the range of values found in the literature for natural cartilage, which range from 0.002 to 0.5 [54], depending not only on the experimental conditions used (e.g., test configuration, lubricant, applied load), but also on the type of cartilage and donor age [7,8].

After the tribological tests, marks were clearly visible on the hydrogels' surface. SEM images of the inside of the wear tracks caused by the tribological tests performed with an applied load of 30 N in SSF are depicted in Figure 10. The non-sterile samples showed some signs of delamination. MW-3 and HHP treatments induced some rippling on the surfaces, which become rougher. According to Dong et al. [55], ripples may be formed due to a stick-slip process that occurs when the hydrogels are pulled and deformed along the sliding direction by the sliding counter-body. This effect was more obvious for the PM samples, suggesting that adhesion was more prominent for this material, which is in agreement with the higher values observed for the coefficient of friction.

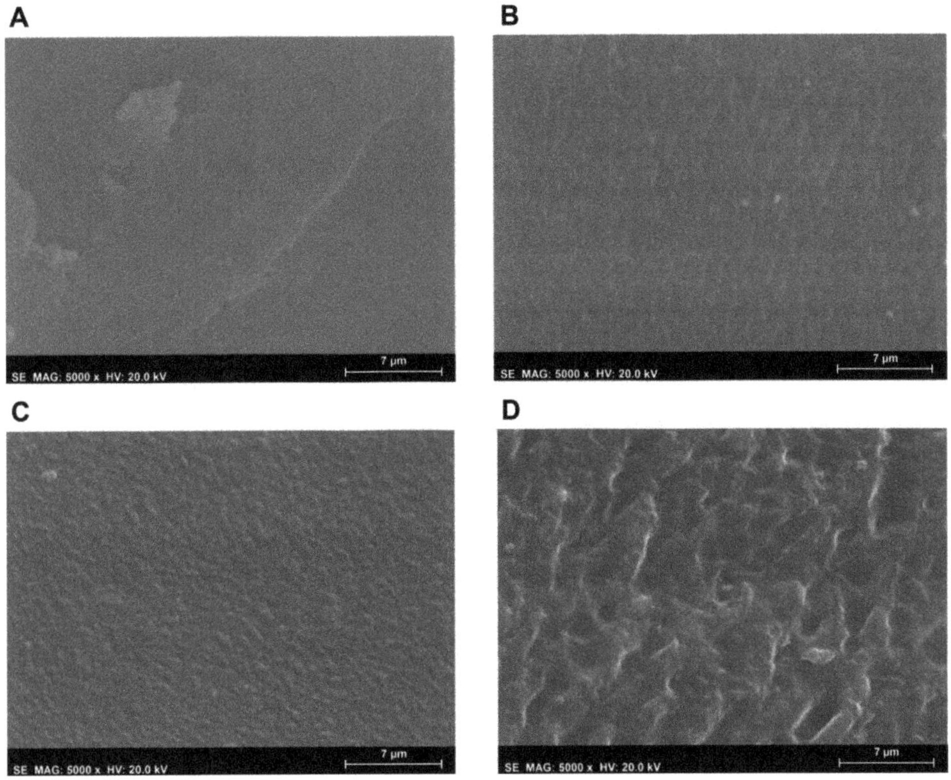

Figure 10. SEM micrographs (5000× magnification, scale bar 7 µm) of the PVA/PBO hydrogels inside the wear tracks, produced in the tribological experiments. Non-sterile (**A**), MW-3 (**B**), HHP (**C**), and PM (**D**) samples.

2.7. HET-CAM Test

The HET-CAM test is a well-established prediction model for tissue irritation due to chemicals, based on the observation of signs of lysis, haemorrhage, or coagulation in the chorioallantoic membrane (CAM) surrounding the chicken embryo in eggs. Although cartilage presents an avascular structure, the surrounding tissues of the joints and bone are vascularized, thereby making the HET-CAM test relevant.

Photographs of the CAM after 5 min of direct contact with the non-sterile and sterile PVA/PBO hydrogels are presented in Figure 11, as well as the positive control (after the addition of NaOH) and the negative control (after the addition of NaCl solution). In the positive control, visible signs of severe irritation can be observed (Figure 11F). Both sterilised and non-sterilised hydrogels induced a behaviour similar to the one observed for the negative control (Figure 11E), i.e., no visual signs of lysis, haemorrhage, or coagulation, were observed. Therefore, the IS was equal to 0, and the hydrogels could be classified as non-irritant.

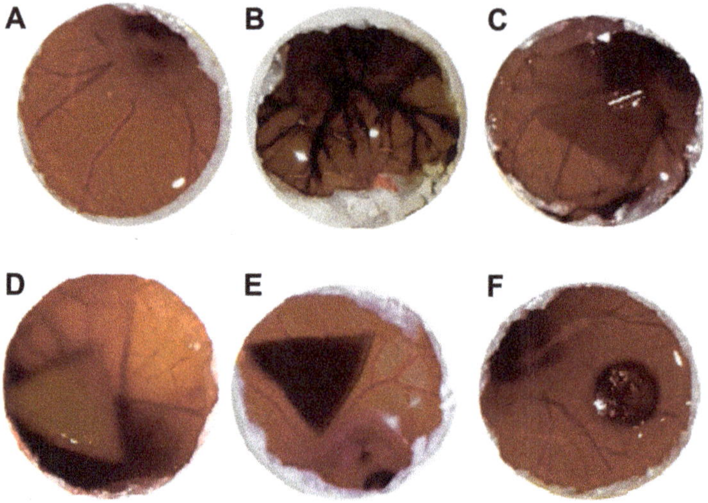

Figure 11. Chorioallantoic membrane images after 5 min of exposure to the negative control (NaCl 0.9%) (**A**), positive control (NaOH, 1 M) (**B**), and to the non-sterile (**C**), MW-3 (**D**), HHP (**E**), and PM (**F**) hydrogel samples.

2.8. Cytotoxicity Analysis

The biocompatibility of the sterile PVA/PBO hydrogels was studied using an MTT assay, after exposure of the chondrocyte cells to the extracts from the samples. The cell viability was not affected by the extracts from PM samples (98.6 ± 6.8%), but MW-3 and HHP hydrogels led to significantly lower viability (68.8 ± 9.3% and 77.6 ± 7.6%, respectively) ($p < 0.001$) (Figure 12A). According to ISO 10993-5:2009, a material is considered non-cytotoxic when the cellular viability ≥70% [56]. While HHP and PM samples led to values above this threshold, MW-3 was slightly below. However, the sensitivity of this assay was significantly higher than in vivo tests, so it is possible to consider the MW-3 samples as within the limit of non-cytotoxicity.

Figure 12B shows optical microscopy images of the incubated chondrocytes. Although the cell proliferation had decreased for MW-3 and HHP, it can be observed that cells displayed the normal elongated chondrocyte morphology, as in the negative control.

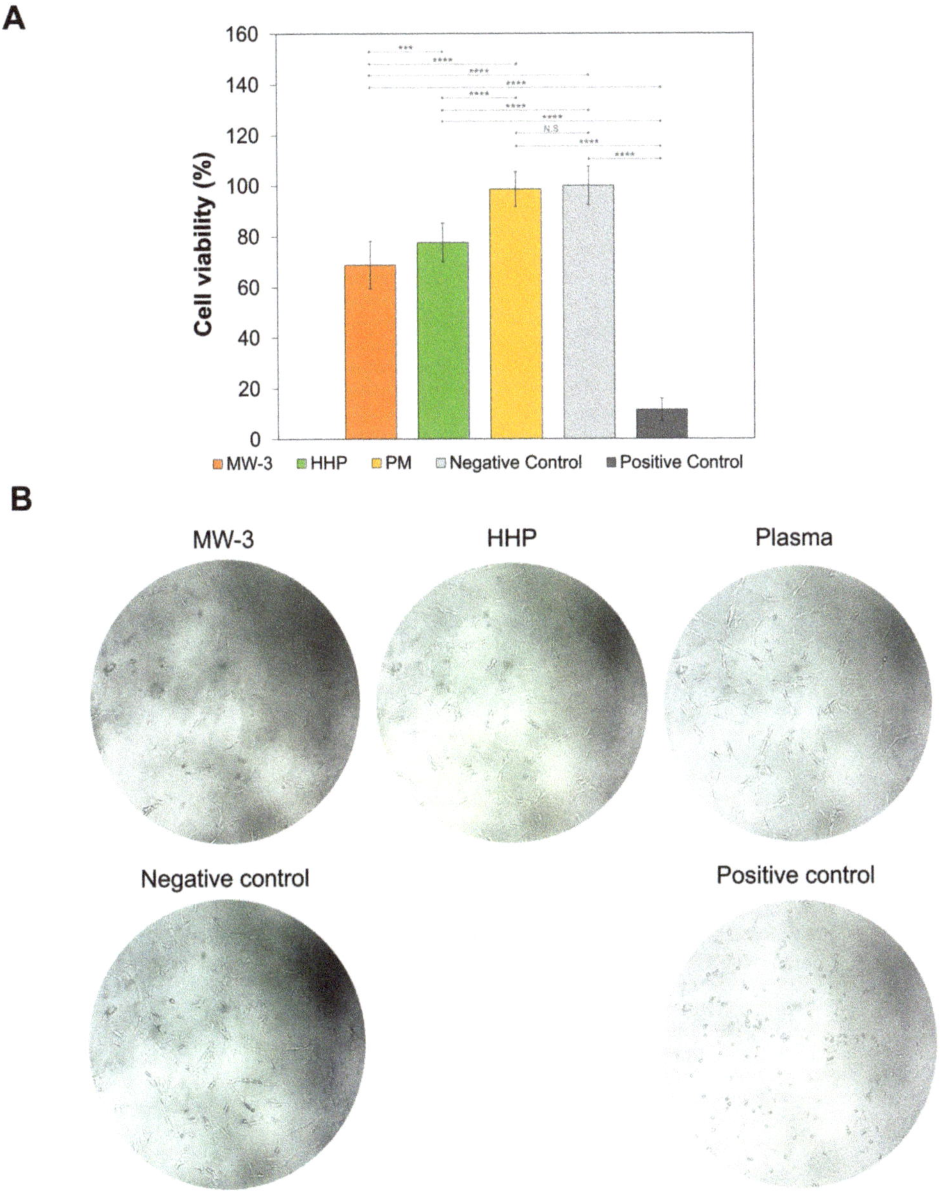

Figure 12. Chondrocyte cell viability (%) determined by the MTT assay (**A**). The relative cell viability is presented as a percentage compared to the negative control. The error bars represent the ±standard deviations (n = 4). Optical microscopy images of the incubated chondrocytes after 48 h exposure to the hydrogels' extracts and controls (**B**). Statistical analysis was performed using a Student-t or Willcoxon test, with significance set at *** $p < 0.005$, **** $p < 0.001$. N.S. = not significant.

3. Conclusions

In this work, the effect of three non-conventional sterilisation methods (i.e., MW, HHP, and PM) on a PVA/PBO hydrogel developed for cartilage replacement was assessed. This hydrogel was able to combine a set of attractive properties that are quite difficult to achieve together in hydrogels, namely a large water content, high mechanical stiffness, low friction, and excellent biocompatibility, which allow mimicking natural cartilage. Since sterilisation is a mandatory step for implantable materials and as hydrogels are generally sensitive materials, finding methods/conditions that ensure the maintenance of hydrogel properties remains a challenge.

The results show that the chosen experimental conditions (MW: 700 W, 3 min; HHP: 600 MPa, 70 °C, 10 min; PM: Ar, 18 W, 5 min) were adequate to ensure sterility in all cases. Sterilisation procedures did not cause significant changes in the surface' morphology, except for PM, which led to an increase in roughness, due to the etching effect. FTIR analysis showed that the treatments induced some changes in the materials' chemical structure, in terms of the degree of crystallinity and crosslinking, explaining some differences in its behaviour. The samples' equilibrium water content did not suffer significant alteration after the treatments, remaining ≈74%. Similarly, the hydrophilicity was also not affected (the water contact angle remained ≈40°). In general, the sterilisation treatments did not negatively impact the materials' mechanical properties. MW-3 stood out due to an increase in hardness, stiffness (both in tensile and compression), and shear modulus. The coefficient of friction obtained against porcine cartilage, using SSF as lubricant, was significantly lower for these samples, when compared with the non-sterile. HHP led to a similar reduction. HET-CAM tests showed that, after the treatments, the samples remained non-irritant. Moreover, contrarily to PM, MW-3 and HHP led to a decrease in the viability of chondrocytes upon exposure to samples' extracts. However, the observed values were still above the cytotoxicity threshold.

Overall, among the three studied methods, MW irradiation proved to be the most adequate to sterilise the samples. Due to its high penetration capacity and by achieving high temperatures in the hydrated materials, it is quite efficient, allowing for terminal sterilisation in exposure times as low as 3 min. Furthermore, it led to a general improvement of the mechanical properties of the hydrogel. In turn, argon plasma had some detrimental effects on the material, but more than that, it presents the drawback of not allowing terminal sterilisation, since the materials cannot be processed in their final packaging (due to the low penetration of plasma). Concerning HHP, this led to effective sterilisation of the hydrogel and minor property changes. However, it is surpassed by MW, since it is necessary to have specialized personnel to operate the expensive equipment required.

It must be stressed that in the choice of the ideal sterilisation method of a hydrogel, besides guaranteeing that it keeps its integrity and key properties essential to ensure its functionality, no toxic or hazardous residues should be left in the material. Moreover, other aspects such as a high penetration capability, short processing time, low cost and simplicity and ease of application are also desirable. MW meets all these requisites.

4. Materials and Methods

4.1. Materials

Polyvinyl alcohol powder (PVA, 99% hydrolysed, average molecular weight 146,000–186,000 g·mol^{-1}), phosphate buffer solution (PBS, pH 7.4, buffer strength 150 mM), 3-(4,5-dimethyl-2-thiazolyl)-2,5-diphenyl-2H-tetrazolium bromide (MTT), soybean casein digest broth medium (CASO), Dulbecco's modified Eagle's medium (DMEM), calf serum, L-glutamine, penicillin-streptomycin solution (10,000 U·mL^{-1} penicillin, 10 mg·mL^{-1} streptomycin), isopropanol, and dimethyl sulfoxide (DMSO) were all obtained from Sigma-Aldrich (Saint Louis, MO, USA). Poly(p-phenylene-2,6-benzobisoxazole) (PBO, Zylon®, AS type) fibres were obtained from Toyobo Co., Ltd. (Osaka, Japan). Thioglycolate liquid medium (TIO) and sodium chloride (purity ≥99%, NaCl) were purchased from PanReac (Barcelona, Spain). Sodium hydroxide (purity ≥99%, NaOH), octylphenoxy-polyethoxyethanol (IGEPAL®) were obtained from Merck (Darmstadt, Germany). Trifluoroacetic acid (TFA) (CF_3COOH) and methanesulfonic acid (99% extra pure MSA, CH_4O_3S), were purchased from CARLO ERBA Reagents (Milano, Italy) and ACROS Organics (Thermo Fisher Scientific, Waltham, MA, USA), respectively. Hyaluronic acid sodium salt (HA, average molecular weight of 1,000,000–2,000,000 g·mol^{-1}) was purchased from Carbosynth (Compton, Berkshire, UK). Lyophilized bovine serum albumin (BSA, Fraction V, pH 7.0) was provided by Serva Electrophoresis GmbH (Heidelberg, Germany). Human chondrocytes were acquired from CELL Applications, Inc. (San Diego, CA, USA). Special sealed bags (polyamide and polyethylene, 90 μm, 10 × 10 cm^2) were purchased from Penta Ibérica (Torres Vedras, Portugal). Distilled and deionized (DD) water (18 MΩ cm, pH 7.7) was obtained with a Millipore water purifying system (Millipore system, Millipore Merck, Darmstadt, Germany). Simulated synovial fluid was formulated with 3 mg·mL^{-1} of HA and 4 mg·mL^{-1} of BSA, dissolved in PBS, and stored in a refrigerator at 4 °C between each use.

4.2. Synthesis of the Polymeric Materials

PVA hydrogels reinforced with PBO fibres were synthesized according to a protocol described previously [20,57]. Briefly, PVA (6% *w*/*v*) was dissolved in pure TFA, while PBO fibres (1% *w*/*v*) were added to a mixture of 80% TFA and 20% MSA solution. Both solutions were left under magnetic stirring until complete dissolution of the solutes. After complete dissolution, the vials were placed at 45 °C for 30 min, to decrease their viscosity, before mixing. The composites were produced by mixing adequate volumes of PVA and PBO solutions, to obtain a formulation with 6:1 PVA:PBO mass ratio. The PVA/PBO solution was then poured into borosilicate petri dishes and, the lid was put on and left for 1 h. Then a small airflow was let into the dishes for 2 h, by opening the lid slightly. Following this, the lid was completely removed, and the gel was left for an additional 21 h. After complete gelation, the hydrogels were removed from the petri dishes and washed in DD water at 50 °C (7 days, changed 2 times a day, stirring at 200 rpm stirring) to remove acidic components (until the pH was neutral). The washed samples were further cross-linked using a thermal process, in an oven at 45 °C for 48 h. All the samples produced had an average thickness of 2.0 ± 0.5 mm in their hydrated state (unless otherwise indicated).

4.3. Sterilisation Processes

4.3.1. Microwave Irradiation (MW)

PVA/PBO hydrogels were individually placed inside special sealed bags, with 3 mL of DD water per 6 g of sample. The sterilisation procedure was carried out using a 2450 MHz microwave oven (Kunft KMW-1698, Worten, Sonae group, Maia, Portugal), at 700 W for 3 min (MW-3 samples) and 5 min (MW-5 samples).

4.3.2. High Hydrostatic Pressure (HHP)

The PVA/PBO samples were packaged for HHP treatment in the same manner as for microwave irradiation. The sealed samples were then submitted to 600 MPa at 70 °C for 10 min in a High Pressure equipment (Hiperbaric 55, Burgos, Spain) (HHP samples) according to [25].

4.3.3. Plasma (PM)

Plasma treatment was carried out using a compact Harrick PDC-32G Plasma Cleaner/Steriliser (115 V, Harrick Plasma, Ithaca, NY, USA). The equipment was connected to a vacuum pump (LVO 100, Leybold, Cologne, Germany) and an argon gas bottle (20 MPa, Alphagaz™, Air Liquid, Lisbon, Portugal). Dry PVA/PBO samples were directly placed into the quartz glass chamber and exposed to ionized argon at 18 W for 5 min (PM samples). Then the samples were placed in sterilised falcon tubes containing 10 mL of DD water per 20 g of sample, in aseptic conditions.

4.4. Bioburden

The microbiological load (bioburden) of the non-sterilised PVA/PBO hydrogels (samples with an 8-mm diameter) was evaluated. Buffered peptone broth with sodium chloride (pH 7.0) was sterilised by steam heat in an autoclave at 10^5 Pa, 121 °C for 20 min, and the samples were immersed in this medium (50 mL) for 2 h at 180 rpm. Afterwards, a pre-selected quantity of the buffer was removed (25 mL) and filtered through cellulose nitrate filters (pore size of 0.45 μm). The used filters were incubated in Sabouraud dextrose agar plates to test for fungi growth (25 °C, 5 days) and in tryptic soy agar (TSA) plates for bacteria growth in aerobic conditions (30 °C, 5 days). The colonies were visually counted to verify the microbiological load of the hydrogels and the mean of triplicates were calculated for each sample.

4.5. Sterility Assessment

The sterility of the treated samples was verified following the Portuguese Pharmacopeia [58] and the European Pharmacopeia 10th Edition [59], using the direct inoculation method. Two culture media, TIO and CASO for bacterial and fungi growth, respectively, were prepared and sterilised according to the manufacturer specifications (autoclave, at 10^5 Pa, 121 °C, for 20 min). Sterilised PVA/PBO samples (8-mm diameter) were placed into the culture medium in a laminar flow cabinet, to ensure aseptic conditions. Two positive controls were performed with bacteria contamination by inoculation of Pseudomonas aeruginosa (ATCC 15442) and Staphylococcus aureus (ATCC 6538) into TIO medium, and a positive control for fungal contamination was carried out by inoculation of the CASO medium with Candida albicans (ATCC 10231). Negative controls (sterile media) were also performed. All flasks were incubated in aerobic conditions for 14 days, at 35 °C (TIO medium) and at 25 °C (CASO medium). Validation of the sterility was performed if no microorganism growth was verified after that period in both media. Triplicates were carried out for each condition.

4.6. Characterisation of the Sterile and Non-Sterile PVA/PBO Hydrogels

4.6.1. Morphology

Surface morphology of the non-sterile and sterile samples was analysed through scanning electron microscopy (SEM, Hitachi S2400, Chiyoda, Tokyo, Japan). Samples (squares with 3 × 3 mm^2) were first dried at 45 °C for 48 h and coated with Au/Pd (100 nm) using a sputter coater and evaporator (Polaron Quorum Technologies, Laughton, East Sussex, UK) for conductivity purposes.

4.6.2. Chemical Structure

The materials' chemical structure was studied through Fourier transform infrared spectroscopy (FTIR), with attenuated total reflectance (ATR). FTIR equipment (model Spectrum Two from PerkinElmer, Waltham, MA, USA) with a lithium tantalate ($LiTaO_3$) mid-infrared (MIR) detector (signal/noise ratio 9300:1) was used. This was equipped with a diamond crystal ATR accessory (model UATR Two). The applied force was controlled, to ensure a good contact between the crystal and samples. Spectra were collected at 4 cm^{-1} resolution and 8 scans of data accumulation and normalized using the OriginPro 8.5 software. Three hydrogel disks (with 14 mm of diameter), non-sterilised and sterilised in the different conditions, were analysed.

4.6.3. Water Content

The non-sterile and sterile samples (6 mm diameter) were dried at 45 °C for 2 days prior to being weighed (dry weight, W_d). Then, they were individually placed in 5 mL of DD water, incubated at 37 °C, and periodically removed and weighed (wet weight, W_w) until a constant value was achieved. The samples were carefully blotted with absorbent paper before each measurement, to remove any remaining water present on their surface. The water content (*WC*) was calculated along the tested time range (120 h), through:

$$WC\ (\%) = \frac{W_w - W_d}{W_w} \times 100 \quad (1)$$

Measurements were performed in triplicate for each condition.

4.6.4. Wettability

The water contact angle on non-sterile and sterilised samples was determined through the captive bubble method (at least one week after treatment). The hydrated hydrogels were fixed to a support and placed inside a quartz glass liquid cell filled with DD water. Air bubbles were formed under the hydrogels' surface using a syringe with an end curved needle. Pictures were taken at predetermined times for 30 s after bubble deposition using a video camera (jAi CV-A50, Spain), connected to an optical microscope (Wild M3Z, Leica Microsystems, Wetzlar, Germany). Images were analysed using the ADSA software (Applied Surface Thermodynamics Research Associates, Toronto, Canada). At least 10 bubbles were captures in 3 different samples for each condition.

4.6.5. Compressive and Tensile Behaviour

The hydrogels' compressive and tensile mechanical behaviour was analysed in a TA.XT Express Texture Analyser (Stable Micro Systems, Godalming, UK) with a load cell of 49 N, using the software Exponent. Uniaxial compression tests were performed in unconfined mode.

For the compression tests, the hydrated samples (discs with 8 mm diameter and 4 mm thickness) were completely immersed in DD water at room temperature and compressed at a strain rate of 0.1 mm·s^{-1} up to a 40% strain.

Tensile tests were carried out with a force of 49 N at a speed of 0.5 mm·s^{-1} using hydrated dumbbell-shaped hydrogel samples (5 mm width, 2.5 mm gauge width, 8 mm gauge length, total length 18 mm).

The tangent compressive/tensile modulus (E_T) were calculated in the strain range of 5–35%, in 5% increments, and using the finite difference method [32], through the following equation:

$$E_T = \frac{(\sigma_\epsilon + \Delta_\epsilon - \sigma_\epsilon - \Delta_\epsilon)}{2\Delta_\epsilon} \quad (2)$$

where $\sigma_\epsilon + \Delta_\epsilon$ and $\sigma_\epsilon - \Delta_\epsilon$ are the stresses correspondent to the strains $\epsilon + \Delta_\epsilon$ and $\epsilon - \Delta_\epsilon$, respectively and Δ_ϵ was 1%. At least, five experiments were carried out for each condition.

4.6.6. Rheological Behaviour

The viscoelastic properties of the hydrogels were studied using a rheometer MCR 92 (Anton Paar, Ashland, VA, USA) with a parallel plate set up (PP25 geometry) and RheoCompass^{TM} software, Ashland, VA, USA, at 37 °C. Hydrated hydrogel samples (25 mm diameter) were placed at the measurement gap, 15 min prior to the measurements, with DD water surrounding the samples, to avoid evaporation artefacts. Amplitude sweep tests were carried out at a fixed angular frequency (1 Hz) in a shear strain range of 0.01–70%, to determine the plateau values of the material's storage (G') and loss (G'') modulus corresponding to the linear viscoelastic range (LVER). Dynamic frequency sweeps, between 0.1 and 100 rad·s^{-1} were performed, with a strain value fixed within the obtained LVER. Furthermore, the samples were also submitted to an isothermal test for 1 h, at a constant strain of 0.1% (within the LVER) and angular frequency 1 Hz. Triplicates of non-sterile and sterile samples were analysed.

The steady shear rate viscosity of the SSF used in the tribology tests was also measured, for a shear rate range of 0–100 s^{-1} at room temperature. This measurement was performed using a cone-plate geometry (CP50).

4.6.7. Hardness

A Shore A durometer (PCE-DD-A, PCE Instruments, Southampton, UK) was used to measure the hardness of the non-sterile and sterile hydrogels in their hydrated state (DD water). The samples were produced with a 4-mm thickness and cut into disks with an 8-mm diameter. Triplicates were analysed for each condition.

4.6.8. Tribological Behaviour

The coefficient of friction between the hydrogel samples and natural cartilage was measured at room temperature, using the lubricants DD water and SSF. Cartilage pins were harvested from porcine cartilage, obtained from a local butcher's shop. Full-depth osteochondral plugs were collected using 6-mm hole punchers at randomly selected sites. The cartilage samples' edges were carefully cut with a scalpel, to prevent sharp edges, washed in DD water, and stored at −20 °C in PBS solution till 2 h before use. The hydrogel samples were pre-hydrated (24 h) in DD water or SSF and placed in a liquid cell with ~25 mL of those lubricant media. The experiments were carried out in a pin-on-disk tribometer TRB3 (Anton Paar, Ashland, VA, USA), in reciprocal oscillating mode (2500 cycles, amplitude 4 mm, sliding velocity 8 $mm \cdot s^{-1}$). The data were collected as a function of time for 1 h, using the InstrumX 9.0.12 software, with an acquisition frequency of 10 Hz. The samples hydrated in water were tested with applied normal loads of 10, 20, and 30 N, corresponding to a Hertzian contact pressure of 0.354, 0.707, and 1.061 MPa, respectively, while the sterile samples were only tested in SSF and for the highest normal load (i.e., 30 N). A minimum of three replicates were analysed for each condition studied.

4.7. Irritability Assay

HET–CAM was carried out, to evaluate if PVA/PBO non-sterile and sterile samples could induce a possible adverse irritation reaction in tissues. Fertilized hen eggs (Sociedade Agrícola da Quinta da Freiria S.A., Roliça, Portugal) were incubated in an egg incubator (Intelligent Incubator 56S, Nanchang Edward Technology Co., Ltd., Nanchang, China) for 8 days at 37.0 ± 0.5 °C with a relative humidity of 60 ± 1%. On the 9th day of incubation, the eggs' shells were cut at the larger far end (which contains the air chamber) using a rotary saw (Dremel 3000 from Breda, Netherlands) to expose the inner membrane. This membrane was hydrated with a NaCl (0.9%) solution, and the egg was placed in the incubator for 30 min.

Afterwards, the inner membrane was removed to expose the CAM. Each sample type was individually placed on top of the CAM for 5 min. The membranes were analysed, to evaluate the appearance of any signals of irritation (i.e., lysis, haemorrhage, and coagulation), and the irritation score was calculated according to [60], using the following equation:

$$IS = \left(\frac{301 - T_H}{300} \times 5\right) + \left(\frac{301 - T_L}{300} \times 7\right) + \left(\frac{301 - T_C}{300} \times 9\right) \quad (3)$$

where T_H, T_L, and T_C represent the time (in seconds) when the first appearance of haemorrhage, lysis, and coagulation occurs, respectively. *IS* classification was done according to the HET-CAM score, ranging from 0–21. An *IS* score of 0–0.9 signifies no signs of irritation, while between 1–4.9 there are signs of slight irritation, from 5–8.9 a moderate irritation occurred, and from 9–21 the eggs show severe irritation [61]. The assay was carried out in triplicate for each hydrogel type. Two controls, a positive, and a negative were utilized, through the addition to the membrane of NaOH (1 M) and NaCl (0.9%), respectively.

4.8. Cell Viability

Cytotoxicity analysis of the sterile samples was performed through cell viability assessment. The assays were carried out using the extract test method, in accordance with ISO 10993-5 (ISO 10993—part 5: tests for in vitro cytotoxicity) [56]. The sterile hydrogels were first placed in DMEM (3 $cm^2 \cdot mL^{-1}$) and left for 24 h at 37 °C, in a humified 5% CO_2 incubator, to obtain the corresponding extract from the hydrogels' leachate. Human chondrocyte cells were cultured in DMEM, supplemented with 2 mM L-glutamine, 10% calf serum, and 1% penicillin-streptomycin solution, and then seeded in 24-well plates to a density of 0.5 × 10^5 cell/well and left to incubate for 24 h at 37 °C, with 5% CO_2 humidity. After that, the cells were inspected using an inverted light optical microscope (Axiovert® 25, ZEISS Microscopy, Jena, Germany), and the culture medium was removed and replaced by the hydrogels' extract in the same amount. Both positive (DMEM + 10% DMSO) and negative (DMEM) controls were prepared simultaneously. An additional incubation of 48 h was carried out in the same conditions. The cells were visualized using an optical microscope, and micrographs were taken.

Quantification of the cell viability was done using an MTT assay. Briefly, 300 µL of MTT (0.5 mg·mL^{-1} in serum-free DMEM) were added to each well and incubated for 3 h. Following this time, 450 µL of MTT solvent (4 mM HCl, 0.1% IGEPAL in isopropanol) was added to each well. The formed MTT formazan crystals were dissolved by agitating in an orbital shaker for 15 min. Finally, the absorbance was measured in a microplate reader (AMP Platos R 496, AMEDA, Labordiagnostik, Graz, Austria), and the relative quantification of cell viability was normalized to the negative control.

4.9. Statistical Analysis

Statistical analysis was carried out, to evaluate the significance of the obtained quantitative data, using the software R Project v. 4.2.1. The normality of the data was verified through the application of a Shapiro–Wilk test. Data with normality were evaluated in terms of similarity of variances with Levene's test, and subsequently, the data' significance was evaluated using the parametric tests, one-way ANOVA test and Student's *t*-test. For the cases where equality of variances did not occur, Welch's *t*-test was carried out. The non-parametric data was evaluated through the application of the Kruskal–Wallis or Willcoxon tests. The level of significance was set to 0.05.

Author Contributions: Conceptualization, D.S. and A.P.S.; methodology, A.S.O., A.C.M. and M.S.-O.; validation, T.P., C.G.F.-P. and A.P.S.; formal analysis, D.S.; investigation, T.P., A.C.M. and M.S.-O.; resources, A.C.M., M.S.-O. and A.P.S.; data curation, T.P., A.C.M. and M.S.-O.; writing—original draft preparation, D.S.; writing—review and editing, A.S.O., A.C.M., M.S.-O., C.G.F.-P., D.S. and A.P.S.; visualization, D.S.; supervision, C.G.F.-P., D.S. and A.P.S.; project administration, A.P.S.; funding acquisition, C.G.F.-P. and A.P.S. All authors have read and agreed to the published version of the manuscript.

Funding: This research was funded by Fundação para a Ciência e a Tecnologia (FCT) through the research project CartHeal—Cartilages for hip prosthesis with controlled drug release ability PTDC/CTM-CTM/29593/2017 and the unit projects CQE—UIDB/00100/2020, UIDP/00100/2020 and IMS—LA/P/0056/2020 (Centro de Química Estrutural, CQE). Andreia Sofia Oliveira acknowledges FCT for the Ph.D. grant PD/BD/128140/2016 (MIT-Portugal Program).

Institutional Review Board Statement: Not applicable.

Informed Consent Statement: Not applicable. Approval by an ethics committee was not required for the HET-CAM assay, since the use of chicken embryos during the first two-thirds of their development is not considered an animal experiment.

Data Availability Statement: Raw data are available upon request.

Acknowledgments: Jorge Saraiva and Carlos Pinto from the University of Aveiro are deeply acknowledged by performing the HHP sterilisations. Toyobo Co. Ltd. (Osaka, Japan) for providing the Zylon® fibres.

Conflicts of Interest: The authors declare that they have no known competing financial interests or personal relationships that could have appeared to influence the work reported in this manuscript.

References

1. Blewis, M.E.; Nugent-Derfus, G.E.; Schimidt, T.A.; Schumacher, B.L.; Sah, R.L. A Model of Synovial Fluid Lubricant Composition in Normal and Injured Joints. *Eur. Cells Mater.* **2007**, *13*, 26–39. [CrossRef] [PubMed]
2. Sophia Fox, A.J.; Bedi, A.; Rodeo, S.A. The Basic Science of Articular Cartilage: Structure, Composition, and Function. *Sports Health* **2009**, *1*, 461–468. [CrossRef]
3. Horvai, A. Cartilage Imaging: Singificance, Techniques, and New Developments. In *Cartilage Imaging: Singificance, Techniques, and New Developments*; Springer: Berlin/Heidelberg, Germany, 2011; pp. 1–10. ISBN 978-1441984371.
4. Krishnan, Y.; Grodzinsky, A.J. Cartilage Diseases. *Matrix Biol.* **2018**, *71*, 51–69. [CrossRef] [PubMed]
5. Dai, W.; Sun, M.; Leng, X.; Hu, X.; Ao, Y. Recent Progress in 3D Printing of Elastic and High-Strength Hydrogels for the Treatment of Osteochondral and Cartilage Diseases. *Front. Bioeng. Biotechnol.* **2020**, *8*, 604814. [CrossRef]

6. Xia, Y.; Momot, K.I.; Chen, Z.; Chen, C.T.; Kahn, D.; Badar, F. Introduction to Cartilage. *R. Soc. Chem.* **2016**, *8*, 1–43. [CrossRef]
7. Khan, M.; Adili, A.; Winemaker, M.; Bhandari, M. Management of Osteoarthritis of the Knee in Younger Patients. *CMAJ* **2018**, *190*, E72–E79. [CrossRef] [PubMed]
8. Milcovich, G.; Antunes, F.E.; Farra, R.; Grassi, G.; Grassi, M.; Asaro, F. Modulating Carbohydrate-Based Hydrogels as Viscoelastic Lubricant Substitute for Articular Cartilages. *Int. J. Biol. Macromol.* **2017**, *102*, 796–804. [CrossRef] [PubMed]
9. DeJulius, C.R.; Gulati, S.; Hasty, K.A.; Crofford, L.J.; Duvall, C.L. Recent Advances in Clinical Translation of Intra-Articular Osteoarthritis Drug Delivery Systems. *Adv. Ther.* **2021**, *4*, 2000088. [CrossRef]
10. Bodugoz-Sentruk, H.; Macias, C.E.; Kung, J.H.; Muratoglu, O.K. Poly(Vinyl Alcohol)–Acrylamide Hydrogels as Load-Bearing Cartilage Substitute. *Biomaterials* **2009**, *30*, 589–596. [CrossRef] [PubMed]
11. Jackson, D.W.; Scheer, M.J.; Simon, T.M. Cartilage Substitutes: Overview of Basic Science and Treatment Options. *J. Am. Acad. Orthop. Surg.* **2001**, *9*, 37–52. [CrossRef]
12. Buwalda, S.J.; Boere, K.W.M.; Dijkstra, P.J.; Feijen, J.; Vermonden, T.; Hennink, W.E. Hydrogels in a Historical Perspective: From Simple Networks to Smart Materials. *J. Control. Release* **2014**, *190*, 254–273. [CrossRef] [PubMed]
13. Kobayashi, M.; Hyu, H.S. Development and Evaluation of Polyvinyl Alcohol-Hydrogels as an Artificial Articular Cartilage for Orthopedic Implants. *Materials* **2010**, *3*, 2753–2771. [CrossRef]
14. Oliveira, A.S.; Seidi, O.; Ribeiro, N.; Colaço, R.; Serro, A.P. Tribomechanical Comparison between PVA Hydrogels Obtained Using Different Processing Conditions and Human Cartilage. *Materials* **2019**, *12*, 3413. [CrossRef] [PubMed]
15. Oliveira, A.S.; Schweizer, S.; Nolasco, P.; Barahona, I.; Saraiva, J.; Colaço, R.; Serro, A.P. Tough and Low Friction Polyvinyl Alcohol Hydrogels Loaded with Anti-Inflammatories for Cartilage Replacement. *Lubricants* **2020**, *8*, 36. [CrossRef]
16. Li, R.; Lin, J.; Fang, Y.; Yu, C.; Zhang, J.; Xue, Y.; Liu, Z.; Zhang, J.; Tang, C.; Huang, Y. Porous Boron Nitride Nanofibers/PVA Hydrogels with Improved Mechanical Property and Thermal Stability. *Ceram. Int.* **2018**, *44*, 22439–22444. [CrossRef]
17. Xu, Z.; Li, J.; Zhou, H.; Jiang, X.; Yang, C.; Wang, F.; Pan, Y.; Li, N.; Li, X.; Shi, L.; et al. Morphological and Swelling Behavior of Cellulose Nanofiber (CNF)/Poly(Vinyl Alcohol) (PVA) Hydrogels: Poly(Ethylene Glycol) (PEG) as Porogen. *RSC Adv.* **2016**, *6*, 43626–43633. [CrossRef]
18. Zhang, J.; Liu, T.; Liu, Z.; Wang, Q. Facile Fabrication of Tough Photocrosslinked Polyvinyl Alcohol Hydrogels with Cellulose Nanofibrils Reinforcement. *Polymer* **2019**, *173*, 103–109. [CrossRef]
19. Rodrigues, A.A.; Batista, N.; Bavaresco, V.P.; Baranauskas, V. In Vivo Evaluation of Hydrogels of Polyvinyl Alcohol with and without Carbon Nanoparticles for Osteochondral Repair. *Carbon* **2012**, *50*, 2091–2099. [CrossRef]
20. Oliveira, A.S.; Silva, J.C.; Loureiro, M.V.; Marques, A.M.; Kotov, N.A.; Colaço, R.; Serro, A.P. Super-Strong Hydrogel Composites Reinforced with PBO Nanofibers for Cartilage Replacement. *Macromol. Biosci.* 2022; *in press*.
21. Chen, M.; Mo, Y.; Li, Z.; Lin, X.; He, Q. Poly(p-Phenylenebenzobisoxazole) Nanofiber Layered Composite Films with High Thermomechanical Performance. *Eur. Polym. J.* **2016**, *86*, 622–630. [CrossRef]
22. Ifuku, S.; Maeta, H.; Izawa, H.; Morimoto, M.; Saimoto, H. Preparation of Polybenzoxazole Nanofibers by a Downsizing Process. *RSC Adv.* **2015**, *5*, 35307–35310. [CrossRef]
23. Karajanagi, S.S.; Yoganathan, R.; Mammucari, R.; Park, H.; Cox, J.; Zeitels, S.M.; Langer, R.; Foster, N.R. Application of a Dense Gas Technique for Sterilizing Soft Biomaterials. *Biotechnol. Bioeng.* **2011**, *108*, 1716–1725. [CrossRef]
24. Qiu, Q.Q.; Sun, W.; Connor, J. Sterilization of Biomaterials of Synthetic and Biological Origin. In *Comprehensive Biomaterials*; Elsevier: Amsterdam, The Netherlands, 2011; pp. 127–144.
25. Jasim Ahmed, H.S.R. Microwave Pasteurization and Sterilization of Foods. In *Handbook of Food Preservation*; Chapter 28; CRC Press: Boca Raton, FL, USA, 2014.
26. Rogers, W.J. Steam and Dry Heat Sterilization of Biomaterials and Medical Devices. In *Sterilisation of Biomaterials and Medical Devices*; Woodhead Publishing Limited: Sawston, UK, 2012; pp. 20–55.
27. Topete, A.; Pinto, C.A.; Barroso, H.; Saraiva, J.A.; Barahonac, I.; Saramago, B.; Serro, A.P. High Hydrostatic Pressure (HHP) as Sterilization Method for Drug-Loaded Intraocular Lenses. *ACS Biomater. Sci. Eng.* **2020**, *7*, 4051–4061. [CrossRef] [PubMed]
28. Silva, D.; de Sousa, H.C.; Gil, M.H.; Santos, L.F.; Amaral, R.A.; Saraiva, J.A.; Salema Oom, M.; Alvarez-Lorenzo, C.; Serro, A.P.; Saramago, B. Imprinted Hydrogels with LbL Coating for Dual Drug Release from Soft Contact Lenses Materials. *Mater. Sci. Eng. C* **2021**, *120*, 111687. [CrossRef]
29. Yaldagard, M.; Mortazavi, S.A.; Tabatabale, F. The Principles of Ultra High Pressure Technology and Its Application in Food Processing/Preservation: A Review of Microbiological and Quality Aspects. *Afr. J. Biotechnol.* **2008**, *7*, 2739–2767.
30. Moisan, M.; Barbeau, J.; Moreau, S.; Pelletier, J.; Tabrizian, M.; Yahia, L. Low-Temperature Sterilization Using Gas Plasmas: A Review of the Experiments and an Analysis of the Inactivation Mechanisms. *Int. J. Pharm.* **2001**, *226*, 1–21. [CrossRef]
31. Yang, L.; Chen, J.; Gao, J. Low Temperature Argon Plasma Sterilization Effect on Pseudomonas Aeruginosa and Its Mechanisms. *J. Electrostat.* **2009**, *67*, 646–651. [CrossRef]
32. Kempson, G.E. The Mechanical Properties of Articular Cartilage. In *The Joints and Synovial Fluid Vol 2*; Academic Press: Cambridge, MA, USA, 1983; pp. 177–238.
33. Chen, Y.; Song, J.; Liu, W. PVA-Based Hydrogels: Promising Candidates for Articular Cartilage Repair. *Macromol. Biosci.* **2021**, *21*, 2100147. [CrossRef]
34. Otsuka, E.; Suzuki, A. A Simple Method to Obtain a Swollen PVA Gel Crosslinked by Hydrogen Bonds. *J. Appl. Polym. Sci.* **2009**, *114*, 10–16. [CrossRef]

35. Galante, R.; Pinto, T.J.A.; Colaço, R.; Serro, A.P. Sterilization of Hydrogels for Biomedical Applications: A Review. *J. Biomed. Mater. Res. B Appl. Biomater.* **2018**, *106B*, 2472–2492. [CrossRef]
36. Srivastava, N.; Wang, C. Effects of Water Addition on OH Radical Generation and Plasma Properties in an Atmospheric Argon Microwave Plasma Jet. *J. Appl. Phys.* **2011**, *110*, 053304. [CrossRef]
37. Paneru, R.; Hoon Ki, S.; Lamichhane, P.; Nguyen, L.N.; Adhikari, B.C.; Jeong, I.J.; Mumtaz, S.; Choi, J.; Kwon, J.S.; Choi, E.H. Enhancement of Antibacterial and Wettability Performances of Polyvinyl Alcohol/Chitosan Film Using Non-Thermal Atmospheric Pressure Plasma. *Appl. Surf. Sci.* **2020**, *532*, 147339. [CrossRef]
38. Coates, J. Interpretation of Infrared Spectra. In *Encyclopedia of Analytical Chemistry*; John Wiley & Sons Ltd.: Hoboken, NJ, USA, 2006; pp. 1–13.
39. Yu, L.; Zhang, Y.; Tang, J.; Gao, J. Friction and Wear Behavior of Polyimide Composites Reinforced by Surface-Modified Poly-p-Phenylenebenzobisoxazole (PBO) Fibers in High Ambient Temperatures. *Polymers* **2019**, *11*, 1805. [CrossRef]
40. Afzal, H.M.; Shehzad, F.; Zubair, M.; Bakather, O.Y.; Al-Harthi, M.A. Influence of Microwave Irradiation on Thermal Properties of Pva and Pva/Graphene Nanocomposites. *J. Therm. Anal. Calorim.* **2020**, *139*, 353–365. [CrossRef]
41. Peppas, N.A. Infrared Spectroscopy of Semicrystalline Poly(Vinyl Alcohol) Networks. *Macromol. Chem. Phys.* **1977**, *178*, 595–601. [CrossRef]
42. Fanetti, S.; Citroni, M.; Dziubek, K.; Nobrega, M.M.; Bini, R. The Role of H-Bond in the High-Pressure Chemistry of Model Molecules. *J. Phys. Condens. Matter* **2018**, *30*, 094001. [CrossRef]
43. Negishi, J.; Nam, K.; Kimura, T.; Kishida, A. High Pressure Technique Is an Effective Method for the Preparation of PVA-Heparin Hybrid Gel. *Eur. J. Pharm. Sci.* **2010**, *41*, 617–622. [CrossRef]
44. Ino, J.M.; Chevallier, P.; Letourneur, D.; Mantovani, D.; Le Visage, C. Plasma Functionalization of Poly(Vinyl Alcohol) Hydrogel for Cell Adhesion Enhancement. *Biomatter* **2013**, *3*, e25414. [CrossRef]
45. Afshari, E.; Mazinani, S.; Ranaei-Siadat, S.-O.; Ghomi, H. Surface Modification of Polyvinyl Alcohol/Malonic Acid Nanofibers by Gaseous Dielectric Barrier Discharge Plasma for Glucose Oxidase Immobilization. *Appl. Surf. Sci.* **2016**, *385*, 349–355. [CrossRef]
46. Grassel, S.; Aszodi, A. *Cartilage Volume 2: Pathophysiology*; Springer: Berlin/Heidelberg, Germany, 2017; Volume 2.
47. Groth, T.; Seifert, B.; Malsch, G.; Albrecht, W.; Paul, D.; Kostadinova, A.; Krasteva, N.; Altankov, G. Interaction of Human Fibroblasts with Moderate Wettable Polyacrylonitrile-Copolymer Membranes. *J. Biomed. Mater. Res.* **2002**, *61*, 290–300. [CrossRef] [PubMed]
48. Mei, Y.; Saha, K.; Bogatyrev, S.R.; Yang, J.; Hook, A.L.; Kalcioglu, Z.I.; Cho, S.-W.; Mitalipova, M.; Pyzocha, N.; Rojas, F.; et al. Combinatorial Development of Biomaterials for Clonal Growth of Human Pluripotent Stem Cells. *Nat. Mater.* **2010**, *9*, 768–778. [CrossRef] [PubMed]
49. Gupta, S.; Webster, T.J.; Sinha, A. Evolution of Pva Gels Prepared without Crosslinking Agents as a Cell Adhesive Surface. *J. Mater. Sci.* **2011**, *22*, 1763–1772. [CrossRef] [PubMed]
50. Salinas, E.Y.; Hu, J.C.; Athanasiou, K. A Guide for Using Mechanical Stimulation to Enhance Tissue-Engineered Articular Cartilage Properties. *Tissue Eng. Part B* **2018**, *24*, 345–358. [CrossRef] [PubMed]
51. Little, C.J.; Bawolin, N.K.; Chen, X. Mechanical Properties of Natural Cartilage and Tissueengineered Constructs. *Tissue Eng. Part B Rev.* **2011**, *17*, 213–227. [CrossRef]
52. Boettcher, K.; Grumbein, S.; Winkler, U.; Nachtsheim, J.; Lieleg, O. Adapting a Commercial Shear Rheometer for Applications in Cartilage Research. *Rev. Sci. Instrum.* **2014**, *85*, 093903. [CrossRef]
53. Gispert, M.; Serro, A.P.; Colaço, R.; Saramago, B. Friction and Wear Mechanisms in Hip Prosthesis: Comparison of Joint Materials Behaviour in Several Lubricants. *Wear* **2006**, *260*, 149–158. [CrossRef]
54. Oungoulian, S.R.; Durney, K.M.; Jones, B.K.; Ahmad, C.S.; Clark, T.; Ateshian, G.A. Wear and Damage of Articular Cartilage with Friction against Orth Implant Materials. *J. Biomech.* **2015**, *48*, 1957–1964. [CrossRef]
55. Dong, C.; Yuan, C.; Xu, A.; Bai, X.; Tian, Y. Rippled Polymer Surface Generated by Stick-Slip Friction. *Langmuir* **2019**, *35*, 2878–2884. [CrossRef]
56. *ISO 10993-5:2009*; Biological Evaluation of Medical Devices—Part 5: Tests for In Vitro Cytotoxicity. ISO: Geneva, Switzerland, 2009.
57. Oliveira, A.S.; Silva, J.C.; Figueiredo, L.; Ferreira, F.C.; Kotov, N.A.; Colaço, R.; Serro, A.P. High-Performance Bilayer Composites for the Replacement of Osteochondral Defects. *Biomater. Sci.* **2022**, *10*, 5856–5875. [CrossRef]
58. INFARMED. *Farmacopeia Portuguesa 9 e Suplementos*; INFARMED: Lisboa, Portugal, 2008.
59. Council of Europe. *European Pharmacopoeia (Ph. Eur.)*, 10th ed.; Council of Europe: Strasbourg, France, 2019.
60. Alvarez-Rivera, F.; Concheiro, A.; Alvarez-Lorenzo, C. Epalrestat-Loaded Silicone Hydrogels as Contact Lenses to Address Diabetic-Eye Complications. *Eur. J. Pharm. Biopharm.* **2018**, *122*, 126–136. [CrossRef]
61. Spielmann, H.; Kalweit, S.; Liebsch, M.; Wirnsberger, G.I.; Bertram-Neis, E.; Krauser, K.; Kreiling, R.; Miltenburger, H.G.; Pape, W.; Steiling, W. Validation Project of Alternatives for the Draize Eye Test. *Toxicol. Vitr.* **1993**, *7*, 505–510. [CrossRef]

gels

MDPI

Article

Pluronic® F127 Hydrogel Containing Silver Nanoparticles in Skin Burn Regeneration: An Experimental Approach from Fundamental to Translational Research

Pedro Francisco [1,†], Mariana Neves Amaral [1,2,†], Afonso Neves [1], Tânia Ferreira-Gonçalves [1,2], Ana S. Viana [3], José Catarino [4], Pedro Faísca [4,5], Sandra Simões [1], João Perdigão [1], Adília J. Charmier [6], M. Manuela Gaspar [1] and Catarina Pinto Reis [1,2,*]

[1] Research Institute for Medicines (iMed.ULisboa), Faculty of Pharmacy, Universidade de Lisboa, 1649-003 Lisbon, Portugal

[2] Instituto de Biofísica e Engenharia Biomédica (IBEB), Faculdade de Ciências, Universidade de Lisboa, Campo Grande, 1749-016 Lisbon, Portugal

[3] Centro de Química Estrutural, Institute of Molecular Sciences, Faculdade de Ciências, Universidade de Lisboa, 1749-016 Lisbon, Portugal

[4] Faculdade de Medicina Veterinária, Universidade Lusoófona de Humanidades e Tecnologias, 1749-024 Lisbon, Portugal

[5] CBIOS—Research Center for Biosciences & Health Technologies, Universidade Lusófona de Humanidades e Tecnologias, Campo Grande 376, 1749-024 Lisbon, Portugal

[6] DREAMS, Universidade Lusófona de Humanidades e Tecnologias, Campo Grande 376, 1749-024 Lisbon, Portugal

* Correspondence: catarinareis@ff.ulisboa.pt; Tel.: +351-217-946-429 (ext. 14244)

† These authors contributed equally to this work.

Citation: Francisco, P.; Neves Amaral, M.; Neves, A.; Ferreira-Gonçalves, T.; Viana, A.S.; Catarino, J.; Faísca, P.; Simões, S.; Perdigão, J.; Charmier, A.J.; et al. Pluronic® F127 Hydrogel Containing Silver Nanoparticles in Skin Burn Regeneration: An Experimental Approach from Fundamental to Translational Research. *Gels* **2023**, *9*, 200. https://doi.org/10.3390/gels9030200

Academic Editor: Kummara Madhusudana Rao

Received: 17 February 2023
Revised: 2 March 2023
Accepted: 3 March 2023
Published: 6 March 2023

Abstract: Presently, skin burns are considered one of the main public health problems and lack therapeutic options. In recent years, silver nanoparticles (AgNPs) have been widely studied, playing an increasingly important role in wound healing due to their antibacterial activity. This work is focused on the production and characterization of AgNPs loaded in a Pluronic® F127 hydrogel, as well as assessing its antimicrobial and wound-healing potential. Pluronic® F127 has been extensively explored for therapeutic applications mainly due to its appealing properties. The developed AgNPs had an average size of 48.04 ± 14.87 nm (when prepared by method C) and a negative surface charge. Macroscopically, the AgNPs solution presented a translucent yellow coloration with a characteristic absorption peak at 407 nm. Microscopically, the AgNPs presented a multiform morphology with small sizes (~50 nm). Skin permeation studies revealed that no AgNPs permeated the skin after 24 h. AgNPs further demonstrated antimicrobial activity against different bacterial species predominant in burns. A chemical burn model was developed to perform preliminary in vivo assays and the results showed that the performance of the developed AgNPs loaded in hydrogel, with smaller silver dose, was comparable with a commercial silver cream using higher doses. In conclusion, hydrogel-loaded AgNPs is potentially an important resource in the treatment of skin burns due to their proven efficacy by topical administration.

Keywords: silver nanoparticles; burns; wound healing; nanotechnology; topical administration

1. Introduction

Skin is responsible for a very different set of functions essential for human survival, including acting as a barrier [1–5]. It is composed of three main layers, the epidermis, dermis and hypodermis, differing in composition and purpose [2,3]. Upon injury, the skin undergoes wound healing, an intricate and dynamic physiological process through which the skin repairs itself, a key process to restore its normal function [6]. To heal, the wound will undergo four stages. The first stage is haemostasias, starting with vasoconstriction and clot formation, acting as a protein reservoir. Next, the wound undergoes an inflammatory process, in which vasodilation increases vascular permeability to promote chemotaxis; consequently, neutrophils and macrophages migrate to the wound site, mediating and ending a debridement process and secreting cytokines and growth factors. The third stage, known as the proliferative stage, consists in epithelization, angiogenesis, granulation tissue formation and collagen deposition, mediated by endothelial cells and fibroblasts. The fourth and last stage is remodeling, with collagen depositing in an organized manner, increasing the tensile strength of the wound. This last stage may last up to one year after wound healing has started [6,7]. In exceptional cases, where the injured tissues are unable to undergo complete regeneration, fibrous tissue will be deposited, creating scars [8].

Burns are one of the most common skin injuries and are the fourth most common type of trauma worldwide [9,10]. Regarding their severity, they can be classified and divided into three degrees, depending on the affected structures and area burned [3,11,12]. The classification of skin burns largely depends on which skin layers have been affected: in first degree burns, only the epidermis is affected; second degree burns usually extend to the dermal layer of the skin; while third degree burns implicate the hypodermis, sometimes completely burning through all the skin layers [13]. Whenever the skin is burned, its barrier function is reduced, making the body much more susceptible to infection [14]. In addition, extensive lesions that progress to adjacent tissue can lead to immunosuppression [12,14]. Moreover, burns can also lead to death by shock or septicemia caused by skin infection [15].

The treatment for burnt skin is described by many guidelines worldwide [16,17]. In general, these guidelines suggest the application of antimicrobial products, such as a silver sulfadiazine cream, followed by the application of a hydrocolloid to promote healing of the affected skin [18,19]. This therapeutic strategy, starting with an antibiotic and followed by a healing agent, is based on the premise that a non-infected burn heals faster.

Furthermore, innovative hydrocolloid dressings impregnated with silver also appear to be effective, increasing the interest in these devices [18]. However, the treatment of an infected burn is more difficult, prolonged and sometimes ineffective [20]. Although healing agents may promote skin regeneration, they simultaneously create a favorable environment for bacterial proliferation, which considerably delays healing time [20,21]. Thus, it is important to first apply an antibiotic or proceed with the concomitant application of an antibiotic agent with regenerating properties for these burns. Moreover, wounds can also be treated using dressings such as silver-based dressings that can release silver ions into the wound, simultaneously presenting antibacterial potential and promoting skin regeneration [22,23]. Currently, silver in topical sulfadiazine-based form is one of the marketed silver formulations for treating burns in the western half of the globe [24].

Nanotechnology emerged in the last century and its application in medicine, referred to as nanomedicine, has emerged mainly in the last three decades [25,26]. Silver nanoparticles (AgNPs) are widely used in several fields [25,27–32], for example, in the treatment of water, textile products, biosensors and storage of food products [33]. In healthcare, AgNPs are essentially used as an antimicrobial agent, but other metallic nanoparticles also present these properties (i.e., gold nanoparticles) [22,25,27–29,34,35]. In addition to antibacterial properties, AgNPs also present antiviral, antifungal, anti-inflammatory, anti-angiogenic, anti-tumoral and antioxidant activity, with applications as a delivery system in imageology and in cosmetics [33,34]. The antimicrobial activity of AgNPs relies on their physicochemical characteristics, such as size, shape, distribution and concentration [35–39]. Due to their high surface to volume ratio, AgNPs will require lower concentration and, consequently, leading to lower toxicity when compared with conventional silver, i.e., silver sulfadiazine or silver nitrate [36–39]. However, it is required that the formulation is maintained in the burn area. With this aim, a thermoreversible hydrogel was prepared.

Gels have a great importance for many applications [40]. The production of polymer-based gels started in the '70s, but in recent years, the interest in the physical gelation of polymers has increased [40,41]. This rise in interest is also explained due to hazard and toxic concerns related to conventional vehicles. In situ thermoreversible gels like hydrogels are alternative vehicle systems with many advantages [42]. Hydrogels are gels in which the dispersed phase is composed by water and the gelling agents are polymers [43]. They are three-dimensional (3D) hydrophilic polymeric networks able to retain large amounts of water or biological fluids and characterized by soft and rubbery consistence in analogy to living tissues. These gels are promising biomaterials due to their interesting properties such as biocompatibility, biodegradability, hydrophilicity and lack of toxicity [44]. These and other properties make hydrogels vehicles for many applications in the medical and pharmaceutical fields [44]. Pluronic® F127 is an amphiphilic block copolymer, a poly(ethylene oxide)/poly(propylene oxide)/poly(ethylene oxide) (PEO–PPO–PEO) triblock copolymers, extensively explored for therapeutic applications mainly due to its appealing properties, such as being non-toxic, bioadhesive and stable, and presenting the ability to transform into a gel at physiologic temperature [45–47]. At low temperatures, Pluronic® F127 is a solution but, as temperature increases, the hydrogen bonds in the hydrophilic chains of the Pluronic® F127 copolymer desolvate, favoring hydrophobic interactions between the polyoxypropylene domains, forming a stable gel [46]. Due to its thermoreversible nature, Pluronic® F127 becomes a gel upon administration, forming a physical and protective barrier [46,47].

This work aims to develop a semi-solid formulation of AgNPs, incorporated in a Pluronic® F127 hydrogel, for topical application with antimicrobial and skin regeneration properties using three different methodological approaches. Compared to a commercial formulation based on silver ions, it is expected to achieve a comparable therapeutic effect. Several techniques and methodologies were applied to evaluate the skin permeation, antibacterial activity, in vitro and in vivo efficiency and safety properties of this new formulation.

2. Results

2.1. Physicochemical Characterization

AgNPs were prepared following three different preparation methods, using the same reagents but differing in the temperature of the reagents used. The three methods (A, B and C) resulted in the successful preparation of AgNPs, but especially the described method C. As presented in Figure 1, the macroscopic appearance of the AgNPs prepared by this method resulted in a yellow translucid dispersion, in contrast to the AgNPs produced by methods A and B. Spectrophotometry analysis of AgNPs prepared by method C showed a characteristic absorption peak at 407 nm.

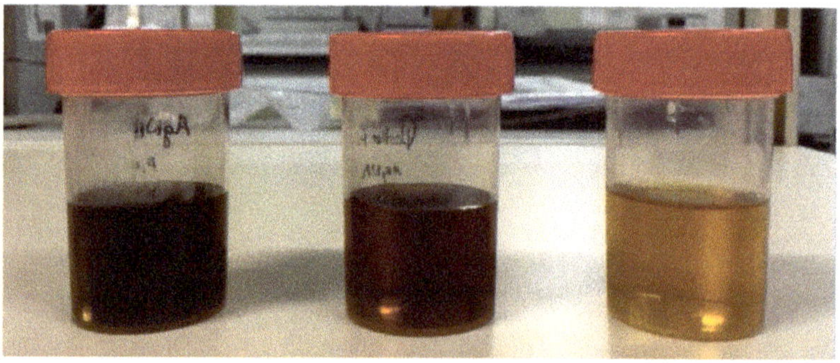

Figure 1. Macroscopic appearance of samples prepared by methods A, B and C (from **left** to **right**).

Table 1 presents results regarding mean size, polydispersity index (PdI) and zeta potential of AgNPs prepared using the different methods (A, B and C). Regarding particle size, the AgNPs prepared following method C presented a similar average size compared to AgNPs prepared following method A. However, the size dispersion of the AgNPs prepared by method C was lower than that of A and B, exhibiting lower PdI values, representing a more monodisperse formulation. Regarding its surface charge, determined by measuring the zeta potential, all of the prepared AgNPs have negative values of surface charge, close to neutrality (−10 to 10 mV), regardless of the preparation method employed, suggesting a potential biocompatibility.

Table 1. Mean size, polydispersity index (PdI) and zeta potential (mean ± SD).

Preparation Method	Size (nm)	PdI	Zeta Potential (mV)
A	47.85 ± 1.01	0.887 ± 0.003	−0.06 ± 1.50
B	18.92 ± 2.61	0.628 ± 0.162	−2.02 ± 0.34
C	48.04 ± 14.87	0.180 ± 0.013	−0.79 ± 2.17

The morphology of the different AgNPs was assessed by AFM (Figure 2). Macroscopically, AgNPs suspension prepared by methods A, B or C were visually different, as previously mentioned, and the observed color changes are closely related to the presence of AgNPs' aggregates. By analyzing the images obtained by AFM (Figure 2), it is possible to conclude that the AgNPs prepared by method A, presenting a greyish tone, contained clusters of much greater dimensions than that expected for free AgNPs, this being in accordance with data obtained by DLS (Table 1). The AFM images (Figure 2) show the presence of these aggregates for the AgNPs synthetized by methods A and B, corroborating the large PdI observed by DLS. In general, AgNPs present a non-spherical shape and sizes lower than 50 nm. Especially for AgNPs prepared by method A, large aggregates of AgNPs were consistently present, with dimensions between 100-200 nm. Taking into account these results, method C was selected for further tests as it yielded more homogeneous AgNPs, with a PdI below 0.2, when compared with the AgNPs prepared following methods A and B.

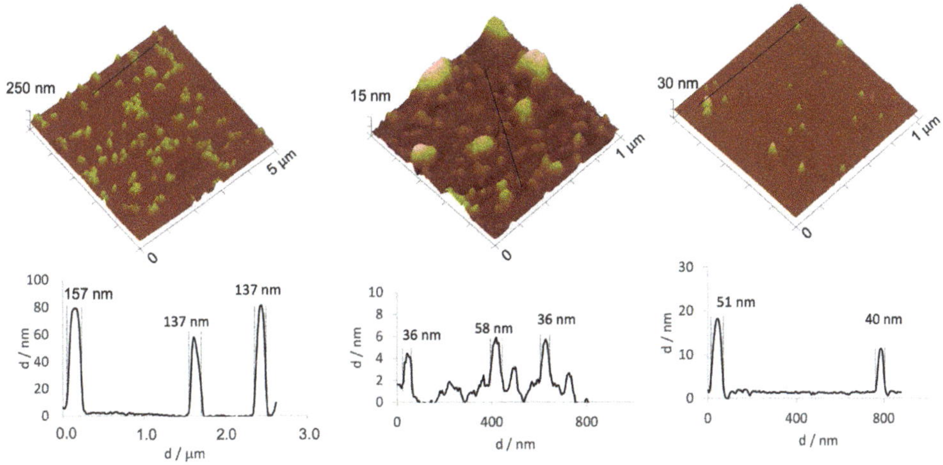

Figure 2. 3D atomic force microscopy (AFM) images and corresponding profiles for AgNPs prepared following (**left**) method A, (**middle**) method B and (**right**) method C.

With method C selected, different recovery processes were also assessed to select the most suitable: centrifugation, lyophilization or solvent evaporation. The macroscopic results of recovering AgNPs using the mentioned processes are shown in Figure 3. AgNPs centrifugation led to the formation of a very dense pellet, impossible to resuspend, leading to the rejection of this recovery method. The same was observed for lyophilization, as the AgNPs were very difficult to resuspend in water. As for the vacuum rotary evaporator recovery method, it was much faster than lyophilization, and the AgNPs were easy to resuspend.

Figure 3. Pellet of the centrifuged sample (**left**), samples prepared for lyophilization (**middle**) and sample after solvent elimination by evaporation (**right**).

To assess if recovering AgNPs by solvent evaporation using a vacuum rotary evaporator influenced AgNPs morphology, AFM images of AgNPs following solvent evaporation were obtained, for the non-diluted and diluted AgNPs (Figure 4). A very dense sample with considerable aggregates was observed in the non-diluted AgNPs recovered by a vacuum rotary evaporator. However, when the sample was diluted, it was very similar to the sample prepared by method C (Figure 4). This fact suggests that temperature has greater influence during the AgNPs formation (i.e., reagents temperature), but does not seem to have influence after AgNPs formation (i.e., AgNPs recovery). Thus, solvent evaporation using a vacuum rotary evaporator was the selected recovery method.

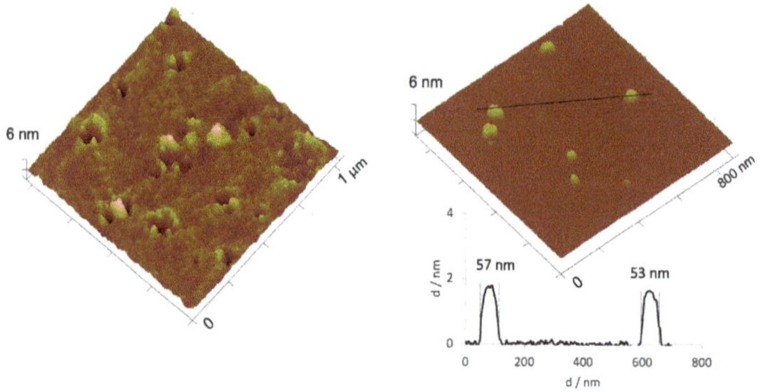

Figure 4. 3D atomic force microscopy (AFM) images of non-diluted (**left**) and diluted (**right**) AgNPs recovered by solvent evaporation. A profile was added for the diluted sample.

2.2. *Antimicrobial Preliminary Efficacy Assessment*

The bacterial strains used in this study were selected according to their tendency to colonize and infect burnt skin (i.e., *Escherichia coli*, *Staphylococcus aureus* and *Pseudomonas aeruginosa*) [48]. Table 2 displays the main results. Minimum inhibitory concentrations (MICs) were determined by broth microdilution for the three bacteria under study and for the different samples. The colloidal dispersion of AgNPs under the conditions of synthesis showed very little or no inhibition for most of the strains. In turn, when concentrated, whether by lyophilization, centrifugation or by evaporation, the bacterial inhibition was more expressive. Though there is inhibitory activity, MIC values are beyond the range of concentrations tested. However, AgNPs concentrated by lyophilization presented a more accentuated inhibitory activity against *P. aeruginosa*, and AgNPs concentrated by evaporation presented a more pronounced inhibitory activity against *E. coli* and *P. aeruginosa*, thus presenting higher efficacy against these strains.

Table 2. Minimum inhibitory concentrations (MIC) for *Escherichia coli*, *Staphylococcus aureus* and *Pseudomonas aeruginosa* (n = 3, presented values are coincident between $n \geq 2$).

Sample	MIC (nM)		
	Escherichia coli	*Staphylococcus aureus*	*Pseudomonas aeruginosa*
AgNPs	>1.65	>1.65	1.65
Centrifuged AgNPs (66 nM)	>33	>33	>33
Lyophilized AgNPs (66 nM) + Pluronic® F127 hydrogel	>33	>33	33
AgNPs concentrated by evaporation (66 nM)	33	N/A	33

2.3. *In Vitro Permeation Studies*

An in vitro skin permeation study was conducted for 24 h using an artificial membrane that mimics the skin. After 24 h, the amount of AgNPs that permeated the membrane was below the limit of detection of the analytical method used. This non-permeation of the AgNPs through the deep layers of the skin can be considered a good safety indicator.

2.4. *Physical Characterization of Pluronic® F127 Hydrogel*

In order for the AgNPs to remain in the burned region of the skin, a thermoreversible Pluronic® F127 hydrogel was prepared and characterized regarding its viscosity and textural properties. Pluronics or Poloxamers are non-toxic FDA-approved poly(ethylene oxide)/poly(propylene oxide)/poly(ethylene oxide) (PEO-PPO-PEO) triblock copolymers. A variety of Pluronics is available on the market, differing for the molecular weight of the building blocks and the ratio between hydrophobic and hydrophilic units. Pictures of the prepared gel were taken at two different temperatures, in order to macroscopically characterize the gel. The obtained images are shown in Figure 5. At 4 °C (storage temperature), the hydrogel is in a liquid form, as can be seen in Figure 5A, and at 37 °C, physiologic temperature and the temperature at which the hydrogel will be applied, Pluronic® F127 is in its gelled form, as shown in Figure 5B.

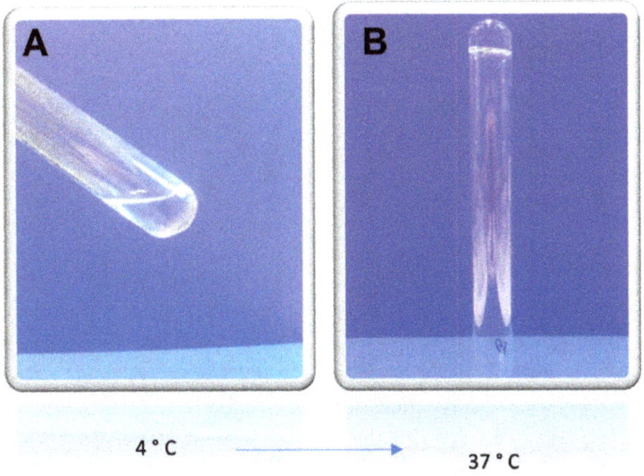

Figure 5. Macroscopic appearance of the thermoreversible Pluronic® F127 hydrogel at (**A**) 4 °C, in liquid state, and (**B**) 37 °C, gelled.

Viscosity is also a very important parameter for any semi-solid formulation. The viscosity was determined at different temperatures and results are shown in Table 3. Looking at the results, the thermoreversible properties of the Pluronic® F127 hydrogel are apparent, as at 4 °C, the gel presented a viscosity of 75.7 ± 0.5 mPas, in its liquid form, and at 37 °C, a notorious higher viscosity, of 7333.3 ± 23.1 mPas.

Table 3. Viscosity of the Pluronic® F127 hydrogel at different temperatures.

Time (min)	Viscosity (mPas)		
	4 °C	25 °C	37 °C
10	76.0	360.0	7320.0
20	76.0	380.0	7360.0
30	75.2	360.0	7320.0
Mean ± SD	75.7 ± 0.5	366.7 ± 11.6	7333.3 ± 23.1

The textural properties of the gel were also evaluated in triplicates, and the maximum peak force of displacement (F_{max}), also denoted hardness, obtained was 0.6 ± 0.01 N. Viscosity and textural properties were also evaluated after the incorporation of the lyophilized AgNPs powder in the Pluronic® F127 hydrogel, and its properties remain stable when compared to the hydrogel without nanoparticles.

2.5. *In Vivo Efficacy and Safety Assessments*

Skin burns started to appear on the second day of SDS application, and all animals completed the assay. Moreover, the animals did not present signs of stress or pain during the duration of the preliminary assessments.

The body weight of the animals was recorded for all groups, and the results are shown in Figure 6. The body weight of the animals decreased for all the groups following the burn induction but recovered after the beginning of treatment. Moreover, the body weight of all groups followed the same trend except for the negative control (Pluronic® F127 hydrogel), in which body weight presented a smaller decrease following the skin burn when compared to the other groups. Furthermore, when comparing the body weights of the different groups on the last day of the assay, the group treated with AgNPs incorporated in Pluronic® F127 hydrogel presented the best results, as the animals in this group recovered 96% of their initial body weight.

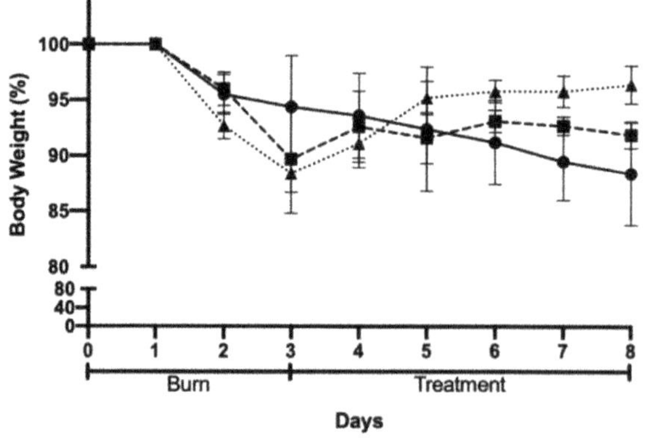

Figure 6. Body weight variation after chemically induced skin burn and treatment: Pluronic® F127 hydrogel (negative control, circles); Positive control (commercial formulation of silver sulfadiazine squares); and test group, AgNPs incorporated in Pluronic® F127 hydrogel (triangles).

Representative images of mice from each group are shown in Figures 7–9 and were taken daily after the beginning of the treatment. By evaluating the photographic records, it is possible to note that the AgNPs incorporated in Pluronic® F127 hydrogel (Figure 7) caused rapid skin regeneration in the test group, with the skin practically healthy at the end of the treatment schedule and without noticeable scarring. In contrast, the negative control (Figure 8), Pluronic® F127 hydrogel, led to a pronounced scar. Regarding the positive control (Figure 9), the commercial topical formulation of silver sulfadiazine led to a complete regeneration of the skin at the same time as the test group, without leaving any noticeable scars on the skin. However, the concentration of silver in the positive control (15.3 μmol of silver per cm^2) was not equivalent, i.e., AgNPs were administered at a very low dose (3.3 pmol of silver per cm^2).

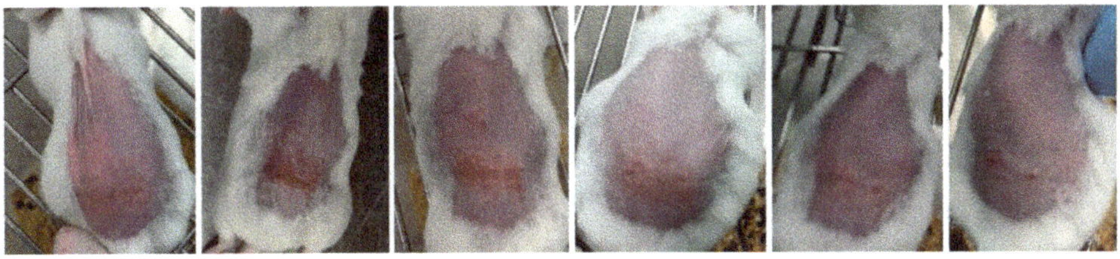

Figure 7. Progress of skin healing in test group treated with AgNPs incorporated in Pluronic® F127 hydrogel (66 nM, from day 3 to day 8, following the course of treatment).

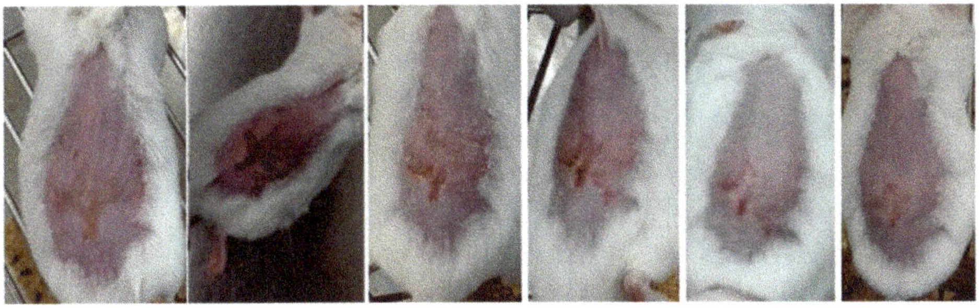

Figure 8. Progress of skin healing in the group treated with Pluronic® F127 hydrogel (from day 3 to day 8, following the course of treatment).

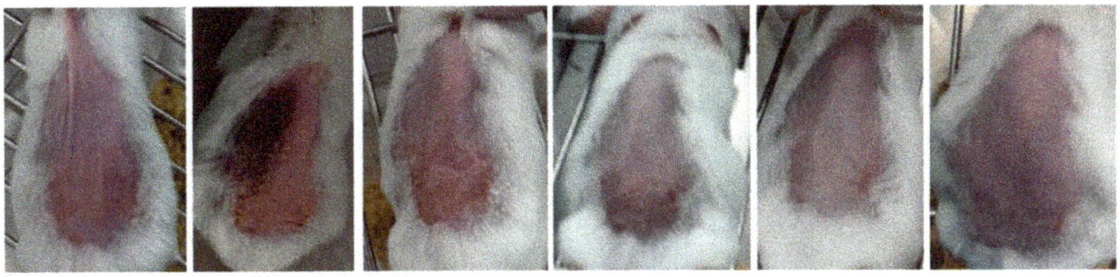

Figure 9. Progress of skin healing in the group treated with the positive control (commercial cream of silver sulfadiazine, 10 mg/g, from day 3 to day 8, following the course of treatment).

Skin thickness was also evaluated and tended to increase with the progression of the injury due to inflammation and subsequent skin regeneration with the formation of crusts. This increase was quite consistent across the different groups (Figure 10). An ideal result would be the achievement of a similar skin thickness the burn and at the end of the protocol.

This did not happen for any of the groups under study. However, results showed a higher tendency of improvement in skin thickness for animals treated with AgNPs incorporated in Pluronic® F127 hydrogel.

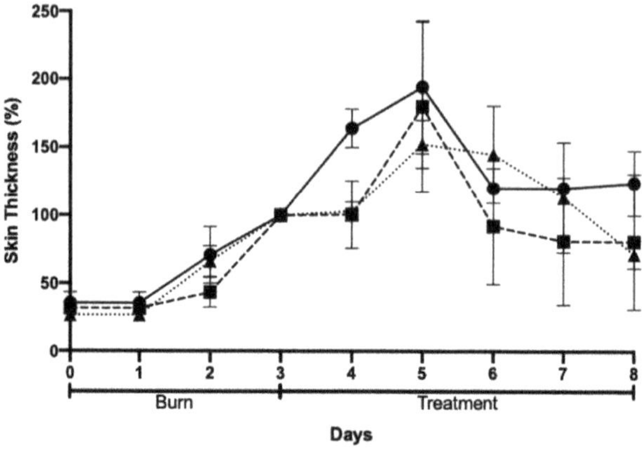

Figure 10. Skin thickness after chemically induced burn and treatment: negative control (Pluronic® F127 hydrogel, circles); positive control (commercial cream of silver sulfadiazine, squares); and AgNPs incorporated in Pluronic® F127 hydrogel (triangles).

2.6. Histopathological Analysis

The burns were histologically analyzed and representative images are presented in Figure 11. Both skin of mice treated with silver sulfadiazine and AgNPs incorporated in Pluronic® F127 presented epidermal closure with full epidermal differentiation, but skin of mice treated with the positive control presented increased thickness and marked hyperkeratosis. Moreover, the skin of mice in the positive control group presented scanty granulation tissue and mixed orientation of the collagen fibers, while granulation tissue was absent in the test group, and the collagen was present in a horizontal pattern. Moreover, a previously described skin regeneration scoring system was used to compare the burns of all animals that participated in this preliminary study (Table 4). The chosen scoring system takes different aspects into account such as the presence of an ulcer or if the wound is completely closed, the degree of epidermis differentiation, the amount of granulation tissue present, and lastly, the collagen fibber orientation and pattern. A score of 16 indicates full regeneration of the skin lesion while lower scores refer to the presence of histologic changes compatible with skin injury. The highest score was of animals treated with AgNPs incorporated in Pluronic® F127 hydrogel (test group, score of 14.60 $\pm$ 3.13). It is to be noted that the lowest score was seen in the animals of the positive control group, treated with a commercial formulation of silver sulfadiazine. This animal model was previously validated by our group, and the same scoring system was used to score the untreated burns of this animal model, obtaining a score of 12.0 $\pm$ 2.8 [9]. Thus, the skin of mice treated with AgNPs incorporated in Pluronic® F127 presented an advanced wound healing.

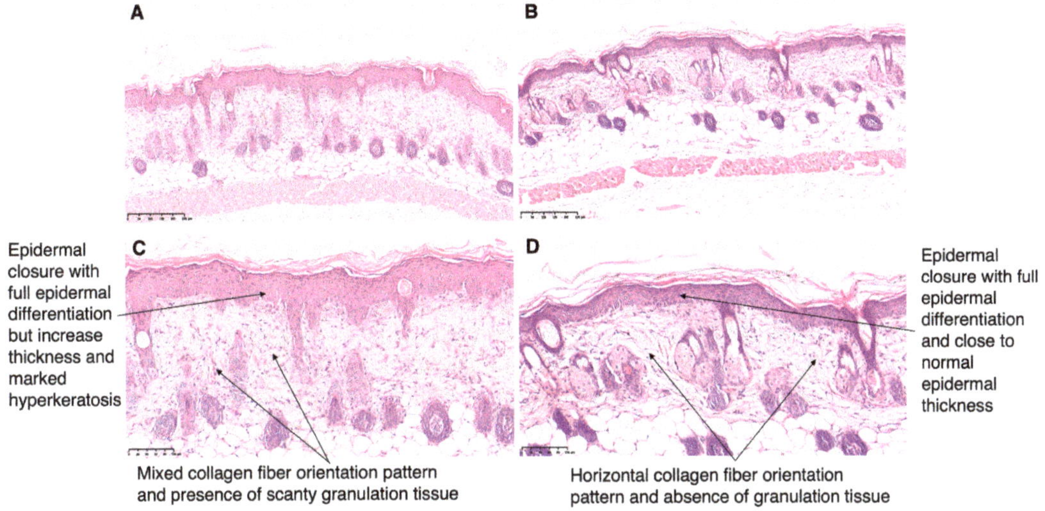

Figure 11. Histological section images of the skin of the mice after 5 days of treatment with (**A**,**C**) commercial cream of silver sulfadiazine (10 mg/g), and (**B**,**D**) AgNPs incorporated in Pluronic® F127 hydrogel (66 nM). (**A**,**B**) Magnification 10× and (**C**,**D**) magnification 20×.

Table 4. Scoring of skin regeneration following 5 days of treatment, as follows: epidermal closure (0—ulcerated skin, 1—closed wound), epidermal differentiation (0—absent, 1—spinous epidermal, 2—granular layer), amount of granulation tissue (1—profound, 2—moderate, 3—scanty, 4—absent), presence of inflammatory infiltrates (1—plenty, 2—moderate, 3—few), the orientation of collagen fibers (1—vertical, 2—mixed, 3—horizontal) and collagen pattern (1—reticular, 2—mixed, 3—fascicle) [9,38].

	Epidermal Closure	Epidermal Differentiation	Amount of Granulation Tissue	Inflammatory Infiltrate	Orientation of Collagen Fibers	Collagen Pattern	Total
Commercial formulation of silver sulfadiazine (n = 3)	0.67 ± 0.58	1.33 ± 1.15	3.33 ± 1.15	2.67 ± 0.58	2.33 ± 1.15	2.33 ± 1.15	12.67 ± 5.77
AgNPs incorporated in Pluronic® F127 hydrogel (n = 4)	1.0 ± 0.0	1.60 ± 0.89	3.33 ± 0.89	2.80 ± 0.45	2.80 ± 0.45	2.80 ± 0.45	14.60 ± 3.13

In the set of seven mice, only two of them did not have a maximum score indicating why the skin was not fully regenerated. One of these animals was allocated to the group treated with the commercial formulation of silver sulfadiazine. The skin was ulcerated and showed no differentiation at the level of the epithelium. The granulation tissue and the inflammatory infiltrate had moderate levels. Collagen appeared in a vertical orientation with a reticular pattern. The total score for this mouse in the group treated with the commercial formulation of silver sulfadiazine was 6. The other mice that did not reach the maximum score were part of the group treated with AgNPs incorporated in Pluronic® F127 hydrogel. In this case, contrary to the situation described above, the lesion was already closed and the collagen was already in a more approximate to normal disposition. However, both inflammatory infiltrate and granular tissue are present, indicating that skin regeneration would not be completed. Thus, the final score for this animal was 9.

Although the skin of most animals, independently of treatment or control groups, had completely recovered, as corroborated by the scoring system, the macroscopic analysis allowed for the classification of animals into two distinct categories: recovered with scarring and recovered without scarring. As seen in Figure 8, the animal representative of the negative control group had regenerated skin but presented a pronounced scar, unlike what happened with the animals of the other groups (Figure 11). The spleen was also subjected to analysis to check for possible toxicity, and all animals in the study showed no changes in this organ.

3. Discussion

Different methods have been described to prepare AgNPs, and these can be of three types: physical (i.e., evaporation–condensation, laser ablation), chemical (i.e., reduction, electrochemical and photochemical methods) or biological (i.e., based on oxidation-reduction reactions mediated by microorganisms such as bacteria, fungi or plant extracts) [49–52]. Amongst these methods, the most commonly used is chemical reduction, as it has a low cost of production, high performance and is fairly simple [49,52]. As described above, this method uses a solution of $NaBH_4$ to reduce Ag^+ to Ag^0, forming a cluster and originating colloidal AgNPs [49,50,52]. To obtain a monodisperse AgNPs, all nuclei must be formed at the same time, to consequently have the same growth, and this is dependent on pH and temperature [53]. In this work, it is worth noting that preparation methods A, B and C differed in the temperature of the reagents, and it seems that the temperature of the reagents used in the preparation of the AgNPs influences their size, dispersion, morphology and the presence of aggregates. On the other hand, AgNPs suspension produced by method C showed a yellow color that previous studies claim to be indicative of the formation of AgNPs without aggregates. In a previous study, AgNPs produced at 50 °C presented a brown color after synthesis. When analyzed by AFM, the produced AgNPs were large (~50 nm), non-spherical and presented several aggregates. When the same preparation method was used but using cooled reagents, at a temperature of 10 °C, the reaction speed was slower and the initial yellow color, seen right after synthesis, only shifted after a few hours [53].

The influence of the particle size on the antimicrobial activity of AgNPs is not consensual [33,54–56]. Although different bacterial species differ in the size ranges of AgNPs that inhibit their bacterial growth, most bacterial growth is inhibited by smaller AgNPs (<50 nm) [57,58]. Martínez-Castañón et al. compared the MIC's of AgNPs with different sizes (7, 29 and 89 nm), demonstrating that smaller AgNPs (7 nm) presented higher antibacterial activity against *E. coli* and *S. aureus* [59]. In another study, Jeong et al. also compared the antimicrobial activity of AgNPs with 10 and 100 nm. AgNPs with the size of 10 nm showed comparable antibacterial activity against *Methylobacterium* spp. with the positive control (methanol) [60]. Skin penetration rates and depth-of-penetration play significant roles in determining the therapeutic potential of topical agents and their systemic toxicity. Theoretically, the smaller the size of the particles, the higher the rate of penetration. In our case, AgNPs produced by method C presented greater homogeneity (PdI of 0.180). Although NPs are mostly preferred for their large surface area, smallness should not be a core goal, as the physicochemical properties of NPs can be efficiently utilized in topical antimicrobial formulations. A study carried out with gold NPs showed that smaller particles with 15 nm reached the deepest layers of the mouse skin while NPs with sizes of 102 and 198 nm only reached the epidermis and the dermis [61]. Another study with polymeric NPs demonstrated that particles with a diameter of approximately 300 nm did not permeate the human skin during the 6 h after their application without mechanical stress (passive permeation) [62]. This was shown by Ezealisiji et al., who tested the skin penetration of AgNPs with 22, 58, 76 and 378 nm [63]. Thus, considering the size of the herein developed AgNPs formulation following method C (ca. 50 nm), our AgNPs should only reach the most external layers of the skin, aiming at the goal of this study and potentially decreasing the probability of the systemic absorption.

Silver itself is non-toxic to humans within the reference dose, i.e., oral reference dose (RfD) = 5×10^{-3} mg/kg-day [64]. Overconsumption of silver, however, may lead to argyria, which results in permanent blue-grayish pigmentation of the skin, eyes and mucous membranes. Systemic toxicity can be caused by rapid accumulation of NPs at capillary/lymphatic junctions in the dermal layer, membrane pores/ligand-mediated endocytosis and physically breached leaky endothelium. Rapidly penetrating NPs could circumvent macrophage-mediated immunological responses and can enter the blood circulatory system. In contrast, slow penetration offers better efficacy of NPs against the infected cells and provides adequate time for the body's immune system to detoxify NPs through phagocytosis. In this work, the skin permeation study did not detect any measurable value of silver, which is a good indicator regarding putative systemic toxic effects.

Besides the particle size, surface charge, determined by zeta potential, is a measure of stability of colloidal dispersion, for which higher values in the module indicate greater physical stability of the dispersion under analysis and less tendency to form aggregates. It is described how negatively charged AgNPs diffuse through the skin at a greater speed [65]. In turn, particles that have a positive zeta potential (above 10 mV) are more likely to bind cells and be recognized by the immune cells [29]. Therefore, particles developed in this work presented a surface charge close to neutrality (between -10 and 10 mV) and would be ideal due to presenting low or no skin penetration and increased biocompatibility. Martínez-Higuera et al. developed negatively charged AgNPs, incorporated in a Carbopol® hydrogel with *Mimosa tenuiflora* extracts, demonstrating the wound healing potential of AgNPs [66].

The condition of the skin (i.e., whether intact or damaged) is another factor that influences the ability of AgNPs to penetrate the skin. A study conducted by Larese et al. showed that AgNPs with a mean size of 25 nm had an increase in skin penetration when the skin was damaged (2.32 ng/cm^2) when compared to intact skin (0.46 ng/cm^2), in vitro [67,68]. When the AgNPs are able to penetrate the skin, several works have shown that these particles usually precipitate at the stratum corneum, preventing AgNPs from precipitating into deeper layers of the skin [68–70]. Contrary to these studies, an in vivo study conducted by George et al. on normal intact skin showed that AgNPs penetrate into deeper layers of the skin, the reticular dermis, and thus, AgNPs are not retained in the stratum corneum. Regardless, none of these studies reported the presence of AgNPs in systemic circulation [68].

In the present study, it was also observed that the therapeutic effect of the resultant AgNPs prepared by method C varied for different tested bacteria, probably due to the disparity in the way AgNPs interact with different bacterial strains [48,71–73]. When compared to the commercialized formulation, the developed AgNPs outperformed the commercial silver sulfadiazine regarding in vitro antimicrobial activity for the tested strains, using lower concentrations of silver.

The exact mechanism of action is not completely understood. After contacting with the skin, one of the main concerns with AgNPs is the possible depletion of mitochondrial function with the production of ROS. Results from an in vitro study carried out on 3D-fibroblast cultures demonstrated that the reduction in mitochondrial activity only occurred temporarily and did not affect their viability [74]. In addition, an in vivo study of a biopsy of AgNPs-treated skin of a single patient showed a large amount of AgNPs without showing signs of apoptosis or necrosis, which corroborates the absence of toxicity [74]. From the results of these studies, it can be concluded that the toxicity conferred by silver is not as accentuated as initially thought. Regarding the possible systemic toxicity of AgNPs, the skin permeation study did not show any silver over time.

Semi-solid formulations have been indicated for better consumer acceptance of the treatment and allow for good skin-spread ability of the formulation [75]. In particular, hydrogels leave a semi-transparent layer over the burn, allowing a burn protection from the external environment and accelerate the wound healing processes [43]. Thus, in order to increase the contact time of AgNPs with the burn site and promote the wound healing properties of the AgNPs, the lyophilized AgNPs were incorporated in a semi-solid formulation, a Pluronic® F127 hydrogel. AgNPs synthesized by other groups have been incorporated into hydrogels based on other polymers, e.g., Masood et al. impregnated a chitosan-PEG hydrogel with AgNPs [76], Nguyen et al. loaded AgNPs into chitosan/Polyvinyl Alcohol hydrogel [77], Ahsan et al. used a PVA hydrogel for AgNPs-hydrogel patches [78], Xie et al. reinforced chitosan hydrogels with AgNPs [79] and Badhwar et al. loaded quercetin hydrogels with AgNPs [80], with wound-healing applications. In fact, Pluronic® F127 presents unique features, such as being thermoreversible, even at low concentrations, being liquid at temperatures lower than the physiological temperature, at which it becomes a gel [47,81–83]. This has led Pluronic® F127 to be vastly researched for dermal and transdermal applications [46,83].

The wound-healing process was treatment dependent. The test group, treated with AgNPs incorporated in Pluronic® F127 hydrogel, had a skin regeneration score below 16. Although this indicates that the wound of the animals in the test group is yet to completely heal, it was still higher than the skin regeneration score of the animals in the positive control group, treated with a commercialized silver sulfadiazine cream. As the AgNPs incorporated in Pluronic® F127 hydrogel presented a much lower concentration of silver in comparison to the commercialized silver sulfadiazine cream, this result is very promising, as the AgNPs were more effective in treating the skin burns in this chemically induced burn in vivo mice model than the positive control. Zhang et al. analyzed the delayed treatment of burns with AgNPs and besides results being also very promising, the wound healing was slower and a higher concentration of silver was used [84]. Comparing our in vivo results with the work developed by Stojkovska and colleagues, this was also the case, in which burns were treated with formulations containing alginate and AgNPs but, again, with a higher concentration of silver [85]. Posteriorly, biochemical analysis, to quantify inflammatory and pro-inflammatory factors in the animals' serum, and identification of the bacterial species colonizing the scarred burn were also performed. Biochemical analysis demonstrated that the animals in the test group did not present an inflammatory response, as the values for IL-6 and TNF-α were 0 pg/mL and <4 pg/mL, respectively. Moreover, the bacterial strains identified at the scarred burn site were common between the different experimental group and were consistent with environment and/or fecal contaminations and/or colonization (i.e., *S. aureus*, *Enterococcus faecalis*, *Staphylococcus xylosus* and *Micrococcus luteus*), as these strains are commonly found in the environment, feces and/or commensal bacteria of the skin, indication that the burns did not become infected. We hope that our proof-of-concept study could facilitate a new paradigm for understanding NPs while developing an ideal antimicrobial topical formulation.

4. Conclusions

The current study explores the application of silver nanoparticles (AgNPs) as a topical treatment of skin burns, one of the most common types of skin injury in the world, with antibacterial and wound-healing properties. In addition, this work explores the use of a thermoreversible hydrogel to deliver those AgNPs.

Skin therapy using hydrogel drug delivery systems has been gaining attention because of its dual functionality to simultaneously supply moisture and loaded actives onto infected sites on the skin. AgNPs were successfully prepared by a chemical reduction method with sodium borohydride at a specific temperature. Different methods were assessed, differing in reagents' temperature only. According to the results obtained in the various assessments, it was observed that temperature differences when producing AgNPs have a significant impact on its physicochemical characteristics. Moreover, the method that yielded AgNPs with more desirable physicochemical characteristics used sodium borohydride at a lower temperature, after being stored at −18 °C for 10 min, allowing the reaction to occur at a slower speed. Of the different recovery processes assessed, solvent elimination by evaporation resulted in AgNPs with efficient bacterial growth inhibition that were easily dispersed in water and maintained AgNPs properties.

In vivo studies using a chemical burn model showed that AgNPs incorporated in Pluronic® F127 hydrogel, at a lower concentration of silver, performed similarly to the positive control, a commercialized formulation of silver sulfadiazine with higher silver concentration, in terms of skin thickness and wound healing. The water-rich structure of the hydrogel and wound-healing properties of AgNPs seem to have the required dual-characteristics for the treatment of skin burns, since they have high efficacy when topically used, since it requires smaller concentrations of silver for the treatment of burns compared to formulations found in the market and skin regeneration was effective using new safe and hydrogel-based materials.

5. Materials and Methods

5.1. Materials

5.1.1. Reagents

Silver nitrate ($AgNO_3$, Cat. No. 209139), Pluronic® F127 (Cat. No. P2443) and sodium borohydride ($NaBH_4$, Cat. No. 71320) were purchased from Sigma Aldrich (Steinheim, Germany). Milli-Q water was obtained in filtration equipment from Millipore Corporation (Burlington, MA, USA). A commercial silver-based formulation in which each gram of cream contains 1 mg of micronized silver sulfadiazine was purchased. Reaction buffer for NZY TaqII DNA polymerase (Cat. No. MB354), NZYTaq II DNA polymerase (Cat. No. MB355) 5 U/µL, magnesium chloride 50 mM, a standard solution, NZYDNA Ladder VI (Cat. No. MB089), agarose powder of routine grade, GreenSafe Premium dye (Cat. No. MB13201) and NZYGelpure (Cat. No. MB011) kit were purchased from Nzytech (Lisbon, Portugal). All other reagents were of analytical grade.

5.1.2. Microbial Strains

The in vitro antimicrobial study was carried out using Gram-positive bacteria (*Staphylococcus aureus*, ATCC 29213) and Gram-negative bacteria (*Escherichia coli*, ATCC 25922 and ATCC 8739, and *Pseudomonas aeruginosa*, ATCC 27853).

5.1.3. Animals

Eight-week-old female CD-1 mice (25–40 g), obtained from Charles River (Barcelona, Spain) were housed in polypropylene cages in a 12–12 h light-dark cycle with a constant temperature environment of 20–24 °C, relative humidity of 55 ± 5% and received standard diet and water *ad libitum*. All animal experiments were conducted according to the recommendations of the Animal Welfare Board (ORBEA) of the Faculty of Pharmacy, *Universidade de Lisboa*, approved by the competent national authority Direção-Geral de Alimentação e Veterinária (DGAV) for project with reference PTDC/BBBBMD/0611/2012, DGAV/2013, and per the EU Directive (2010/63/EU), the Portuguese laws (DL 113/2013, 2880/2015, 260/2016 and 1/2019), and all relevant legislation.

5.2. Methods

5.2.1. Preparation of AgNPs

A solution of $AgNO_3$ (1 mM) was added dropwise to a NaBH4 (2 mM) solution under constant stirring. The optimization of the synthesis protocol was achieved by varying the temperature of the solutions used for preparation (Table 5) and thus, different batches of AgNPs were prepared (method A, B and C).

Table 5. Temperature of the solutions used for AgNPs production following methods A, B and C.

Method	Conditions
A	$AgNO_3$ was cooled prior to the preparation of the $AgNO_3$ solution, and the solutions used to prepare the AgNPs were at room temperature.
B	The $AgNO_3$ and $NaBH_4$ solutions were prepared and stored at 4 °C prior to AgNPs production.
C	$AgNO_3$ solution was produced and used at room temperature, while $NaBH_4$ solution was cooled prior to being used in AgNPs production.

The AgNPs suspension was protected from light with aluminum foil, at 4 °C. Synthesis of AgNPs was confirmed by spectrophotometry (UV-Vis Spectrophotometer, Hitachi U-2000 Dual-Beam UV-Vis, Oxford, United Kingdom) in which an absorption peak around 400 nm should be present.

Recovery of the AgNPs was performed by three different methods. The first one consists of centrifuging the prepared AgNPs (Sigma 3-30KS, Sigma Zentrifugen, Osterode am Harz, Germany) at 60,000× *g* for 20 min, followed by 40 min at 40,000× *g* at 4 °C; the solvent elimination by rotary vacuum evaporator method was adapted from a previous study in which a vacuum rotary evaporator (Butchi RE 111, Butchi, Switzerland) in a hot bath (70 °C, Butchi 461 Water Bath, Butchi, Switzerland) was used; lastly, lyophilization was carried out by leaving AgNPs in a freeze dryer (Modulyo, Edwards, CO, USA) at 102 mbar for 24 h.

5.2.2. AgNPs Characterization

- Mean Size and Surface Charge

AgNPs diluted in Milli-Q water (1:10, pH 7) were analyzed in terms of size and polydispersity index (PdI) by Dynamic Laser Scattering (DLS) (Nano Z Zetasizer, Malvern Instruments, Malvern, United Kingdom).

The AgNPs' surface charge was measured through Laser Doppler Anemometry (Nano Z Zetasizer, Malvern Instruments, Malvern, United Kingdom). For this measurement, samples were diluted in NaCl 0.1 M solution (1:10, pH 7).

- Morphology Analysis

The morphology of the nanoparticles was assessed by atomic force microscopy (AFM). Briefly, 40 µL of the sample was placed on a freshly cleaved mica surface. The mica was left to dry overnight and analysis was performed the following day. Images were acquired by Multimode 8 HR coupled to Nanoscope V (Bruker, Billerica, MA, USA) using Peak Force Tapping and ScanAssist AFM mode and silicon nitride ScanAsyst-Air probes (spring constant of ca.0.4 N/m, Bruker).

- Quantification of AgNPs

Finally, AgNPs were quantified using spectrophotometry. A calibration curve was prepared, and the following equation was obtained (R^2 = 0.998):

$$\text{Abs} = 0.198\text{C} + 9.209 \times 10^{-3} \quad (1)$$

where Abs is absorbance, and C is AgNPs concentration.

5.2.3. Antimicrobial Preliminary Efficacy Studies

All strains used were kept stored at −80 °C and were previously grown overnight at 37 °C in Muller–Hinton Agar plates. All antimicrobial assays were carried out by broth microdilution in Muller–Hinton broth in non-treated 96-well plates containing two-fold dilutions of the compound/formulation tested. The inoculum was prepared by suspending overnight bacterial growth in sterile distilled water as to obtain a bacterial suspension adjusted to a 0.5 McFarland standard, followed by 1:100 dilution in Muller Hinton broth. An equal volume to that present in each well was used to inoculate the plates to obtain a final inoculum concentration of ca. 5.0×10^5 colony-forming units (CFU)/mL. Compound-free and non-inoculated wells were included in each plate as positive and negative controls, respectively. The plates were incubated overnight at 37 °C and the Minimum Inhibitory Concentration (MIC) determined as the lowest concentration that inhibited visual growth by each strain.

5.2.4. In Vitro Skin Permeation Studies

Permeation studies were conducted with Franz cells using a silicon membrane, in a water bath at 32 °C for 24 h. The donor chamber was filled with 300 µL of AgNPs

(2 mg/mL) using Tween®80 (0.04%, *v*/*v*) as a dispersing agent, and the receptor chamber was filled with PBS at pH 7.4 (USP39) under constant stirring (100 rpm). Samples were collected every hour for the first eight hours, as well as all of the cell content at the end of the study being finished. The concentration of AgNPs was accessed by spectrophotometry, following the above-described method.

5.2.5. Preparation and Physical Characterization of Pluronic® F127 Hydrogel

For in vivo assessment of AgNPs and to deliver these nanoparticles to the target (burn) area, a thermoreversible hydrogel was prepared. Twenty-eight grams of Pluronic® F127 were added to 100 mL of phosphate buffer solution (PBS pH 7.4, USP32) in a beaker, according to a previous study [86]. The solubilization was then carried out using a magnetic plate (100 rpm, 2 h). The prepared Pluronic® F127 hydrogel was stored at 4 °C. Viscosity was determined using 100 rpm with needle n.° 3, in triplicate ($n = 3$), at different temperatures (4, 25 and 37 °C) using a Brookfield® Rotational Viscometer (Middleborough, MA, USA). Texture analysis (firmness and adhesiveness) was performed using the Stable Micro Systems TA-XT2i Texturometer (Godalming, United Kingdom). A test probe P/25P (25 mm/s) was used, with a test speed of 3 mm/s, distance 5 mm, load cell with 5 k and Trigger Force of 0.049.

5.2.6. In Vivo Efficacy and Safety Assays: Proof of Concept

Animals were randomly allocated into three experimental groups: a group dosed with the vehicle of the test formulation ($n = 3$); a group dosed with a commercial cream of silver sulfadiazine (10 mg/g of cream) ($n = 3$); a test group dosed with AgNPs (66 nM) dispersed in Pluronic® F127 hydrogel ($n = 5$). Each application of commercial cream of silver sulfadiazine corresponds to 15.3 µmol of silver per cm^2 and each application of AgNPs dispersed in Pluronic® F127 hydrogel corresponds to 3.3 pmol of silver per cm^2.

Before experimentation began, animals were lightly anaesthetized with isoflurane and an area of 2 cm^2 of the back of each mouse was shaved with a commercial depilatory cream to expose the skin. Then, a chemical burn was induced by topical application of 100 µL Carbopol 940® gel containing 40% of SDS for two consecutive days.

All formulations were administered topically (100 µL) with a syringe and performed daily, during 5 days of protocol, after light sedation with isoflurane. After administration, the water provided to the animals contained codeine (30 mg/500 mL) to reduce any pain and ensure welfare. During the 8-day experiment (since time zero), the body weight, skin thickness (Fisherbrand™ Traceable™ Carbon Fiber Calipers 6", FisherScientific, Hampton, NH, USA) and welfare of all animal groups were monitored. The burns were photographed each day to record burn evolution. A sterile swab was used to collect the skin flora of animals in each group for bacterial strain identification.

After 8 days, the animals were sacrificed and approximately 1 cm^2 of the burn area was harvested from each mouse, along with the spleen, and stored in formalin for histological analysis. Biochemical analysis was performed on the serum of all animals in each group to quantify IL-6 and TNF-α.

5.2.7. Histology

Specimens of skin and spleen were excised and fixed in 10% buffered formalin for a minimum period of 48 h and were routinely processed, embedded and sectioned into 3 µm thick sections, and stained with H&E. Slides were analyzed with a CX31 microscope (Olympus Corporation, Tokyo, Japan), and images were acquired with the NanoZoomer-SQ Digital slide scanner C13140-01 (Hamamatsu Photonics, Shizuoka, Japan).

A scoring system for wound healing was developed by adaptation of previously published scoring systems [9,87]. Briefly, skins were scored for epidermal closure (0—ulcerated skin, 1—closed wound); epidermal differentiation (0—absent, 1—spinous epidermal, 2—granular layer); amount of granulation tissue (1—profound, 2—moderate, 3—scanty, 4—absent); inflammatory infiltrate (1—plenty, 2—moderate, 3—few); collagen fiber orientation

(1—vertical, 2—mixed, 3—horizontal) and pattern of collagen (1—reticular, 2—mixed, 3—fascicle).

5.2.8. Statistical Analysis

Results were expressed as mean ± standard deviation (SD) for in vitro studies. For biological assays, results were expressed as mean ± standard error of the mean (SEM).

Author Contributions: Conceptualization, C.P.R.; methodology, C.P.R., M.M.G., S.S., P.F. (Pedro Francisco) and M.N.A.; formal analysis, P.F. (Pedro Francisco), M.N.A., A.N., T.F.-G., A.S.V., J.C., P.F. (Pedro Faísca), J.P., M.M.G., S.S. and C.P.R.; investigation, P.F. (Pedro Francisco), M.N.A., A.N., A.S.V., J.C., P.F. (Pedro Faísca), J.P., M.M.G., S.S. and C.P.R.; writing—original draft preparation, P.F. (Pedro Francisco) and M.N.A.; writing—review and editing, T.F.-G., A.S.V., J.C., P.F. (Pedro Faísca), J.P., M.M.G., S.S. and C.P.R.; supervision, C.P.R. and S.S.; project administration, C.P.R.; funding acquisition, A.J.C., T.F.-G. and C.P.R. All authors have read and agreed to the published version of the manuscript.

Funding: The authors gratefully acknowledge Fundação para a Ciência e a Tecnologia (FCT) through projects UIDB/04138/2020, UIDP/04138/2020, UIDB/00645/2020, UIDB/00100/2020 and UIDP/00100/2020. M.N.A. would like to thank FCT for the PhD Fellowship SFRH/BD/05377/2021, and T.F.-G. for the PhD Fellowship SFRH/BD/147306/2019. J.P. is supported by FCT through Estímulo Individual ao Emprego Científico [CEECIND/00394/2017].

Institutional Review Board Statement: All animal studies were performed in compliance with the guidelines outlined in the Guide for the Care and Use of Laboratory Animals, in accordance with the national (DL 113/2013, 2880/2015, 260/2016 and 1/2019) and international (Directive 2010/63/EU) accepted principles for laboratory animals' use (3 R's principles). All animal experiments were reviewed and approved by the DGAV (national authority) and by the Animal Experiment Ethics Committee of University of Coimbra (ORBEA) (project with reference PTDC/BBBBMD/0611/2012, DGAV/2013).

Informed Consent Statement: Not applicable.

Data Availability Statement: Not applicable.

Acknowledgments: The authors thank Maria Lídia Palma from ULHT for contributing to this work.

Conflicts of Interest: The authors declare no conflict of interest.

References

1. Reis, C.P.; Gomes, A.; Rijo, P.; Candeias, S.; Pinto, P.; Baptista, M.; Martinho, N.; Ascensão, L. Development and Evaluation of a Novel Topical Treatment for Acne with Azelaic Acid-Loaded Nanoparticles. *Microsc. Microanal.* **2013**, *19*, 1141–1150. [CrossRef] [PubMed]
2. Proksch, E.; Brandner, J.M.; Jensen, J.-M. The skin: An indispensable barrier. *Exp. Dermatol.* **2008**, *17*, 1063–1072. [CrossRef] [PubMed]
3. Kalantari, K.; Mostafavi, E.; Afifi, A.M.; Izadiyan, Z.; Jahangirian, H.; Rafiee-Moghaddam, R.; Webster, T.J. Wound dressings functionalized with silver nanoparticles: Promises and pitfalls. *Nanoscale* **2020**, *12*, 2268–2291. [CrossRef] [PubMed]
4. Mota, A.H.; Rijo, P.; Molpeceres, J.; Reis, C.P. Broad overview of engineering of functional nanosystems for skin delivery. *Int. J. Pharm.* **2017**, *532*, 710–728. [CrossRef]
5. Reis, C.P.; Damgé, C. Nanotechnology as a Promising Strategy for Alternative Routes of Insulin Delivery. *Methods Enzymol.* **2012**, *508*, 271–294.
6. Politano, A.D.; Campbell, K.T.; Rosenberger, L.H.; Sawyer, R.G. Use of Silver in the Prevention and Treatment of Infections: Silver Review. *Surg. Infect.* **2013**, *14*, 8–20. [CrossRef]
7. Ovais, M.; Ahmad, I.; Khalil, A.T.; Mukherjee, S.; Javed, R.; Ayaz, M.; Raza, A.; Shinwari, Z.K. Wound healing applications of biogenic colloidal silver and gold nanoparticles: Recent trends and future prospects. *Appl. Microbiol. Biotechnol.* **2018**, *102*, 4305–4318. [CrossRef]
8. Guillamat-Prats, R. The Role of MSC in Wound Healing, Scarring and Regeneration. *Cells* **2021**, *10*, 1729. [CrossRef]
9. Quitério, M.; Simões, S.; Ascenso, A.; Carvalheiro, M.; Leandro, A.P.; Correia, I.; Viana, A.S.; Faísca, P.; Ascensão, L.; Molpeceres, J.; et al. Development of a Topical Insulin Polymeric Nanoformulation for Skin Burn Regeneration: An Experimental Approach. *Int. J. Mol. Sci.* **2021**, *22*, 4087. [CrossRef]
10. Greenhalgh, D.G. Management of Burns. *N. Engl. J. Med.* **2019**, *380*, 2349–2359. [CrossRef]
11. Reinke, J.M.; Sorg, H. Wound Repair and Regeneration. *Eur. Surg. Res.* **2012**, *49*, 35–43. [CrossRef] [PubMed]

12. Liu, S.-H.; Huang, Y.-C.; Chen, L.Y.; Yu, S.-C.; Yu, H.-Y.; Chuang, S.-S. The skin microbiome of wound scars and unaffected skin in patients with moderate to severe burns in the subacute phase. *Wound Repair Regen.* **2018**, *26*, 182–191. [CrossRef]
13. Abraham, J.P.; Plourde, B.D.; Vallez, L.J.; Nelson-Cheeseman, B.B.; Stark, J.R.; Sparrow, E.M.; Gorman, J.M. Skin Burns. In *Theory and Applications of Heat Transfer in Humans*; John Wiley & Sons Ltd.: Chichester, UK, 2018; pp. 723–739.
14. Norman, G.; Christie, J.; Liu, Z.; Westby, M.J.; Jefferies, J.M.; Hudson, T.; Edwards, J.; Mohapatra, D.P.; Hassan, I.A.; Dumville, J.C. Antiseptics for burns. *Cochrane Database Syst. Rev.* **2017**, *7*. [CrossRef] [PubMed]
15. Zhang, P.; Zou, B.; Liou, Y.-C.; Huang, C. The pathogenesis and diagnosis of sepsis post burn injury. *Burn. Trauma* **2021**, *9*, tkaa047. [CrossRef] [PubMed]
16. Yoshino, Y.; Ohtsuka, M.; Kawaguchi, M.; Sakai, K.; Hashimoto, A.; Hayashi, M.; Madokoro, N.; Asano, Y.; Abe, M.; Ishii, T.; et al. The wound/burn guidelines—6: Guidelines for the management of burns. *J. Dermatol.* **2016**, *43*, 989–1010. [CrossRef]
17. Ahuja, R.B.; Gibran, N.; Greenhalgh, D.; Jeng, J.; Mackie, D.; Moghazy, A.; Moiemen, N.; Palmieri, T.; Peck, M.; et al.; ISBI Practice Guidelines Committee. ISBI Practice Guidelines for Burn Care. *Burns* **2016**, *42*, 953–1021. [CrossRef]
18. Wiktor, A.; Richards, D.; Torrey, S.B. Treatment of Minor Thermal Burns. 2017. Available online: https://www.medilib.ir/uptodate/show/349 (accessed on 2 March 2023).
19. Lloyd, E.C.O.; Rodgers, B.C.; Michener, M.; Williams, M.S. Outpatient burns: Prevention and care. *Am. Fam. Physician* **2012**, *85*, 25–32.
20. Palmieri, T.L. Infection Prevention: Unique Aspects of Burn Units. *Surg. Infect.* **2019**, *20*, 111–114. [CrossRef]
21. Oryan, A.; Alemzadeh, E.; Moshiri, A. Burn wound healing: Present concepts, treatment strategies and future directions. *J. Wound Care* **2017**, *26*, 5–19. [CrossRef]
22. Khansa, I.; Schoenbrunner, A.R.; Kraft, C.T.; Janis, J.E. Silver in Wound Care—Friend or Foe? A Comprehensive Review. *Plast. Reconstr. Surg.—Glob. Open* **2019**, *7*, e2390. [CrossRef] [PubMed]
23. Negut, I.; Grumezescu, V.; Grumezescu, A. Treatment Strategies for Infected Wounds. *Molecules* **2018**, *23*, 2392. [CrossRef] [PubMed]
24. Poon, V.K.M.; Burd, A. In vitro cytotoxity of silver: Implication for clinical wound care. *Burns* **2004**, *30*, 140–147. [CrossRef]
25. Abou El-Nour, K.M.M.; Eftaiha, A.; Al-Warthan, A.; Ammar, R.A.A. Synthesis and applications of silver nanoparticles. *Arab. J. Chem.* **2010**, *3*, 135–140. [CrossRef]
26. Reis, C.P.; Neufeld, R.J.; Veiga, F.; Ribeirod, A.J. Preparation of drug-loaded polymeric nanoparticles. In *Nanomedicine in Cancer*; Pan Stanford: Singapore, Singapore, 2017; pp. 171–214.
27. Bastos, V.; Ferreira de Oliveira, J.M.P.; Brown, D.; Johnston, H.; Malheiro, E.; Daniel-da-Silva, A.L.; Duarte, I.F.; Santos, C.; Oliveira, H. Corrigendum to "The influence of Citrate or PEG coating on silver nanoparticle toxicity to a human keratinocyte cell line" [Toxicol. Lett. 249 (2016) 29–41]. *Toxicol. Lett.* **2016**, *257*, 97. [CrossRef]
28. De Matteis, V.; Cascione, M.; Toma, C.; Leporatti, S. Silver Nanoparticles: Synthetic Routes, In Vitro Toxicity and Theranostic Applications for Cancer Disease. *Nanomaterials* **2018**, *8*, 319. [CrossRef]
29. Rai, M.; Deshmukh, S.D.; Ingle, A.P.; Gupta, I.R.; Galdiero, M.; Galdiero, S. Metal nanoparticles: The protective nanoshield against virus infection. *Crit. Rev. Microbiol.* **2016**, *42*, 46–56. [CrossRef]
30. Palza, H. Antimicrobial Polymers with Metal Nanoparticles. *Int. J. Mol. Sci.* **2015**, *16*, 2099–2116. [CrossRef]
31. Liao, C.; Li, Y.; Tjong, S. Bactericidal and Cytotoxic Properties of Silver Nanoparticles. *Int. J. Mol. Sci.* **2019**, *20*, 449. [CrossRef]
32. Lee, S.; Jun, B.-H. Silver Nanoparticles: Synthesis and Application for Nanomedicine. *Int. J. Mol. Sci.* **2019**, *20*, 865. [CrossRef]
33. Zhang, X.-F.; Liu, Z.-G.; Shen, W.; Gurunathan, S. Silver Nanoparticles: Synthesis, Characterization, Properties, Applications, and Therapeutic Approaches. *Int. J. Mol. Sci.* **2016**, *17*, 1534. [CrossRef] [PubMed]
34. Cadinoiu, A.N.; Rata, D.M.; Daraba, O.M.; Ichim, D.L.; Popescu, I.; Solcan, C.; Solcan, G. Silver Nanoparticles Biocomposite Films with Antimicrobial Activity: In Vitro and In Vivo Tests. *Int. J. Mol. Sci.* **2022**, *23*, 10671. [CrossRef]
35. Al-Musawi, S.; Albukhaty, S.; Al-Karagoly, H.; Sulaiman, G.M.; Alwahibi, M.S.; Dewir, Y.H.; Soliman, D.A.; Rizwana, H. Antibacterial Activity of Honey/Chitosan Nanofibers Loaded with Capsaicin and Gold Nanoparticles for Wound Dressing. *Molecules* **2020**, *25*, 4770. [CrossRef]
36. Mihai, M.M.; Dima, M.B.; Dima, B.; Holban, A.M. Nanomaterials for Wound Healing and Infection Control. *Materials* **2019**, *12*, 2176. [CrossRef] [PubMed]
37. Bruna, T.; Maldonado-Bravo, F.; Jara, P.; Caro, N. Silver Nanoparticles and Their Antibacterial Applications. *Int. J. Mol. Sci.* **2021**, *22*, 7202. [CrossRef] [PubMed]
38. Nam, G.; Rangasamy, S.; Purushothaman, B.; Song, J.M. The Application of Bactericidal Silver Nanoparticles in Wound Treatment. *Nanomater. Nanotechnol.* **2015**, *5*, 23. [CrossRef]
39. Wong, K.K.Y.; Liu, X. Silver nanoparticles—The real "silver bullet" in clinical medicine? *Medchemcomm* **2010**, *1*, 125. [CrossRef]
40. Bercea, M.; Darie, R.N.; Nit, L.E.; Morariu, S. Temperature Responsive Gels Based on Pluronic F127 and Poly (vinyl alcohol). *Ind. Eng. Chem. Res.* **2011**, *50*, 4199–4206. [CrossRef]
41. Moreno, E.; Schwartz, J.; Larrañeta, E.; Nguewa, P.A.; Sanmartín, C.; Agüeros, M.; Irache, J.M.; Espuelas, S. Thermosensitive hydrogels of poly(methyl vinyl ether-co-maleic anhydride)—Pluronic® F127 copolymers for controlled protein release. *Int. J. Pharm.* **2014**, *459*, 1–9. [CrossRef]

42. Dewan, M.; Sarkar, G.; Bhowmik, M.; Das, B.; Chattoapadhyay, A.K.; Rana, D.; Chattopadhyay, D. Effect of gellan gum on the thermogelation property and drug release profile of Poloxamer 407 based ophthalmic formulation. *Int. J. Biol. Macromol.* **2017**, *102*, 258–265. [CrossRef]
43. Miastkowska, M.; Kulawik-Pióro, A.; Szczurek, M. Nanoemulsion Gel Formulation Optimization for Burn Wounds: Analysis of Rheological and Sensory Properties. *Processes* **2020**, *8*, 1416. [CrossRef]
44. Parhi, R. Cross-linked hydrogel for pharmaceutical applications: A review. *Adv. Pharm. Bull.* **2017**, *7*, 515–530. [CrossRef] [PubMed]
45. Jaquilin, P.J.R.; Oluwafemi, O.S.; Thomas, S.; Oyedeji, A.O. Recent advances in drug delivery nanocarriers incorporated in temperature-sensitive Pluronic F-127–A critical review. *J. Drug Deliv. Sci. Technol.* **2022**, *72*, 103390. [CrossRef]
46. Escobar-Chávez, J.J.; López-Cervantes, M.; Naïk, A.; Kalia, Y.N.; Quintanar-Guerrero, D.; Ganem-Quintanar, A. Applications of thermo-reversible pluronic F-127 gels in pharmaceutical formulations. *J. Pharm. Pharm. Sci.* **2006**, *9*, 339–358.
47. Diniz, I.M.A.; Chen, C.; Xu, X.; Ansari, S.; Zadeh, H.H.; Marques, M.M.; Shi, S.; Moshaverinia, A. Pluronic F-127 hydrogel as a promising scaffold for encapsulation of dental-derived mesenchymal stem cells. *J. Mater. Sci. Mater. Med.* **2015**, *26*, 153. [CrossRef] [PubMed]
48. Park, H.-S.; Pham, C.; Paul, E.; Padiglione, A.; Lo, C.; Cleland, H. Early pathogenic colonisers of acute burn wounds: A retrospective review. *Burns* **2017**, *43*, 1757–1765. [CrossRef]
49. Iravani, S.; Korbekandi, H.; Mirmohammadi, S.V.; Zolfaghari, B. Synthesis of silver nanoparticles: Chemical, physical and biological methods. *Res. Pharm. Sci.* **2014**, *9*, 385–406. [PubMed]
50. Guilger-Casagrande, M.; de Lima, R. Synthesis of Silver Nanoparticles Mediated by Fungi: A Review. *Front. Bioeng. Biotechnol.* **2019**, *7*, 287. [CrossRef]
51. Haider, A.; Kang, I.-K. Preparation of Silver Nanoparticles and Their Industrial and Biomedical Applications: A Comprehensive Review. *Adv. Mater. Sci. Eng.* **2015**, *2015*, 165257. [CrossRef]
52. Abbasi, E.; Milani, M.; Fekri Aval, S.; Kouhi, M.; Akbarzadeh, A.; Tayefi Nasrabadi, H.; Nikasa, P.; Joo, S.W.; Hanifehpour, Y.; Nejati-Koshki, K.; et al. Silver nanoparticles: Synthesis methods, bio-applications and properties. *Crit. Rev. Microbiol.* **2016**, *42*, 173–180. [CrossRef]
53. Natsuki, J. A Review of Silver Nanoparticles: Synthesis Methods, Properties and Applications. *Int. J. Mater. Sci. Appl.* **2015**, *4*, 325. [CrossRef]
54. Moreno-Martin, G.; León-González, M.E.; Madrid, Y. Simultaneous determination of the size and concentration of AgNPs in water samples by UV–vis spectrophotometry and chemometrics tools. *Talanta* **2018**, *188*, 393–403. [CrossRef] [PubMed]
55. Singh, R.; Shedbalkar, U.U.; Wadhwani, S.A.; Chopade, B.A. Bacteriagenic silver nanoparticles: Synthesis, mechanism, and applications. *Appl. Microbiol. Biotechnol.* **2015**, *99*, 4579–4593. [CrossRef] [PubMed]
56. Patil, M.P.; Kim, G.-D. Eco-friendly approach for nanoparticles synthesis and mechanism behind antibacterial activity of silver and anticancer activity of gold nanoparticles. *Appl. Microbiol. Biotechnol.* **2017**, *101*, 79–92. [CrossRef]
57. Tang, S.; Zheng, J. Antibacterial Activity of Silver Nanoparticles: Structural Effects. *Adv. Healthc. Mater.* **2018**, *7*, 1701503. [CrossRef]
58. Bélteky, P.; Rónavári, A.; Zakupszky, D.; Boka, E.; Igaz, N.; Szerencsés, B.; Pfeiffer, I.; Vágvölgyi, C.; Kiricsi, M.; Kónya, Z. Are Smaller Nanoparticles Always Better? Understanding the Biological Effect of Size-Dependent Silver Nanoparticle Aggregation Under Biorelevant Conditions. *Int. J. Nanomed.* **2021**, *16*, 3021–3040. [CrossRef] [PubMed]
59. Martínez-Castañón, G.A.; Niño-Martínez, N.; Martínez-Gutierrez, F.; Martínez-Mendoza, J.R.; Ruiz, F. Synthesis and antibacterial activity of silver nanoparticles with different sizes. *J. Nanoparticle Res.* **2008**, *10*, 1343–1348. [CrossRef]
60. Jeong, Y.; Lim, D.W.; Choi, J. Assessment of Size-Dependent Antimicrobial and Cytotoxic Properties of Silver Nanoparticles. *Adv. Mater. Sci. Eng.* **2014**, *2014*, 763807. [CrossRef]
61. Sonavane, G.; Tomoda, K.; Sano, A.; Ohshima, H.; Terada, H.; Makino, K. In vitro permeation of gold nanoparticles through rat skin and rat intestine: Effect of particle size. *Colloids Surfaces B Biointerfaces* **2008**, *65*, 1–10. [CrossRef]
62. Schneider, M.; Stracke, F.; Hansen, S.; Schaefer, U.F. Nanoparticles and their interactions with the dermal barrier. *Dermatoendocrinol* **2009**, *1*, 197–206. [CrossRef]
63. Ezealisiji, K.M.; Okorie, H.N. Size-dependent skin penetration of silver nanoparticles: Effect of penetration enhancers. *Appl. Nanosci.* **2018**, *8*, 2039–2046.
64. Tak, Y.K.; Pal, S.; Naoghare, P.K.; Rangasamy, S.; Song, J.M. Shape-Dependent Skin Penetration of Silver Nanoparticles: Does It Really Matter? *Sci. Rep.* **2015**, *5*, 16908. [CrossRef] [PubMed]
65. Kraeling, M.E.K.; Topping, V.D.; Keltner, Z.M.; Belgrave, K.R.; Bailey, K.D.; Gao, X.; Yourick, J.J. In vitro percutaneous penetration of silver nanoparticles in pig and human skin. *Regul. Toxicol. Pharmacol.* **2018**, *95*, 314–322. [PubMed]
66. Martínez-Higuera, A.; Rodríguez-Beas, C.; Villalobos-Noriega, J.M.A.; Arizmendi-Grijalva, A.; Ochoa-Sánchez, C.; Larios-Rodríguez, E.; Martínez-Soto, J.M.; Rodríguez-León, E.; Ibarra-Zazueta, C.; Mora-Monroy, R.; et al. Hydrogel with silver nanoparticles synthesized by Mimosa tenuiflora for second-degree burns treatment. *Sci. Rep.* **2021**, *11*, 11312. [CrossRef] [PubMed]
67. Larese, F.F.; D'Agostin, F.; Crosera, M.; Adami, G.; Renzi, N.; Bovenzi, M.; Maina, G. Human skin penetration of silver nanoparticles through intact and damaged skin. *Toxicology* **2009**, *255*, 33–37. [CrossRef]

68. Ong, W.T.J.; Nyam, K.L. Evaluation of silver nanoparticles in cosmeceutical and potential biosafety complications. *Saudi J. Biol. Sci.* **2022**, *29*, 2085–2094. [CrossRef]
69. Bianco, C.; Visser, M.J.; Pluut, O.A.; Svetličić, V.; Pletikapić, G.; Jakasa, I.; Riethmuller, C.; Adami, G.; Larese Filon, F.; Schwegler-Berry, D.; et al. Characterization of silver particles in the stratum corneum of healthy subjects and atopic dermatitis patients dermally exposed to a silver-containing garment. *Nanotoxicology* **2016**, *10*, 1480–1491. [CrossRef]
70. Wang, M.; Marepally, S.K.; Vemula, P.K.; Xu, C. Inorganic Nanoparticles for Transdermal Drug Delivery and Topical Application. In *Nanoscience in Dermatology*; Elsevier: Amsterdam, The Netherlands, 2016; pp. 57–72.
71. Kim, J.S.; Kuk, E.; Yu, K.N.; Kim, J.-H.; Park, S.J.; Lee, H.J.; Kim, S.H.; Park, Y.K.; Park, Y.H.; Hwang, C.-Y.; et al. Antimicrobial effects of silver nanoparticles. *Nanomed. Nanotechnol. Biol. Med.* **2007**, *3*, 95–101. [CrossRef]
72. Ahmadi, M.; Adibhesami, M. The Effect of Silver Nanoparticles on Wounds Contaminated with Pseudomonas aeruginosa in Mice: An Experimental Study. *Iran. J. Pharm. Res. IJPR* **2017**, *16*, 661–669.
73. Latifi, N.A.; Karimi, H. Correlation of occurrence of infection in burn patients. *Ann. Burn. Fire Disasters* **2017**, *30*, 172–176.
74. Rigo, C.; Ferroni, L.; Tocco, I.; Roman, M.; Munivrana, I.; Gardin, C.; Cairns, W.; Vindigni, V.; Azzena, B.; Barbante, C.; et al. Active Silver Nanoparticles for Wound Healing. *Int. J. Mol. Sci.* **2013**, *14*, 4817–4840. [CrossRef] [PubMed]
75. Mota, A.H.; Prazeres, I.; Mestre, H.; Bento-Silva, A.; Rodrigues, M.J.; Duarte, N.; Serra, A.T.; Bronze, M.R.; Rijo, P.; Gaspar, M.M.; et al. A Newfangled Collagenase Inhibitor Topical Formulation Based on Ethosomes with *Sambucus nigra* L. Extract. *Pharmaceuticals* **2021**, *14*, 467. [CrossRef]
76. Masood, N.; Ahmed, R.; Tariq, M.; Ahmed, Z.; Masoud, M.S.; Ali, I.; Asghar, R.; Andleeb, A.; Hasan, A. Silver nanoparticle impregnated chitosan-PEG hydrogel enhances wound healing in diabetes induced rabbits. *Int. J. Pharm.* **2019**, *559*, 23–36. [CrossRef] [PubMed]
77. Nguyen, T.D.; Nguyen, T.T.; Ly, K.L.; Tran, A.H.; Nguyen, T.T.N.; Vo, M.T.; Ho, H.M.; Dang, N.T.N.; Vo, V.T.; Nguyen, D.H.; et al. In Vivo Study of the Antibacterial Chitosan/Polyvinyl Alcohol Loaded with Silver Nanoparticle Hydrogel for Wound Healing Applications. *Int. J. Polym. Sci.* **2019**, *2019*, 7382717. [CrossRef]
78. Ahsan, A.; Farooq, M.A. Therapeutic potential of green synthesized silver nanoparticles loaded PVA hydrogel patches for wound healing. *J. Drug Deliv. Sci. Technol.* **2019**, *54*, 101308. [CrossRef]
79. Xie, Y.; Liao, X.; Zhang, J.; Yang, F.; Fan, Z. Novel chitosan hydrogels reinforced by silver nanoparticles with ultrahigh mechanical and high antibacterial properties for accelerating wound healing. *Int. J. Biol. Macromol.* **2018**, *119*, 402–412. [CrossRef]
80. Badhwar, R.; Mangla, B.; Neupane, Y.R.; Khanna, K.; Popli, H. Quercetin loaded silver nanoparticles in hydrogel matrices for diabetic wound healing. *Nanotechnology* **2021**, *32*, 505102. [CrossRef] [PubMed]
81. Faris Taufeq, F.Y.; Habideen, N.H.; Rao, L.N.; Podder, P.K.; Katas, H. Potential Hemostatic and Wound Healing Effects of Thermoresponsive Wound Dressing Gel Loaded with *Lignosus rhinocerotis* and *Punica granatum* Extracts. *Gels* **2023**, *9*, 48. [CrossRef]
82. Shriky, B.; Kelly, A.; Isreb, M.; Babenko, M.; Mahmoudi, N.; Rogers, S.; Shebanova, O.; Snow, T.; Gough, T. Pluronic F127 thermosensitive injectable smart hydrogels for controlled drug delivery system development. *J. Colloid Interface Sci.* **2020**, *565*, 119–130. [CrossRef]
83. Gioffredi, E.; Boffito, M.; Calzone, S.; Giannitelli, S.M.; Rainer, A.; Trombetta, M.; Mozetic, P.; Chiono, V. Pluronic F127 Hydrogel Characterization and Biofabrication in Cellularized Constructs for Tissue Engineering Applications. *Procedia CIRP* **2016**, *49*, 125–132. [CrossRef]
84. Zhang, K.; Lui, V.C.H.; Chen, Y.; Lok, C.N.; Wong, K.K.Y. Delayed application of silver nanoparticles reveals the role of early inflammation in burn wound healing. *Sci. Rep.* **2020**, *10*, 6338. [CrossRef] [PubMed]
85. Stojkovska, J.; Djurdjevic, Z.; Jancic, I.; Bufan, B.; Milenkovic, M.; Jankovic, R.; Miskovic-Stankovic, V.; Obradovic, B. Comparative in vivo evaluation of novel formulations based on alginate and silver nanoparticles for wound treatments. *J. Biomater. Appl.* **2018**, *32*, 1197–1211. [CrossRef] [PubMed]
86. Yang, Z.; Nie, S.; Hsiao, W.W.; Pam, W. Thermoreversible Pluronic® F127-based hydrogel containing liposomes for the controlled delivery of paclitaxel: In vitro drug release, cell cytotoxicity, and uptake studies. *Int. J. Nanomed.* **2011**, 151. [CrossRef] [PubMed]
87. Braiman-Wiksman, L.; Solomonik, I.; Spira, R.; Tennenbaum, T. Novel Insights into Wound Healing Sequence of Events. *Toxicol. Pathol.* **2007**, *35*, 767–779. [CrossRef]

gels

MDPI

Article

Development of Natural Active Agent-Containing Porous Hydrogel Sheets with High Water Content for Wound Dressings

Thanyaporn Pinthong [1], Maytinee Yooyod [1], Jinjutha Daengmankhong [1], Nantaprapa Tuancharoensri [1], Sararat Mahasaranon [2], Jarupa Viyoch [3], Jirapas Jongjitwimol [4], Sukunya Ross [2] and Gareth M. Ross [2,*]

[1] Biopolymer Group, Department of Chemistry, Faculty of Science, Naresuan University, Phitsanulok 65000, Thailand; thanyapornpi62@nu.ac.th (T.P.)
[2] Biopolymer Group, Department of Chemistry, Center of Excellence in Biomaterials, Faculty of Science, Naresuan University, Phitsanulok 65000, Thailand
[3] Department of Pharmaceutical Technology, Faculty of Pharmaceutical Sciences and Center of Excellence for Innovation in Chemistry, Naresuan University, Phitsanulok 65000, Thailand; jarupav@nu.ac.th
[4] Department of Medical Technology, Faculty of Allied Health Sciences and Center of Excellence in Biomaterials, Faculty of Science, Naresuan University, Phitsanulok 65000, Thailand
* Correspondence: gareth@nu.ac.th

Abstract: This work was concerned with the fabrication of a porous hydrogel system suitable for medium to heavy-exudating wounds where traditional hydrogels cannot be used. The hydrogels were based on 2-acrylamido-2-methyl-1-propane sulfonic acid (AMPs). In order to produce the porous structure, additional components were added (acid, blowing agent, foam stabilizer). Manuka honey (MH) was also incorporated at concentrations of 1 and 10% *w*/*w*. The hydrogel samples were characterized for morphology via scanning electron microscopy, mechanical rheology, swelling using a gravimetric method, surface absorption, and cell cytotoxicity. The results confirmed the formation of porous hydrogels (PH) with pore sizes ranging from ~50–110 μm. The swelling performance showed that the non-porous hydrogel (NPH) swelled to ~2000%, while PH weight increased ~5000%. Additionally, the use of a surface absorption technique showed that the PH absorbed 10 μL in <3000 ms, and NPH absorbed <1 μL over the same time. Incorporating MH the enhanced gel appearance and mechanical properties, including smaller pores and linear swelling. In summary, the PH produced in this study had excellent swelling performance with rapid absorption of surface liquid. Therefore, these materials have the potential to expand the applicability of hydrogels to a range of wound types, as they can both donate and absorb fluid.

Keywords: porous hydrogels; wound dressings; Manuka honey; surface absorption; 2-acrylamido-2-methyl-1-propane sulphonic acid

check for updates

Citation: Pinthong, T.; Yooyod, M.; Daengmankhong, J.; Tuancharoensri, N.; Mahasaranon, S.; Viyoch, J.; Jongjitwimol, J.; Ross, S.; Ross, G.M. Development of Natural Active Agent-Containing Porous Hydrogel Sheets with High Water Content for Wound Dressings. *Gels* **2023**, *9*, 459. https://doi.org/10.3390/gels9060459

Academic Editors: Ana Paula Serro, Ana Isabel Fernandes and Diana Silva

Received: 9 May 2023
Revised: 25 May 2023
Accepted: 31 May 2023
Published: 3 June 2023

1. Introduction

Hydrogels are chemically or physically cross-linked hydrophilic polymer networks that are able to absorb and retain a large amount of water or biological fluids. They are often used in tissue engineering [1], drug delivery [2–4], and biomedical devices (e.g., wound dressings and contact lenses) [5,6] due to their biomimetic properties and high water content. Hydrogel wound dressings are important medical devices that can be used for a wide range of wounds such as shallow and deep open wounds (pressure sores, leg ulcers, surgical and malignant wounds, partial thickness burns, scalds, and lacerations) that may be located on hard-to-fit locations on the body (such as joints, hands, and face) [7,8].

However, one area where hydrogel wound dressings cannot currently be used is medium and heavy-exudating wounds, which usually require foam dressings (e.g., polyurethane) [9]. The reason for this is that hydrogels, despite their capacity to absorb substantial amounts of liquid, typically experience a delay in swelling at the outset of application. During this delay, the hydrogel network undergoes a process of structural reorganization where the polymer chains separate from one another, generating gaps that permit water to enter the hydrogel [3,10]. This limitation means that traditional hydrogel wound dressings cannot be used for medium and heavy-exudating wounds. Nevertheless, by incorporating a porous structure into the hydrogel, it is thought that the hydrogel can become capable of effectively absorbing higher amounts of exudate from wounds with medium and heavy exudation. This is demonstrated by a group of hydrogels termed "superporous hydrogels" (SPH) that have been shown to rapidly absorb large amounts of water due to the presence of interconnected microscopic pores [11].

SPHs exhibit this behavior because of their porous configuration, which provides a significantly larger surface area and shorter diffusion distance compared to conventional hydrogels. These gels can have a similar composition to conventional hydrogels but contain a system for producing a porous structure during the polymerization process. A variety of methods exist to generate macroporous hydrogel structures, such as emulsion [12], freeze drying [13–15], high internal phase emulsion (HIPE) [16], water in oil emulsion templates [17], and the gas blowing technique [18].

Numerous polymers can be utilized to create hydrogel wound dressings, such as poly(vinyl pyrrolidone), poly(vinyl alcohol), poly(acrylic acid), polyesters, 2-acrylamido-2-methylpropane sulfonic acid (AMPS), and its sodium salt [19]. AMPS-based polymers are highly desirable due to their inherent advantages. The presence of a sulfonate group in AMPS resembles the glycosaminoglycan present in the skin's extracellular matrix, which plays a key role in maintaining and providing moisture to the body. This property makes AMPS hydrogels act as synthetic counterparts of proteoglycans. Moreover, AMPS has been proven to accelerate epithelialization, alleviate pain, and stimulate bioactivity in ulcerated wounds [19–22]. This has enabled AMPS to be used for several hydrogel wound dressings and injectable hydrogels. For example, AMPS has been combined with other polymers such as poly(ε-caprolactone) diacrylate and carboxymethyl chitosan to form AMPS-containing wound dressing hydrogels [23]. Injectable sulfonate-containing hydrogels with AMPS have also been produced from thiol-containing copolymers reacted with a four-arm acrylamide-terminated poly(ethylene glycol) via a thiol-ene click reaction [24]. The aforementioned benefits serve as the underlying justification for the utilization of AMPs in this study.

Many hydrogel wound dressing materials also benefit from the incorporation of natural healing aids. There are several options to consider, such as Aloe vera [25], Centella asiatica [26], Echinacea purpurea [27], and Manuka honey [28–30]. Manuka honey (MH) is a mono-floral honey obtained from the Leptospermum scoparium tree native to New Zealand. It has been shown to stimulate angiogenesis, macrophages, and wound epithelialization [31]. In addition, it can provide nutrition components during the wound healing process [32,33] and exhibits inflammation modulation, thereby reducing the inflammation phase and promoting wound healing [33]. Interestingly, MH possesses antibacterial and antibiofilm affects, a low pH range (3.2–4.5), and degradation to hydrogen peroxide which contributes to bacterial death [34,35]. In terms of manufacturing, MH also influences the viscosity of the system, which can help control how the components are mixed.

In this study, we investigated the potential of gas-blown porous hydrogel sheets, which is a novel approach to enhance the absorption properties of the hydrogels for treating medium to heavy-exudating wounds. The primary goal was to develop hydrogels that exhibit not only high absorption capacity but also excellent mechanical stability. The gas-blown porous hydrogel sheets were fabricated through the synthesis of hydrophilic monomer and 2-acrylamido-2-methyl-1-propane sulfonic acid sodium salt (AMPs), a foaming agent, and with the incorporation of MH. Non-porous hydrogels (NPH) were compared to porous hydrogels (PH) by observing the appearance and rheological mechanical properties of the gels. Then, the absorption properties of the gels were assessed in terms of bulk swelling capacity and a novel technique developed to measure the real-time surface absorption. Two different concentrations of MH were added to the system (1 and 10% *w/w*), and a range of material characterization techniques was used to study how the incorporation of MH affected the performance of the gels. Additional assessments were conducted to examine the morphology of the gels and cell viability to ensure that the materials were non-toxic. These properties are crucial for ensuring that the material can effectively as a wound dressing for all wound types, including those with heavy exudate.

2. Results and Discussion

The fabrication of porous hydrogels is achieved when the foaming agent (sodium bicarbonate (BA)) is decomposed by an acid (methacrylic acid (MAA)) during gelation to produce CO_2. The hydrogels are polymerized around the gas bubbles, producing a porous structure. Methacrylic acid (MAA) was selected to be incorporated into the hydrogel structure during the polymerization step. This fabrication of porous hydrogels (PH) requires the balance of gelation time and gas blowing formation to obtain the optimal porous structure. Figure 1 demonstrates that when gelation occurs before or after the maximum foam height, the resulting hydrogels exhibit a two-layered structure. For example, for the gel in Zone A, the concentration of redox initiators was increased to 2 M, resulting in rapid gelation. This caused the gel to form in two layers because the foam reaction was still at the early stages when the gelation was completed. This shows that an excessively fast gelation time leads to the formation of a two-layered system, which is not desirable for creating homogeneously porous hydrogels. In Zone C, the concentration of redox initiators was decreased to 0.5 M. This resulted in a slower gelation time, which allowed for the foam to reach its maximum height prior to dissipating before complete gelation. Consequently, since the polymer matrix could not trap the foam before the gelation was complete, the resulting gel also did not possess a homogeneously porous structure. This suggests that gelation times that are too slow to occur during the maximum foam height may also not be suitable for creating homogeneously porous hydrogels. However, when gelation takes place at the ideal time (Zone B), which corresponds to the peak foam height, the resulting hydrogel exhibits a homogeneous porous structure.

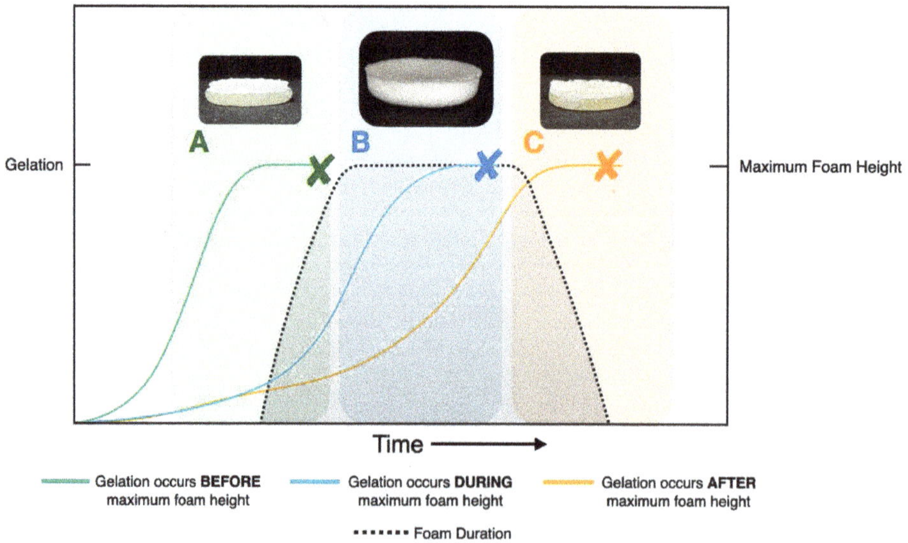

Figure 1. Relationship between gelation time and maximum foam height. (**A**) Gelation occurs before foam reaches maximum height, (**B**) gelation occurs during maximum foam height, and (**C**) gelation occurs after maximum foam height.

Initially, we compared the properties and differences between non-porous hydrogels (NPH) and porous hydrogels (PH) without honey (Section 2.1). The confirmation of the porous structure was examined, as well as how it influences the mechanical strength and swelling behavior of the gels. Additionally, a surface absorption method was employed that was more closely related to the future application of the material in absorbing wound fluid at the surface. After confirming that the porous hydrogel system was viable, the addition of MH was studied. In preliminary studies, the %MH was varied at the following compositions: −1%, 2%, 5%, 10%, 15%, and 25%, with 1 and 10% exhibiting homogeneous porous structures, while the others altered the balance between gelation and foam quality to produce dual-layered systems. Therefore, concentrations of 1 and 10% *w*/*w* were studied further (Section 2.2).

2.1. Comparisons of Non-Porous Hydrogels (NPH) and Porous Hydrogels (PH)

Figure 2 shows a graphical representation (Figure 2A) and visualization/scanning electron microscopy (SEM) micrographs (Figure 2B) of the NPH and PH. There was a marked difference between the two hydrogels, with the NPH presenting a clear appearance, while the PH was opaque. This opaque appearance was due to the porous structure of the PH, which is schematically show in Figure 2A. To validate the morphology depicted in the schematic, SEM was employed to examine the morphologies of both samples. Upon analyzing the SEM samples, it was observed that the NPH had a smooth and unblemished surface. On the other hand, the PH sample exhibited both macro structures formed by "polymer droplets" and microstructures resulting from the gas blowing process.

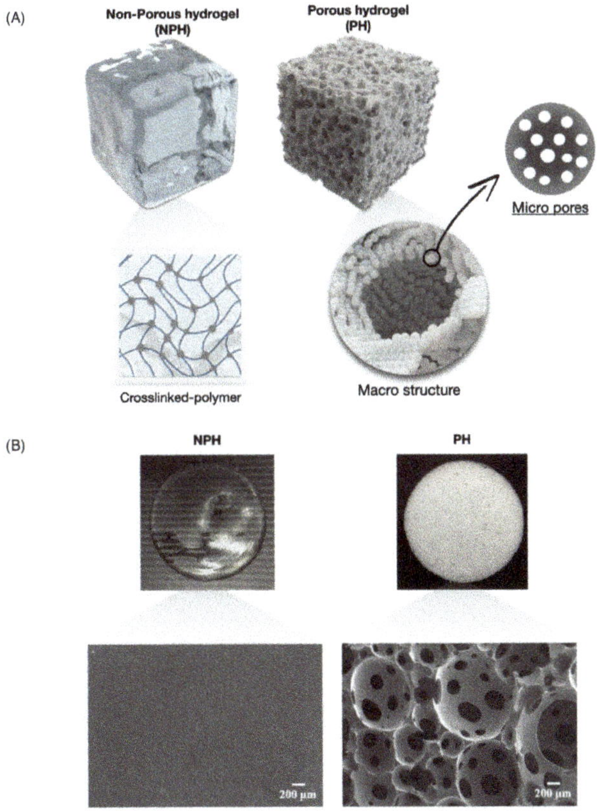

Figure 2. A schematic of three-dimensional non-porous hydrogel (NPH) and porous hydrogel (PH) (**A**). Visualization and scanning electron microscopy micrographs of non-porous hydrogel (NPH) and porous hydrogel (PH) structures (**B**).

The bulk swelling behavior of NPH and PH samples is presented in Figure 3. Figure 3A shows the appearance of the gels before immersion in water at 0 min, after 1 min, and after 5 min. Both the NPH and PH displayed a significant increase in size, swelling from their original 10 mm diameter to 24 mm and 34 mm, respectively, after 5 min. Figure 3B shows the swelling ratios calculated using Equation (1) over a period 30 min. The weight increase for both samples was greater than 1000%, with the PH reaching over 5000%. This value was reached in less than 2 min, with a more consistent % swelling reached after only 5 min. In contrast, the swelling of the NPH samples showed a more linear increase over the entire 30 min. The main difference between the two systems was that the pores present in the PH facilitated faster initial water absorption through capillary action. Another notable observation was that the NPH sample still exhibited considerable swelling, reaching a value of 2–3000% after 24 h. However, this was still considerably lower than the swelling observed in the PH sample. Therefore, the formation of pores in the system led to a twofold increase in swelling.

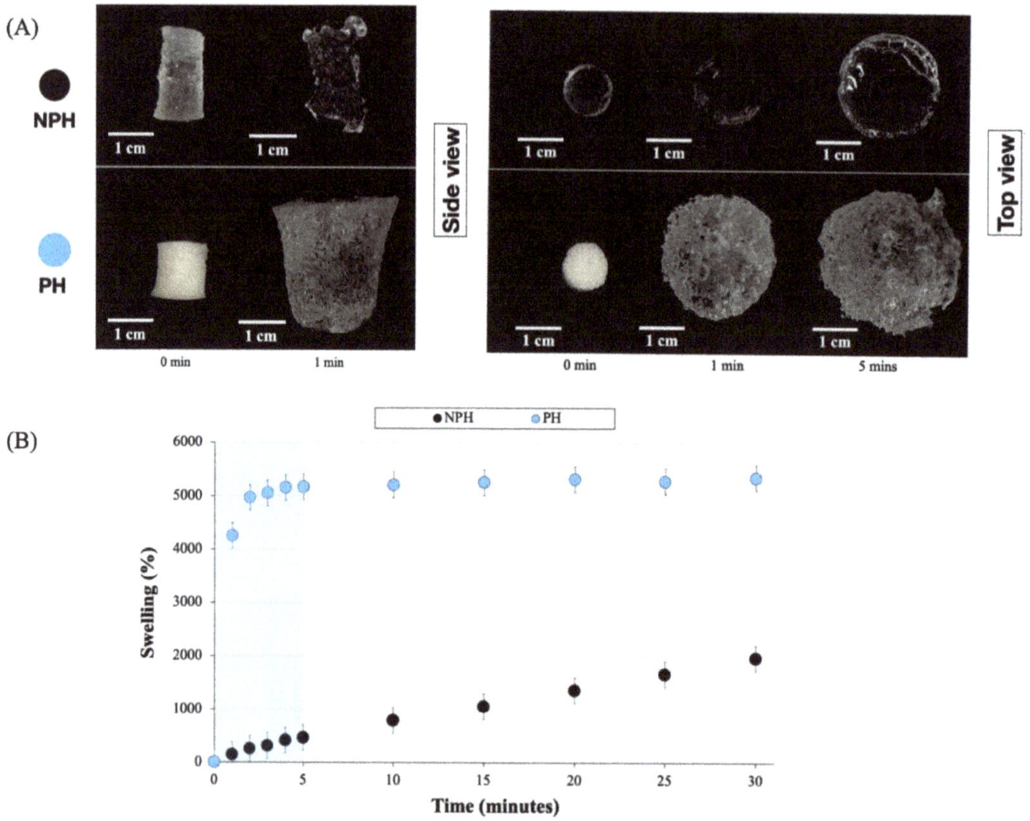

Figure 3. Bulk swelling in water and appearance of NPH and PH at 0 and 1 min respectively. (**A**) Side view and top view. (**B**) Percentage swelling of NPH and PH.

Based on the results of the swelling ratio, further investigation was conducted to examine the initial swelling or surface absorption of these gels. Plotting the drop volume (μL) against drop age (milliseconds) allowed for the measurement of surface absorption. The camera recorded at a rate of 36 frames per second, facilitating precise monitoring of the drop volume throughout the absorption process. This enabled a comprehensive real-time assessment of surface absorption for each sample. Figure 4 illustrates the real-time surface absorption of the NPH and PH gels. The PH gel completely absorbed the initial drop volume of 10 μL in less than 3000 ms (3 s), while the NPH gel absorbed water at a considerably slower rate compared to the porous sample. Over the 4000 ms period shown, the NPH sample only absorbed approximately 1 μL. This once again highlights that the NPH experienced a lag in swelling as the polymer chains rearranged themselves to allow water to enter the gel, whereas the pores in the PH enabled water to enter the system immediately upon contact.

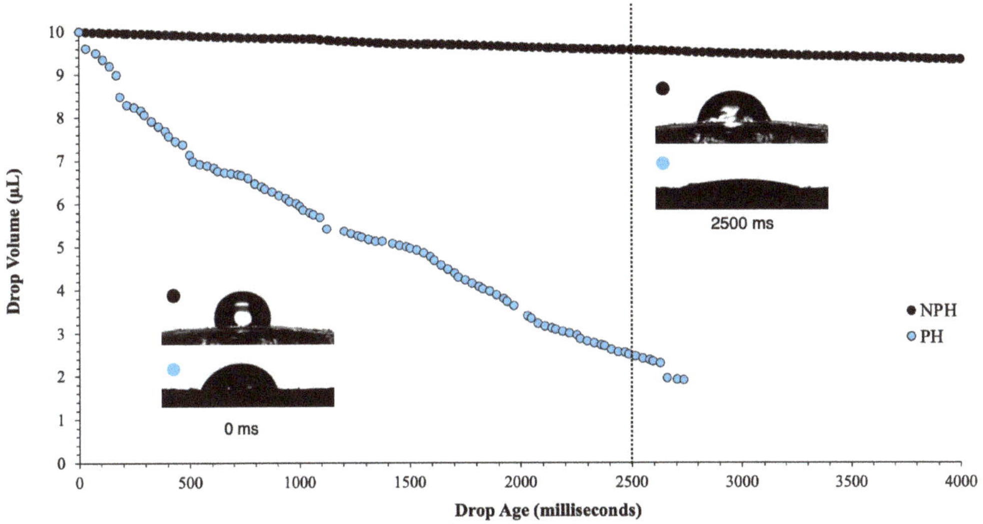

Figure 4. Surface absorption of non-porous hydrogel (NPH) and porous hydrogel (PH). Inset: appearances of droplet at instant of drop placement (0 ms) and after 2500 ms.

2.2. Comparisons of Porous Hydrogels (PH) and Porous Hydrogels with Manuka Honey (PH_MH)

The mechanical properties of the NPH and PH samples were assessed by measuring the rheological properties, which are presented in Figure 5. The mechanical properties of sheet hydrogels are important in many applications that require handling, such as wound dressings. In regards to wound dressing function, two key requirements can be identified: (1) the material should be able to withstand the expected forces during treatment, and (2) the parameters should not impair normal skin function and preferably promote healing [36]. The samples underwent two different testing regimes: a frequency sweep and a strain sweep. The results from the frequency sweep in Figure 5A show that the NPH had an average storage modulus value of ~17,000 Pa, while the PH has an average value of ~1700 Pa. The frequency-sweep test did not show a crossover between G′ and G″ in ether system, and each gel exhibited a modulus was independent of frequency, which is conventional in gel-like systems. The strain sweep in Figure 5B was used to assess the extent of the linear viscoelastic region for each gel. The results indicate linear behavior up to approximately 20% strain (γL) for PH and approximately 30% for NPH, before G′ started to decrease. As the strain increased, crossovers between G′ and G″ occurred at 100% (NPH) and 125% (PH) strain (γF), indicating a transition towards a liquid-like response. A material with good cohesive strength should have a high G′ and a low tan delta. The tan delta values of the samples are also presented in Figure 5C, and the results show that both samples have tan delta values below 0.1 at strains below 20%, with the NPH sample exhibiting a lower tan delta value than the PH sample. When the tan delta value is less than 0.1, it indicates that there is good cohesive force. In Figure 5D, the complex viscosity vs. angular frequency plot shows a decreasing linear relationship for both samples, with the NPH having a higher complex viscosity compared to the PH sample. This indicates that there was yield stress in both systems, and that both hydrogels were viscoelastic solids, as they did not flow at rest.

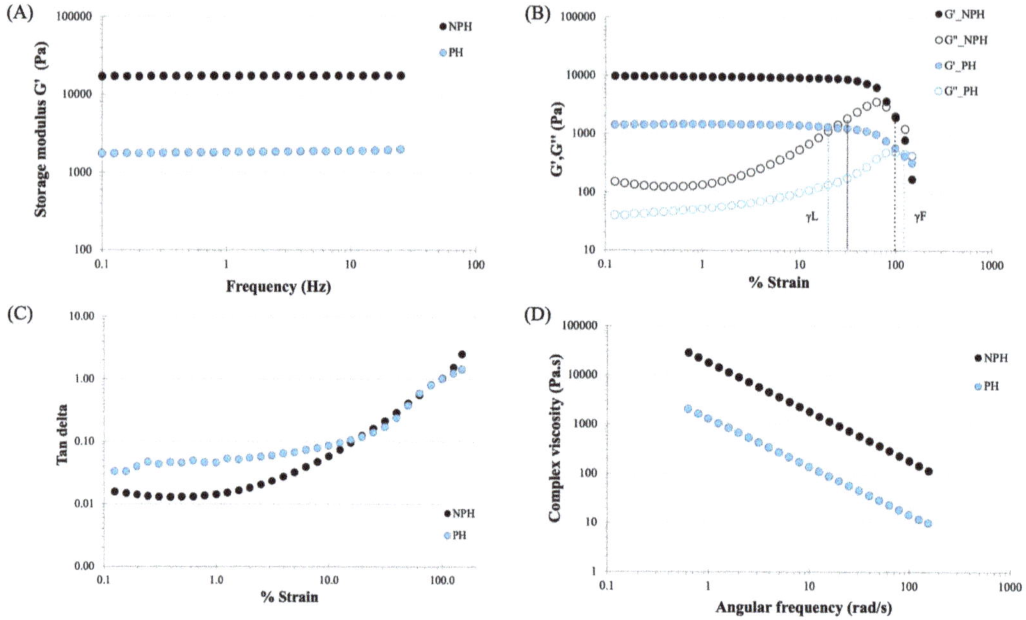

Figure 5. Rheological behavior of NPH and PH under (**A**) frequency sweep—Storage modulus G′ vs. Frequency. (**B**) Strain sweep—Storage and loss modulus G′ and G″ vs. % Strain. (**C**) Strain sweep—Tan delta vs. % Strain. (**D**) Complex viscosity vs. Angular frequency.

The cross-section structural morphology of the hydrogel samples was observed using SEM. The SEM micrographs showed that when MH was incorporated in the gels, there was a reduction in the average pore size. Figure 6A,B, shows the presence of 'polymer droplets' and pores in the structure. In contrast, PH_10% MH (Figure 6C) exhibits a more homogenous polymer structure with similar-sized pores to those in 1% MH. Table 1 presents the averaged pores size and % porosity values. The average pores sizes for PH, PH_1% MH, and PH_10% MH were 108.5 µm ± 46.0, 51.5 ± 24.2, and 50.5 µm ± 11.2, respectively. These pore size values are in line with values observed for other porous hydrogels fabricated using gas foaming techniques, which have been reported within the range of 10–500 µm [37]. The impact of MH concentration on the % porosity within the hydrogel structure shows that the incorporation of 1% MH into the PH resulted in a decrease in porosity from 37.10 ± 34.5% to 21.73 ± 15.5%, whereas the addition of 10% MH increased the porosity to 42.22 ± 11.1%. This indicates that the incorporation of 10% MH caused an increase in the number of pores within the hydrogel structure, leading to a higher overall porosity. The smaller pore sizes and increased porosity resulting from the addition of 10% MH were attributed to the influence of MH on the system. Firstly the viscosity of the system was altered, and secondly, MH is slightly acidic, which affected the size and duration of the gas bubbles formed. Another noteworthy observation was the shape of the PH_10% MH pores, which were highly spherical. This characteristic highlights the enhanced stability of the foam produced by this particular sample.

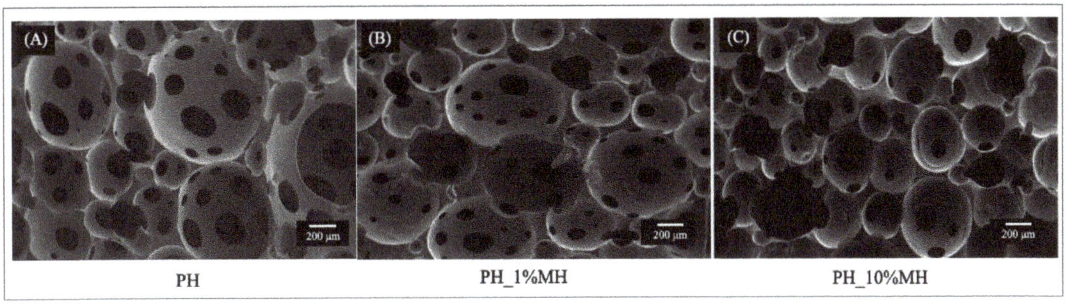

Figure 6. SEM micrographs at 50× magnification of (**A**) PH, (**B**) PH_1% MH, and (**C**) PH_1% MH.

Table 1. Average pore size and % porosity.

Samples	Average Pore Size (µm)	% Porosity
Porous hydrogel (PH)	108.5 ± 46.0	37.10 ± 34.5
Porous hydrogel with 1% Manuka Honey (PH_1% MH)	51.5 ± 24.2	21.73 ± 15.5
Porous hydrogel with 10% Manuka Honey (PH_10% MH)	50.5 ± 11.2	42.22 ± 11.1

Figure 7A,B shows the bulk swelling behavior of PH and PH with MH. The results show that the addition of MH resulted in a reduction in the swelling ratio, with 1% MH resulting in a larger decrease compared to 10% MH. The majority of this additional swelling capacity visibly occurred in the initial period (<3 min). After this initial period, the rate at which the gels swelled was similar for all samples. A noteworthy finding was that the gels containing MH seemed to retain their original shape better than the hydrogels that did not contain MH. This was especially visible in the PH_10% MH after swelling for 5 min.

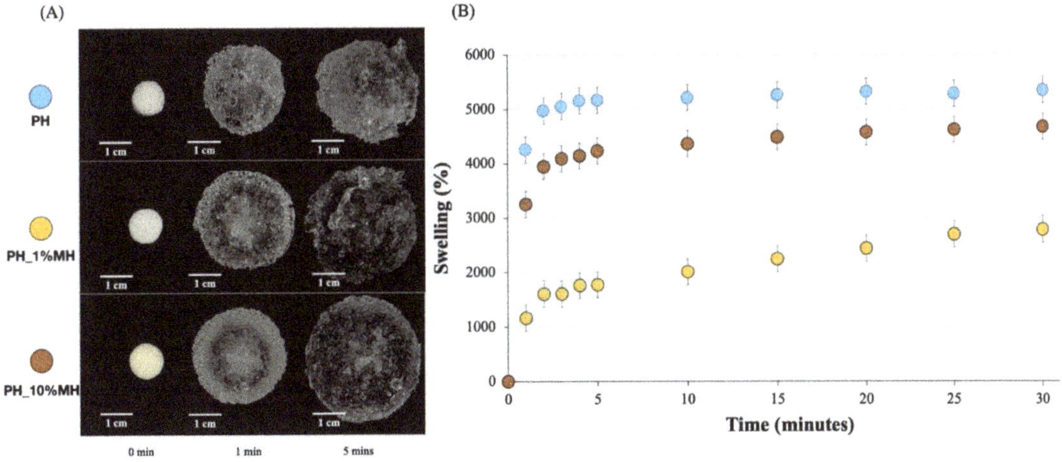

Figure 7. Bulk swelling in water. (**A**) The appearance of PH and PH with 1% and 10% MH at 0, 1, and 5 min, respectively. (**B**) Percentage swelling of PH and PH with 1% and 10% MH.

PH_10% MH exhibited a higher percentage of swelling due to its larger surface area and interconnected network within the structure, which contributed to its higher swelling capacity compared to PH_1% MH. Although PH had the most swelling capacity due to its much larger pore size, when considering mechanical properties after swelling, PH with MH had better appearances when compared to PH without MH.

Figure 8 compares the surface absorption PH samples with 1% and 10% MH. During the initial period (0 to 500 ms), all hydrogels exhibited a very similar and fast absorption rate. After this initial period, the PH samples continued to absorb at a similar rate. However, after 500 ms, the PH_1% MH and PH_10% MH absorption rates decreased to 2700 ms and 3000 ms, respectively. After this time period, the MH samples absorbed the rest of the droplet. One limitation of this technique is that the software could not accurately measure the last microliter of solution. Hence, all sample traces finished before reaching zero. The properties of the hydrogels can be affected by pore size and porosity, with pore size having a significant impact on the movement of water into the gels [38]. In the case of hydrogels containing MH, the pore size was approximately half that of hydrogels without MH, resulting in the distinct absorption behavior observed in these samples. The size of the pores in the hydrogel plays a crucial role in determining the surface absorption capacity. Generally, larger pores have a higher capacity for absorption compared to smaller pores. This is because larger pores provide more surface area and volume for the absorption of fluids or molecules. However, there is an optimal pore size range that balances surface area with the diffusion distance to facilitate efficient absorption. The shape of the pores also affects the surface absorption; pores with irregular shapes or tortuous pathways may hinder the diffusion of fluids or molecules into the hydrogel matrix. On the other hand, well-defined and interconnected pore structures can facilitate the flow and penetration of substances, leading to improved surface absorption. The surface properties of hydrogels significantly impact their ability to absorb substances. These properties include surface charge, surface energy, and surface roughness.

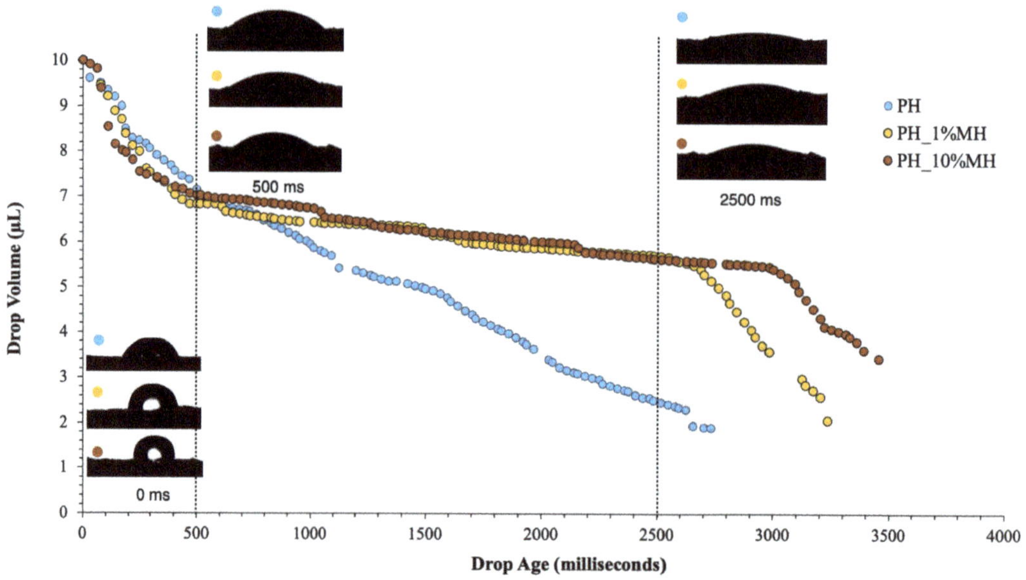

Figure 8. Surface absorption of PH and PH with 1% and 10% MH. Inset: the appearances of droplet at the instant of drop placement (0 ms), after 500 ms, and after 2500 ms.

Figure 9A,D presents the rheology properties of PH and PH with 1% and 10% MH. These samples were tested using the same parameters as shown in Figure 5. Based on the results shown in Figure 9A, the incorporation of 1% and 10% Manuka honey did not significantly affect the storage modulus during the frequency sweep testing, as all samples exhibited storage modulus values between 1000 and 2000 Pa. Figure 9B illustrates the behavior of PH and PH with 1% and 10% MH during the strain sweep. The results showed that the addition of Manuka honey did not alter the storage modulus of the gels (G′), but there was an increase in the loss modulus (G″) observed in the PH and honey-containing gels (PH_1% MH and PH_10% MH) at 34.66 (PH), 58.62 (PH_1% MH), and 120.09 (PH_10% MH) Pa, respectively. The linear viscoelastic region of the hydrogel (γL) was identical to that of PH, but the crossover of G′ and G″ (γF) occurred at approximately 80% strain for the samples containing honey and approximately 100% for the PH samples without honey. The examination of tan delta (Figure 9C), comparing PH with MH at concentrations of 1 and 10% *w*/*w*, revealed a similar pattern. The addition of honey into the PH matrices affected the resulting material's tan delta values, resulting in a marginally reduced tan delta as the honey concentration increased, while the trend in tan delta remained comparable. Figure 9D shows the complex viscosity plotted against angular frequency for all the samples. The results showed that all three samples exhibited a linear response, further confirming that the samples behaved as viscoelastic solids. The rheological performance of the samples showed that the inclusion of MH into the hydrogel matrix did not reduce the mechanical properties of the hydrogel. The minimal impact of honey on the mechanical properties can be attributed to the fact that honey does not undergo considerable physical crosslinking and exists as an interpenetrating network associated with the water and polymer molecules. However, it is important to note that certain components present in honey, such as 1,2 dicarbonyl compounds (Glyoxal and 3-deoxyglucosulose) and phenolic acids (Gallic acid and 4-methoxyphenylactic acid) [39], have the ability to form hydrogen bonds with the hydrogel network. These components play a role in preserving the strength of the gels and contribute to the decomposition of the sodium bicarbonate blowing agent, ultimately leading to an improved porous structure.

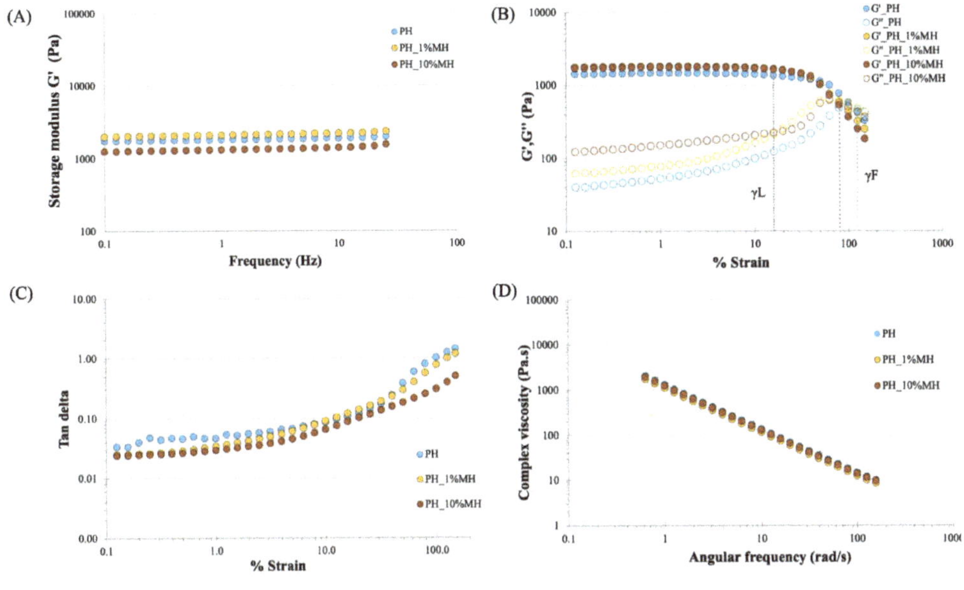

Figure 9. Rheological behavior of PH and PH with 1 and 10% w/w MH. (**A**) Frequency sweep—Storage modulus G′ vs. Frequency. (**B**) Strain sweep—Storage and loss modulus G′ and G″ vs. % Strain. (**C**) Strain sweep—Tan delta vs. % Strain. (**D**) Complex viscosity vs. Angular frequency.

The cytotoxicity of NPH, PH, and PH with 1% and 10% MH was tested using the XTT assay. Fibroblast cells were used to evaluate cell viability over 24 h. Statistical analyses were conducted to indicate significant differences ($p > 0.05$) in cell viability between the samples and the control. Figure 10 shows that all samples exhibited cell viability higher than

80%, with only PH_10% MH demonstrating a cell viability <90% and showing statistical significance compared to the control sample. High Manuka honey content is known to result in a decrease in cell viability due to its hydrogen peroxide and flavonoid content. Moreover, the pH of Manuka honey is acidic, ranging between pH 3.2 and 4.5, which can decrease cell viability. Previous studies have indicated that cytotoxic effects start at 3–5% MH [40]. International guidelines (ISO10993-5, 2009. Biological evaluation of medical Devices, in: Standardization, I.O.f. (Ed.) Part 5: Test for in vitro cytotoxicity, 3 ed. International Organization for Standardization, Geneva, Switzerland) state that a substance is cytotoxic only if it reduces cell viability to less than 70%. All the tested samples exhibited cell viabilities higher than 70% and can therefore be classified as non-toxic.

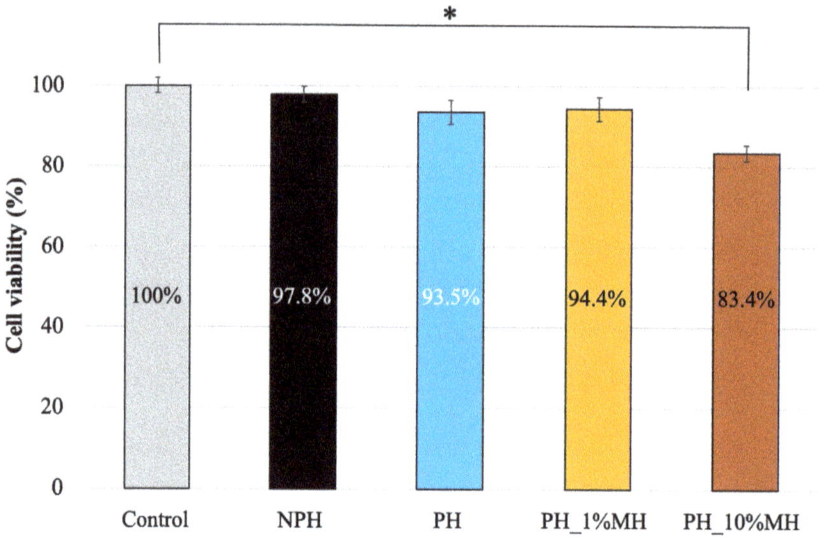

Figure 10. Cytotoxicity test to evaluate cell viability of NPH, PH, PH_1% MH, and PH_10%MH (* $p < 0.05$).

3. Conclusions

This work was concerned with the fabrication method for sheet porous hydrogels (PH) that exhibit exceptional absorption capacity and rapid fluid uptake. The porous hydrogels were synthesized through a delicate balance of three pre-mixtures, resulting in reproducible and reliable production of the gels. Porous hydrogels exhibited higher swelling ratios than non-porous hydrogels (NPH). Additionally, real-time surface absorption analysis revealed significant differences, with porous hydrogels absorbing 10 μL of fluid in 3000 ms, while NPH hydrogels absorbed only 1 μL. To enhance the properties of PH hydrogels for use in wound dressings, Manuka honey (MH) was incorporated into the gel structure at concentrations of 1 and 10% *w*/*w*. The results indicated that MH improved the appearance of the gel, with smaller pores and more linear swelling behavior observed over the first 5 min. The incorporation of MH also improved the mechanical properties of the hydrogels. These improvements were attributed to the inherent properties of MH, which altered the system by increasing viscosity and lowering the pH, enhancing foam production during gelation. Overall, the findings of this study suggest that the synthesized hydrogels have potential for use in wound dressings for medium and heavy-exudating wounds, where conventional hydrogels may not be suitable. When combined with the beneficial properties of MH, the improved properties of PH hydrogels offer a promising avenue for the development of advanced wound dressings with rapid absorption. Future research should focus on modifying the composition in order to achieve the desired pore size of the fabricated porous hydrogels. This can be accomplished by adjusting the balance of surfactants, blowing agents, and acids, while also considering the inclusion of additional components to modify the viscosity. Furthermore, the system can be enhanced by incorporating other active agents, such as silver nanoparticles.

4. Materials and Methods

4.1. Materials

2–acrylamido-2-methyl-1-propanesulfonic acid sodium salt (AMPs) (Mw = 229.23; 50% wt. in water) (monomer), Methacrylic acid 99% (MAA) (monomer), Di(ethylene glycol) diacrylate 99% (XL) (cross-linker), N,N,N′,N′-Tetramethylethylenediamine (TEMED) (initiator), Ammonium persulphate (APS) (initiator), Poloxamer 407 (Pluronic® F-127) (F127) (surfactant), and Sodium hydrogen carbonate (BA) (blowing agent) were all purchased from Sigma-Aldrich Co. Inc, Singapore, Singapore. Manuka Honey (MH) (86% Manuka pollen content) was purchased from Airborne Honey Ltd., Canterbury, New Zealand. For the cell culture studies, Dulbecco's modified Eagle's medium (DMEM), fetal bovine serum (FBS), penicillin–streptomycin, amphotericin B, and 0.25% trypsin–ethylenediaminetetraacetic acid were purchased from Gibco (Grand Island, NY, USA). The XTT solution (Cell Proliferation Kit II) was supplied by Roche Diagnostics GmbH (Mannheim, Germany). Phosphate-buffered saline (PBS) of pH 7.4 was supplied by KEMAUS, Cherrybrook N.S.W., Australia. Normal human dermal fibroblast (NHDF) cells (Lot no. C-12302, Promocell, Eppelheim, Germany) (1×10^5 cells/well, passage number 6) were provided by the Faculty of Pharmacy, Naresuan University, Phitsanulok, Thailand.

4.2. Synthesis of Non-Porous Hydrogels (NPH), Porous Hydrogels (PH), and Porous Hydrogels with Manuka Honey (PH_MH)

Synthesis of Hydrogels

Hydrogel samples were prepared using a redox-initiated free-radical polymerization procedure. Three different hydrogel samples were synthesized, including non-porous hydrogels (NPH), porous hydrogels (PH), and porous hydrogels with Manuka honey (PH_MH).

All the hydrogels were prepared using an Ammonium Persulphate (APS) and N, N, N, N,-tetramethyl ethylenediamine (TEMED) redox pair at a concentration of 1M. For the redox pair, APS was added to vial A and TEMED was added to vial B for all samples.

Non-porous hydrogels (NPH) were synthesized by following the composition in Table 2. Briefly, all components were split between two vials (A and B). The NPH gels were comprised of AMPs, deionized water, XL solution, and a redox initiator pair (APS and TEMED). Both vials A and B contained equal amounts of monomer, cross-linker, and deionized water, with 1 M of APS and 1 M TEMED prepared and added in separate vials. Then, each vial was mixed using an orbital shaker for 30 min until a homogenous solution was obtained. In the final step, vials A and B were poured together into a mold, and the mixture was mixed using an overhead mechanical stirrer for 15 s.

Table 2. Chemical compositions of the hydrogel samples.

Sample	AMPs (g)	DI Water/F127 (g)	DI Water (g)	XL (g)	TEMED (g)	APS (g)	BA (g)	MAA (g)	MH (g)
Non-Porous hydrogel (NPH)	5.00	-	4.00	0.20	-	-	-	-	-
Porous hydrogel (PH)	5.00	4.00	-	0.20	0.65	0.65	0.50	0.25	-
Porous hydrogel with 1% Manuka honey (PH_1% MH)	5.00	3.90	-	0.20	0.65	0.65	0.50	0.25	0.1040
Porous hydrogel with 10% Manuka honey (PH_10% MH)	5.00	1.96	-	0.20	0.65	0.65	0.50	0.25	1.0400

For porous hydrogel samples, the monomers (AMPs/MAA), surfactant solutions, cross-linker, blowing agent, and redox pairs were added in three separate sample vials (A, B, and C). For vials A and B, the blowing agent (Sodium hydrogen carbonate) was dissolved in the pre-prepared stock surfactant solution (Poloxamer 407—0.5% F127) and split equally between vials A and B. The monomer AMPs were added to both vials A and B, along with the cross-linker (Di(ethylene glycol) diacrylate). The final component, methacrylic acid, was prepared in vial C by mixing the remainder of the extra added water/surfactant solution. All the vials were then shaken for 30 min to allow the solutions to mix thoroughly. Finally, vial C, followed by B and then A, were poured into a mold. Using an overhead mechanical stirrer, the solution was rapidly stirred for a controlled time period (20 s) in order to combine all the components before foaming and gelation occurred. The amounts of each component used are listed in Table 2. The porous hydrogels with Manuka honey (PH_MH) were prepared using the same method, with the Manuka honey added to the surfactant solution in vials A and B at two different concentrations (1 and 10% *w*/*w*).

4.3. Swelling Test

The swelling behaviour of the samples was investigated by completely immersing them in deionized water at room temperature. Next, the swollen hydrogels were removed and weighed at selected time intervals ranging from 1 to 30 min. Upon removal from the deionized water, the hydrogels were blotted to remove excess surface water before weighing. The swelling ratio (% swelling) was calculated based on the change in weight using the following equation:

$$\%\ \text{Swelling} = \frac{W_f - W_i}{W_i} \times 100\% \tag{1}$$

where W_i and W_f are initial weight and final weight at different times, respectively. The measurements were conducted three times for each sample and reported as the average % swelling percentage, with the standard deviation reported to indicate the level of uncertainty.

4.4. Surface Absorption

The surface absorption properties of all hydrogel samples were evaluated using a Dataphysics Model OCA20 (Filderstadt, Germany) contact angle apparatus. The surface absorption was found by plotting the drop volume (μL) vs. drop age (milliseconds). The OCA20 camera recorded at 36 frames per second, which allowed the drop volume to be accurately recorded during the absorption process, enabling the detailed, real-time surface absorption to be measured for each sample. The hydrogel samples were first cut into 10 mm sizes using a cork borer and placed on a glass sample holder. Then, a droplet of 10 μL of deionized water was deposited onto the surface of each hydrogel sample, and the software recorded the volume of liquid vs. drop age.

4.5. Morphological Observations

The morphology of the hydrogels was measured using scanning electron microscopy (SEM) (LEO Co., Cambridge, England, Model: 1455VP). The preparation of the samples for SEM were as follows: the hydrogel samples were cut into a diameter of 10 mm and placed on an aluminum stub. The hydrogels were then dehydrated in a desiccator in order to remove the moisture present in the hydrogel before coating it with gold. At this point, the hydrogels were ready for testing. The SEM images were used to measure the pore size, which was calculated using ImageJ software (version 2.3.0) and presented as average value.

The porosity of the hydrogel samples was assessed using SEM images and analyzed with the ImageJ software. The thresholding operation was utilized to distinguish between pores and solid material based on their pixel intensities. Pixels with intensities below a specified threshold value were identified as pores, while pixels with intensities above the threshold were categorized as solid materials. The porosity was calculated by dividing the area of the pores by the total area of the image, and the resulting value was multiplied by 100% to express it as a percentage. The porosity calculation formula can be expressed as:

$$\% \text{ Porosity} = \frac{\text{Area of pores}}{\text{Total area}} \times 100 \quad (2)$$

4.6. Rheological Measurement

A rotational rheometer, ARES G2 (advanced rheometrics expansion system), TA Instrument, New Castle, DE, U.S.A was used to measure the viscoelastic behaviour of all the hydrogel samples. The samples were cut to a diameter of 25 mm using a cork borer and placed between two 25 mm serrated parallel plates with a 2–4 mm gap. The two test parameters consisted of a stain sweep from 0.01% to 150% at a constant frequency of 1 Hz and a frequency sweep from 0.1 to 25 Hz with 1% constant strain at 25 °C.

4.7. Cytotoxicity Test

The samples were cut into a cylinder shape with a diameter of 6 mm. Then, the samples were impregnated in 1 mL of serum-free Dulbecco's Modified Eagle's Medium (DMEM) Grand Island, NY, USA at room temperature for 24 h. At that time, the impregnated medium was sterilized using a syringe filter cap (0.2 μm). A suspension of normal human dermal fibroblast cells (NHDF cells) was placed in 96-well plates at a density of 1×10^4 cells/well and incubated in DMEM containing 10% FBS, 1% penicillin, 1% streptomycin, and 0.1% amphotericin B at 37 °C in a CO_2 incubator for 24 h. The medium was discarded and the NHDF cells were washed with PBS. Then, the cells were treated with the sterilized medium of the impregnated samples in each well. The untreated control was prepared using the NHDF cells with free-serum medium without the impregnation of samples. The cells were incubated at 37 °C in a CO_2 incubator for 24 h and compared with the control group (untreated NHDF). After treatment, the cells were washed with PBS. Then, both groups were replaced with 200 μL of new free-serum medium. Then, 50 μL of XTT solution was added to each well and incubated for 4 h. The cell viability was determined by measuring the optical density (*OD*) at 490 nm using a microplate reader (EonTM, BioTek instrument, Winooski, VT, USA.) and equation 3. The *OD* values of each sample were calculated as the % viability and compared with 100% viability of the untreated cells:

$$\text{Cell Viability } (\%) = \frac{OD_S}{OD_C} \times 100 \quad (3)$$

where OD_S is the absorbance of the samples and OD_C is the control.

Author Contributions: Conceptualization, T.P., J.V., J.J. and G.M.R.; Methodology, T.P., M.Y., J.D., N.T., J.J. and G.M.R.; Validation, T.P. and G.M.R.; Formal analysis, T.P.; Investigation, T.P. and M.Y.; Resources, S.R.; Data curation, T.P.; Writing – original draft, T.P., M.Y., J.D., S.R. and G.M.R.; Writing—review & editing, T.P., N.T., S.M., J.V., J.J., S.R. and G.M.R.; Visualization, T.P. and G.M.R.; Supervision, S.M., J.V., J.J., S.R. and G.M.R.; Project administration, S.R. and G.M.R.; Funding acquisition, S.R. and G.M.R. All authors have read and agreed to the published version of the manuscript.

Funding: This research was funded by Global and Frontier Research University Fund, Naresuan University (Grant number R2566C052) and Thailand Science Research and Innovation (TSRI) (Grant Number FRB660001/0179, Contract Number R2566B004) And The APC was funded by Global and Frontier Research University Fund, Naresuan University (Grant number R2566C052) and Naresuan University.

Institutional Review Board Statement: Not applicable.

Informed Consent Statement: Not applicable.

Data Availability Statement: The raw/processed data required to reproduce these findings cannot be shared at this time as the data also forms part of an ongoing study.

Acknowledgments: We would like to acknowledge the Science Lab Centre, Faculty of Science, Naresuan University for supporting CA and SEM measurements. We are also thankful for the Center of Excellence Excellence in Biomaterials, Faculty of Science, Naresuan University for supporting facilities.

Conflicts of Interest: The authors declare no conflict of interest.

References

1. Rosiak, J.M.; Yoshii, F. Hydrogels and Their Medical Applications. *Nucl. Instrum. Methods Phys. Res. Sect. B Beam Interact. Mater. At.* **1999**, *151*, 56–64. [CrossRef]
2. Hicyilmaz, A.S.; Seckin, A.K.; Cerkez, I. Synthesis, Characterization and Chlorination of 2-Acrylamido-2-Methylpropane Sulfonic Acid Sodium Salt-Based Antibacterial Hydrogels. *React. Funct. Polym.* **2017**, *115*, 109–116. [CrossRef]
3. Makino, K.; Hiyoshi, J.; Ohshima, H. Effects of Thermosensitivity of Poly (N-Isopropylacrylamide) Hydrogel upon the Duration of a Lag Phase at the Beginning of Drug Release from the Hydrogel. *Colloids Surf. B Biointerfaces* **2001**, *20*, 341–346. [CrossRef]
4. Yooyod, M.; Ross, S.; Phewchan, P.; Daengmankhong, J.; Pinthong, T.; Tuancharoensri, N.; Mahasaranon, S.; Viyoch, J.; Ross, G.M. Homo- and Copolymer Hydrogels Based on N-Vinylformamide: An Investigation of the Impact of Water Structure on Controlled Release. *Gels* **2023**, *9*, 333. [CrossRef] [PubMed]
5. Kongprayoon, A.; Ross, G.; Limpeanchob, N.; Mahasaranon, S.; Punyodom, W.; Topham, P.D.; Ross, S. Bio-Derived and Biocompatible Poly(Lactic Acid)/Silk Sericin Nanogels and Their Incorporation within Poly(Lactide-Co-Glycolide) Electrospun Nanofibers. *Polym. Chem.* **2022**, *13*, 3343–3357. [CrossRef]
6. Ross, S.; Yooyod, M.; Limpeanchob, N.; Mahasaranon, S.; Suphrom, N.; Ross, G.M. Novel 3D Porous Semi-IPN Hydrogel Scaffolds of Silk Sericin and Poly(N-Hydroxyethyl Acrylamide) for Dermal Reconstruction. *Express Polym. Lett.* **2017**, *11*, 719–730. [CrossRef]
7. Kamoun, E.A.; Kenawy, E.R.S.; Chen, X. A Review on Polymeric Hydrogel Membranes for Wound Dressing Applications: PVA-Based Hydrogel Dressings. *J. Adv. Res.* **2017**, *8*, 217–233. [CrossRef]
8. Khurana, B.; Gierlich, P.; Meindl, A.; Gomes-Da-Silva, L.C.; Senge, M.O. Hydrogels: Soft Matters in Photomedicine. *Photochem. Photobiol. Sci.* **2019**, *18*, 2613–2656. [CrossRef]
9. Brumberg, V.; Astrelina, T.; Malivanova, T.; Samoilov, A. Modern Wound Dressings: Hydrogel Dressings. *Biomedicines* **2021**, *9*, 1235. [CrossRef] [PubMed]
10. Bajpai, S.K.; Johnson, S. Superabsorbent Hydrogels for Removal of Divalent Toxic Ions. Part I: Synthesis and Swelling Characterization. *React. Funct. Polym.* **2005**, *62*, 271–283. [CrossRef]
11. Patel, P.K.; Mistry, S.N.; Patel, G.J.; Bharadia, P.D.; Pandya, V.M.; Modi, D.A. Recent in Controlled Drug Delivery System: Superporous Hydrogels. *IJPI's J. Pharm. Cosmetol.* **2011**, *1*, 53–65.
12. Zhang, T.; Sanguramath, R.A.; Israel, S.; Silverstein, M.S. Emulsion Templating: Porous Polymers and Beyond. *Macromolecules* **2019**, *52*, 5445–5479. [CrossRef]
13. Whang, K.; Goldstick, T.K.; Healy, K.E. Control of Protein Release from Emulsion Freeze-Dried Scaffolds with Unique Microarchitecture. *Scopus* **1996**, *2*, 774.
14. Baker, S.C.; Rohman, G.; Southgate, J.; Cameron, N.R. The Relationship between the Mechanical Properties and Cell Behaviour on PLGA and PCL Scaffolds for Bladder Tissue Engineering. *Biomaterials* **2009**, *30*, 1321–1328. [CrossRef]
15. Pramanik, R.; Narayanan, A.; Rajan, A.; Konar, S.; Arockiarajan, A. Transversely Isotropic Freeze-Dried PVA Hydrogels: Theoretical Modelling and Experimental Characterization. *Int. J. Eng. Sci.* **2019**, *144*, 103144. [CrossRef]
16. Gong, X.; Rohm, K.; Su, Z.; Zhao, B.; Renner, J.; Manas-Zloczower, I.; Feke, D.L. Porous Hydrogels Templated from Soy-Protein-Stabilized High Internal Phase Emulsions. *J. Mater. Sci.* **2020**, *55*, 17284–17301. [CrossRef]
17. Hori, K.; Sano, M.; Suzuki, M.; Hanabusa, K. Preparation of Porous Polymer Materials Using Water-in-Oil Gel Emulsions as Templates. *Polym. Int.* **2018**, *67*, 909–916. [CrossRef]
18. Omidian, H.; Rocca, J.G.; Park, K. Advances in Superporous Hydrogels. *J. Control. Release* **2005**, *102*, 3–12. [CrossRef]
19. Rungrod, A.; Kapanya, A.; Punyodom, W.; Molloy, R.; Meerak, J.; Somsunan, R. Synthesis of Poly(ε-Caprolactone) Diacrylate for Micelle-Cross-Linked Sodium AMPS Hydrogel for Use as Controlled Drug Delivery Wound Dressing. *Biomacromolecules* **2021**, *22*, 3839–3859. [CrossRef] [PubMed]
20. Sivan, S.S.; Roberts, S.; Urban, J.P.G.; Menage, J.; Bramhill, J.; Campbell, D.; Franklin, V.J.; Lydon, F.; Merkher, Y.; Maroudas, A.; et al. Injectable Hydrogels with High Fixed Charge Density and Swelling Pressure for Nucleus Pulposus Repair: Biomimetic Glycosaminoglycan Analogues. *Acta Biomater.* **2014**, *10*, 1124–1133. [CrossRef]
21. Ghatak, S.; Maytin, E.V.; MacK, J.A.; Hascall, V.C.; Atanelishvili, I.; Moreno Rodriguez, R.; Markwald, R.R.; Misra, S. Roles of Proteoglycans and Glycosaminoglycans in Wound Healing and Fibrosis. *Int. J. Cell Biol.* **2015**, *2015*, 834893. [CrossRef] [PubMed]
22. Bialik-Was, K.; Pluta, K.; Malina, D.; Barczewski, M.; Malarz, K.; Mrozek-Wilczkiewicz, A. The Effect of Glycerin Content in Sodium Alginate/Poly(Vinyl Alcohol)-Based Hydrogels for Wound Dressing Application. *Int. J. Mol. Sci.* **2021**, *22*, 12022. [CrossRef] [PubMed]
23. Kalaithong, W.; Molloy, R.; Nalampang, K.; Somsunan, R. Design and Optimization of Polymerization Parameters of Carboxymethyl Chitosan and Sodium 2-Acrylamido-2-Methylpropane Sulfonate Hydrogels as Wound Dressing Materials. *Eur. Polym. J.* **2021**, *143*, 110186. [CrossRef]

24. Liang, J.; Karakoçak, B.B.; Struckhoff, J.J.; Ravi, N. Synthesis and Characterization of Injectable Sulfonate-Containing Hydrogels. *Biomacromolecules* **2016**, *17*, 4064–4074. [CrossRef]
25. Singh, S.; Gupta, A.; Gupta, B. Scar Free Healing Mediated by the Release of Aloe Vera and Manuka Honey from Dextran Bionanocomposite Wound Dressings. *Int. J. Biol. Macromol.* **2018**, *120*, 1581–1590. [CrossRef]
26. Sh Ahmed, A.; Taher, M.; Mandal, U.K.; Jaffri, J.M.; Susanti, D.; Mahmood, S.; Zakaria, Z.A. Pharmacological Properties of Centella Asiatica Hydrogel in Accelerating Wound Healing in Rabbits. *BMC Complement. Altern. Med.* **2019**, *19*, 213. [CrossRef]
27. Yotsawimonwat, S.; Rattanadechsakul, J.; Rattanadechsakul, P.; Okonogi, S. Skin Improvement and Stability of Echinacea Purpurea Dermatological Formulations. *Int. J. Cosmet. Sci.* **2010**, *32*, 340–346. [CrossRef]
28. Frydman, G.H.; Olaleye, D.; Annamalai, D.; Layne, K.; Yang, I.; Kaafarani, H.M.A.; Fox, J.G. Manuka Honey Microneedles for Enhanced Wound Healing and the Prevention and/or Treatment of Methicillin-Resistant Staphylococcus Aureus (MRSA) Surgical Site Infection. *Sci. Rep.* **2020**, *10*, 13229. [CrossRef]
29. Oryan, A.; Alemzadeh, E.; Moshiri, A. Biological Properties and Therapeutic Activities of Honey in Wound Healing: A Narrative Review and Meta-Analysis. *J. Tissue Viability* **2016**, *25*, 98–118. [CrossRef]
30. Nolan, V.C.; Harrison, J.; Cox, J.A.G. Dissecting the Antimicrobial Composition of Honey. *Antibiotics* **2019**, *8*, 251. [CrossRef]
31. Hixon, K.R.; Bogner, S.J.; Ronning-Arnesen, G.; Janowiak, B.E.; Sell, S.A. Investigating Manuka Honey Antibacterial Properties When Incorporated into Cryogel, Hydrogel, and Electrospun Tissue Engineering Scaffolds. *Gels* **2019**, *5*, 7–9. [CrossRef] [PubMed]
32. Harding, K.; Carville, K.; Chadwick, P.; Moore, Z.; Nicodème, M.; Percival, S.L.; Romanelli, M.; Schultz, G.; Tariq, G. Wuwhs Consensus Document: Executive Summary Wound Exudate: Effective Assessment and Management Introduction. *Wound Int.* **2019**.
33. Hilliard, G.; DeClue, C.E.; Minden-Birkenmaier, B.A.; Dunn, A.J.; Sell, S.A.; Shornick, L.P. Preliminary Investigation of Honey-Doped Electrospun Scaffolds to Delay Wound Closure. *J. Biomed. Mater. Res.-Part B Appl. Biomater.* **2019**, *107*, 2620–2628. [CrossRef] [PubMed]
34. Pormohammad, A.; Monych, N.K.; Ghosh, S.; Turner, D.L.; Turner, R.J. Nanomaterials in Wound Healing and Infection Control. *Antibiotics* **2021**, *10*, 473. [CrossRef]
35. Sung, H.W.; Huang, R.N.; Huang, L.L.; Tsai, C.C. In Vitro Evaluation of Cytotoxicity of a Naturally Occurring Cross-Linking Reagent for Biological Tissue Fixation. *J. Biomater. Sci. Polym. Ed.* **1999**, *10*, 63–78. [CrossRef]
36. Opt Veld, R.C.; Walboomers, X.F.; Jansen, J.A.; Wagener, F.A.D.T.G. Design Considerations for Hydrogel Wound Dressings: Strategic and Molecular Advances. *Tissue Eng.-Part B Rev.* **2020**, *26*, 230–248. [CrossRef]
37. Sergeeva, A.; Vikulina, A.S.; Volodkin, D. Porous Alginate Scaffolds Assembled Using Vaterite $CaCO_3$ Crystals. *Micromachines* **2019**, *10*, 357. [CrossRef]
38. Foudazi, R.; Zowada, R.; Manas-Zloczower, I.; Feke, D.L. Porous Hydrogels: Present Challenges and Future Opportunities. *Langmuir* **2023**, *39*, 2092–2111. [CrossRef]
39. El-Senduny, F.F.; Hegazi, N.M.; Abd Elghani, G.E.; Farag, M.A. Manuka Honey, a Unique Mono-Floral Honey. A Comprehensive Review of Its Bioactives, Metabolism, Action Mechanisms, and Therapeutic Merits. *Food Biosci.* **2021**, *42*, 101038. [CrossRef]
40. Minden-Birkenmaier, B.A.; Cherukuri, K.; Smith, R.A.; Radic, M.Z.; Bowlin, G.L. Manuka Honey Modulates the Inflammatory Behavior of a DHL-60 Neutrophil Model under the Cytotoxic Limit. *Int. J. Biomater.* **2019**, *11*, 6132581. [CrossRef]

gels

MDPI

Article

Formulation and Evaluation of Diclofenac Potassium Gel in Sports Injuries with and without Phonophoresis

Komal Ammar Bukhari [1], Imran Ahmad Khan [1,2,*], Shahid Ishaq [3], Muhammad Omer Iqbal [4,5,*], Ali M. Alqahtani [6], Taha Alqahtani [6] and Farid Menaa [7,*]

[1] Ali-Ul-Murtaza, Department of Rehabilitation Sciences, Muhammad Institute of Medical and Allied Sciences, Multan 60000, Pakistan
[2] Department of Pharmacology and Physiology, MNS University of Agriculture, Multan 60000, Pakistan
[3] Department of Rehabilitation, Bakhtawar Amin Medical and Dental College, Multan 60000, Pakistan
[4] Shandong Provincial Key Laboratory of Glycoscience and Glycoengineering, School of Medicine and Pharmacy, Ocean University of China, Qingdao 266100, China
[5] Royal Institute of Medical Sciences (RIMS), Multan 60000, Pakistan
[6] Department of Pharmacology, College of Pharmacy, King Khalid University, Abha 62529, Saudi Arabia
[7] Departments of Internal Medicine and Nanomedicine, California Innovations Corporation, San Diego, CA 92037, USA
* Correspondence: imranahmadkhandurrani@gmail.com (I.A.K.); oiqbal133@gmail.com (M.O.I.); menaateam@gmail.com (F.M.)

Citation: Bukhari, K.A.; Khan, I.A.; Ishaq, S.; Iqbal, M.O.; Alqahtani, A.M.; Alqahtani, T.; Menaa, F. Formulation and Evaluation of Diclofenac Potassium Gel in Sports Injuries with and without Phonophoresis. *Gels* **2022**, *8*, 612. https://doi.org/10.3390/gels8100612

Academic Editors: Ana Paula Serro, Ana Isabel Fernandes and Diana Silva

Received: 14 August 2022
Accepted: 20 September 2022
Published: 26 September 2022

Abstract: *Background:* Pain remains a global public heath priority. *Phonophoresis*, also known as sonophoresis or ultrasonophoresis, is when an ultrasound is used to maximize the effects of a topical drug. *Purpose:* The objective of this study was to test, in patients injured in sports or accidents (N = 200), the efficacy of diclofenac potassium (DK) 6%, 4%, and 2% formulated gels with and without phonophoresis in comparison with market available standard diclofenac sodium (DS or DN) gel. *Methods:* The patients were enrolled after informed consent. By using the lottery method, 100 patients were randomly segregated into five groups without phonophoresis and repeated similarly with phonophoresis at a frequency of 0.8 MHz, an intensity of about 1.5 W/cm^2, and at continuous mode (2:1). Group-1 was treated with 6% DK gel, group-2 was treated with 4% DK gel, group-3 was treated with 2% DK gel, group-4 was treated with 4% DS gel and group-5 was given control gel three to four times a week for 4 weeks. The patients were screened by using NPRS and WOMAC scales. They were assessed on the baseline, 4th session, 8th session, 12th session, and 16th session. *Results:* Significant dose-dependently relief was observed in NPRS (Numeric Pain Rating Scale) and the WOMAC (Western Ontario McMaster Osteo-Arthritis) index for pain in disability and stiffness for each group treated with DK gel compared to DS gel. Phonophoresis increased these benefits significantly when used after topical application of DK gel or DS gel, and the dose-dependent effects of DK gel plus phonophoresis were stronger than the dose-dependent effects of DS gel plus phonophoresis. The faster and profounder relief was due to phonophoresis, which allows more penetration of the DK gel into the skin as compared to the direct application of DK gel in acute, uncomplicated soft tissue injury, such as plantar fasciitis, bursitis stress injuries, and tendinitis. In addition, DK gel with phonophoresis was well tolerated. Thus, in this personalized clinical setting, according to the degree of inflammation or injured-induced pain, disability, and stiffness, DK gel 6% with phonophoresis appeared more effective and thus more recommendable than DS gel 6% alone or DS gel 6% combined to phonophoresis.

Keywords: diclofenac potassium gel; diclofenac sodium gel; phonophoresis; sport injuries; pain

1. Introduction

Sports injuries (e.g., acute ankle sprain, plantar fasciitis, bursitis) are commonly treated with non-steroidal anti-inflammatory drugs (NSAIDs) to minimize discomfort, swelling, and inflammation [1,2]. They have been shown in clinical trials to be effective for the long-term treatment of recurring or chronic illnesses (e.g., back pain, arthritis) as well as for the quick control of severe and acute mild-to-moderate pain and inflammation resulting from injuries such as ankle sprains, musculoskeletal pain, soft tissue, and/or joint injuries [1,2]. Over the counter, topical NSAIDs (1 to 10%, according to the clinical case, and commonly 1 to 2%) are used as an alternative to oral versions [3–5]. Topical diclofenac is a typical NSAID that can enter joints, muscles, and synovial fluid through the skin. It spreads and lingers primarily in the target tissues that are inflamed [6–9]. Clinical trials have also demonstrated that NSAIDs are potent and well-tolerated anxiolytic and anti-inflammatory pain killers for the treatment of both acute and chronic pain conditions, resulting in improved mobility and recovery [3,10–12]. The ability of a pertinent NSAID to permeate thoroughly into damaged tissue determines its effectiveness [13]. Efficient permeation and therapeutic response require a delicate balance of lipid and water solubility. The formulation can influence local permeation and pharmacokinetic profile [14–16].

Topical diclofenac, a phenyl acetic acid derivative, which is frequently formulated as sodium salt (DS), has been demonstrated through pharmacokinetic investigations to swiftly enter the skin and reach the underlying tissues (e.g., joints, muscles, synovial) [16–20]. Lecithin, a component that forms micelles, and an aqueous-alcohol micro emulsion serve as the foundation of the 4% (*w*/*w*) diclofenac spray gel formulation (MIKA Pharma GmbH, Speyer, Germany) [16]. Low systemic availability and positive cutaneous penetration were observed [16]. Diclofenac is detected in plasma within 15 min, and due to analgesic and antipyretic properties, efficiency (rapid and targeted actions) and security, diclofenac spray's is a viable model in severe painful situations [16–18]. As per its main molecular mechanism, diclofenac blocks cyclooxygenase-2 (COX-2) [3,10], inhibiting the production of prostaglandins (PGs) from arachidonic acid (AA) (Figure 1). In addition to DS spray gel formulations, research has developed diclofenac patches [17,21–23]. Thereby, a semi-occlusive, bioadhesive patch containing 180 mg of diclofenac epolamine (hydroxyethylpyrrolidine), which is comparable to 129.7 mg of diclofenac acid and equivalent to 140 mg of diclofenac sodium, has been prepared [17]. Epolamine (50.3 mg), a pharmacologically inert is used to salify Diclofenac and increase its hydrophilic and lipophilic potencies, making it a viable chemical for topical administration [17]. Unlike DS, DK has only been fabricated for oral administration in the form of a fast-release tablet to deliver better results than oral DS in terms of bioavailability [24–26].

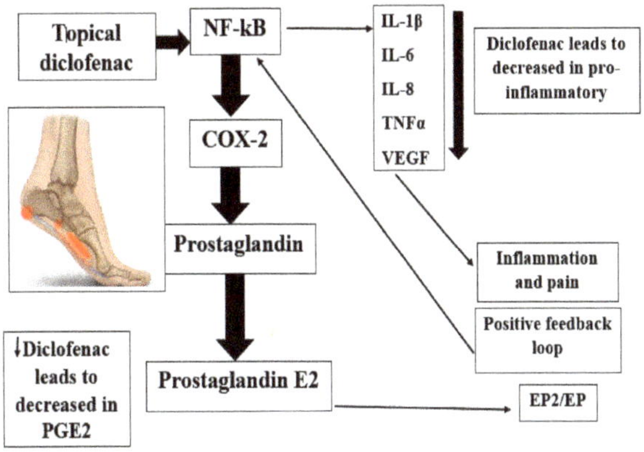

Figure 1. Mechanism of action of diclofenac potassium. **NF-κb:** Nuclear Factor-Kappa B; **COX-2:** Cyclooxygenase-2; **PGE2:** Prostaglandin E2 (dinoprostone); **IL-1β:** Interleukin-1beta (lymphocyte activating factor (LAF)); **IL-6:** Interleukin-6 (B-cell stimulatory factor -2 (BSF-2)); **IL-8:** Interleukin-8 (α chimiokine); **TNF-α:** Tumor Necrosis Factor-alpha; **VEGF:** Vascular Endothelial Growth Factor.

A musculoskeletal ultrasound is recommended to diagnose sport injuries-induced heel pain, such plantar fasciitis [27–29]. Physiotherapists and orthopedic surgeons often prescribe phonophoresis for the treatment of plantar fasciitis grade I and II [28,29]. Based on the defined test, it incorporates various physiological, physiochemical, and chemical effects, concluding that heat induced by sound waves plays a significant role in the management of different medical conditions; it also produced mechanical results [30–32]. Technological advances and high-frequency transducers have made ultrasound technology more desirable than the imaging of abnormal magnetic resonance imaging (MRI) due to its high surface area [30], and recent studies showed that on average, phonophoresis with analgesic gel for the treatment of sports injuries is more effective than local ultrasound alone [32–36]. Although, topical analgesic delivered through phonophoresis in deeper tissues is advised for the treatment of chronic muscular injuries [35,36]and that oral DK displayed a more effective action than oral DS [24–26], we failed to find any previous study using DK gel through phonophoresis for the treatment of sports injuries. To the best of our knowledge, DK gel is not available on the market worldwide yet. Herein, we then decided, for the first time, to test the efficacy of various concentrations (2–6%) of DK gels combined or not with phonophoresis in patients suffering from injury-induced pain. The comparison has been carried out with market-available standard DS gel at averaged use concentration (4%).

2. Results and Discussion

In total, 200 patients suffering from sport injuries were enrolled in this study. The patients with planter fasciitis were the most frequent (16%), whereas bursitis (4%) and capsulitis (4%) were the least frequent sport-related injuries (Figure 2). They were followed up on the baseline, at the 4th session, 8th session, 12th session, and at the 16th session of treatment. The quantitative data were presented for mean ± S.D. Statistical insignificance was considered if the *p*-value was less than 0.05. One-way ANOVA was used to determine whether there are any statistically significant differences between the means of two or more independent (unrelated) groups.

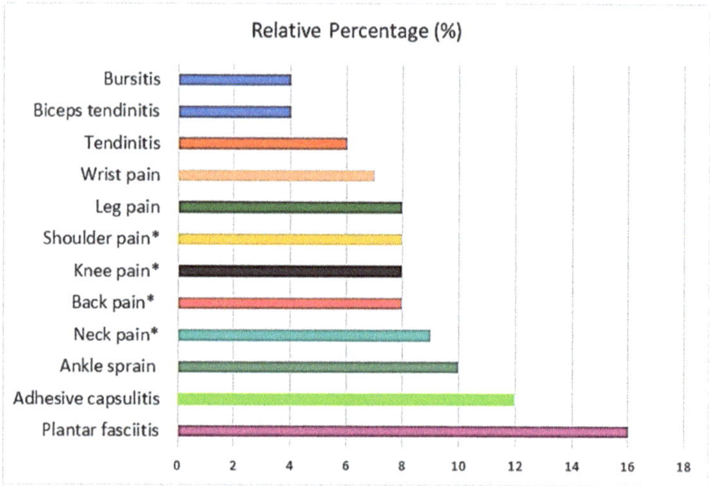

Figure 2. Incidence (%) of patients with sport-related injuries (N = 200) before treatment (which were then subsequently treated). * Means pain in muscles and soft tissues.

Swelling in patients treated with gel combined with phonophoresis was significantly decreasing in a dose-dependent manner (Figure 3a): 6% DK gel (p = 0.09–0.02), 4% DK gel (p = 0.09–0.03), 2% DK gel (p = 0.09–0.05), 4% DS gel (p = 0.08–0.02), and for placebo (p = 0.09–0.08). Comparatively to patients receiving gel but without phonophoresis, swelling was also significantly decreasing in a dose-dependent manner (Figure 3b): 6% DK gel (p = 0.08–0.03), 4% DK gel (p = 0.09–0.04), 2% DK gel (p = 0.09–0.05), 4% DS gel (p = 0.08–0.04), and for placebo (p = 0.09–0.08).

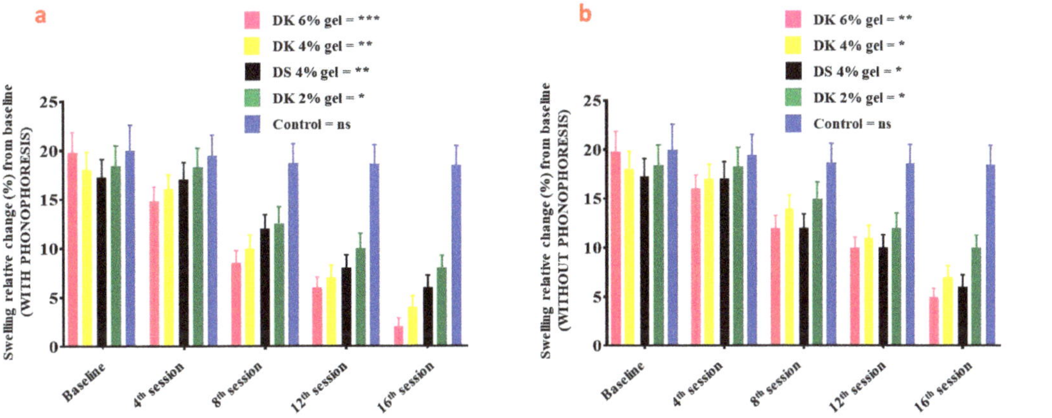

Figure 3. Swelling-related changes on the baseline, 4th session, 8th session, 12th session, and 16th session of Diclofenac potassium (DK) treatment (**a**) with phonophoresis (**b**) without phonophoresis, in sports-injured patients. Diclofenac sodium (DS) was used as a standard control. Control group is the placebo. Statistical significance (p-values) is detailed in the main text. ns = no significant; $p < 0.05$ *, $p < 0.01$ **, $p < 0.001$ ***.

NPRS in patients treated with gel combined with phonophoresis was significantly decreasing in a dose-dependent manner (Figure 4a): 6% DK gel (p = 0.009–0.000), 4% DK gel (p = 0.08–0.02), 2% DK gel (p = 0.08–0.03), 4% DS gel (p = 0.08–0.02) and for placebo (p = 0.09–0.07). Comparatively to patients receiving gel but without phonophoresis, NPRS was also significantly decreasing in a dose-dependent manner (Figure 4b): 6% DK gel (p = 0.009–0.03), 4% DK gel (p = 0.09–0.04), 2% DK gel (p = 0.08–0.04), 4% DS gel (p = 0.08–0.04), and for placebo (p = 0.09–0.08).

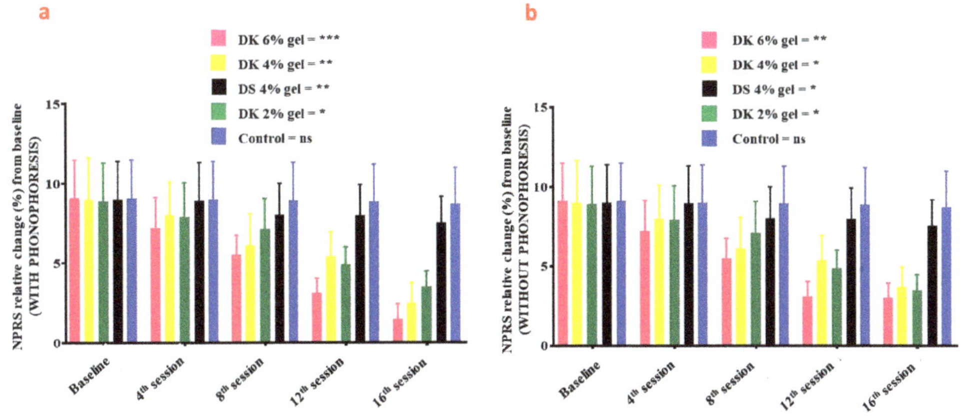

Figure 4. NPRS-related changes on the baseline, 4th session, 8th session, 12th session, and 16th session of Diclofenac potassium (DK) treatment (**a**) with phonophoresis (**b**) without phonophoresis, in sports-injured patients. Diclofenac sodium (DS) was used as a standard control. Control group is the placebo. Statistical significance (p-values) is detailed in the main text. ns = no significant; $p < 0.05$ *, $p < 0.01$ **, $p < 0.001$ ***. NPRS: Numerical Pain Rating Scale.

WOMAC ADLs (activities of daily living) in patients treated with gel combined with phonophoresis was significantly decreasing in a dose-dependent manner (Figure 5a): 6% DK gel (p = 0.009–0.001), 4% DK gel (p = 0.08–0.03), DK gel (p = 0.08–0.04), 4% DN gel (p = 0.09–0.03), and for placebo (p = 0.09–0.07). Comparatively to patients receiving gel but without phonophoresis, WOMAC ADLs was also significantly decreasing in a dose-dependent manner (Figure 5b): 6% DK gel (p = 0.009–0.04), 4% DK gel (p = 0.08–0.03), 2% DK gel (p = 0.09–0.04), 4% DS gel (p = 0.08–0.03), and for placebo (p = 0.09–0.08).

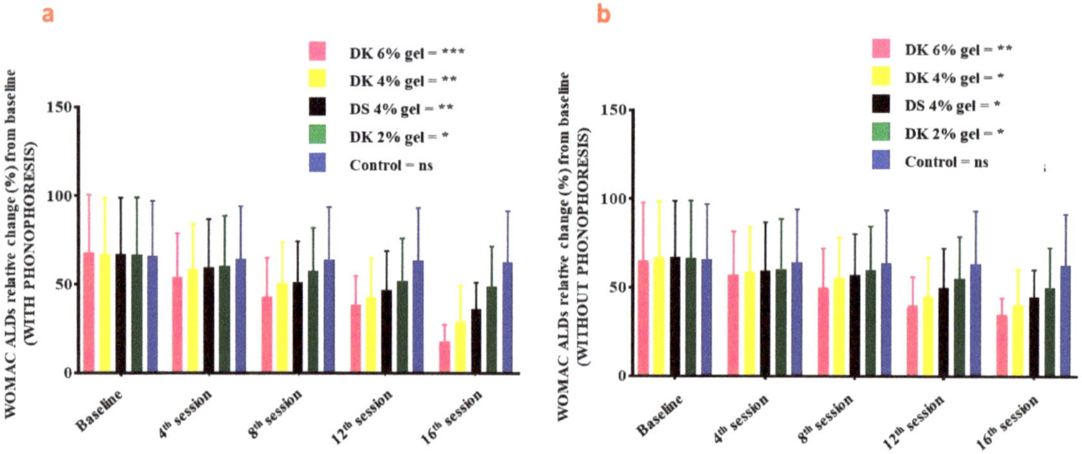

Figure 5. Index of WOMAC ADLs on the baseline, 4th session, 8th session, 12th session, and 16th session of Diclofenac potassium (DK) treatment (**a**) with phonophoresis (**b**) without phonophoresis, in sports-injured patients. Diclofenac sodium (DS) was used as a standard control. Control group is the placebo. Statistical significance (p-values) is detailed in the main text. ns = no significant; $p < 0.05$ *, $p < 0.01$ **, $p < 0.001$ ***. WOMAC: Western Ontario and McMaster Universities Arthritis Index, ADLs: activities of daily living.

WOMAC pain in patients treated with gel combined with phonophoresis was significantly decreasing in a dose-dependent manner (Figure 6a): 6% DK gel (p = 0.007–0.001), 4% DK gel (p = 0.08–0.02), 2% DK gel (p = 0.09–0.05), 4% DS gel (p = 0.08–0.03), and for placebo (p = 0.09–0.07). Comparatively to patients receiving gel but without phonophoresis, WOMAC pain was also significantly decreasing in a dose-dependent manner (Figure 6b): 6% DK gel (p = 0.008–0.04), 4% DK gel (p = 0.09–0.04), 2% DK gel (p = 0.09–0.05), 4% DN gel (p = 0.08–0.04), and for placebo (p = 0.09–0.08).

WOMAC stiffness in patients treated with gel combined with phonophoresis was significantly decreasing in a dose-dependent manner (Figure 7a): 6% DK gel (p = 0.008–0.001), 4% DK gel (p = 0.09–0.03), 2% DK gel (p = 0.08–0.05), 4% DS gel (p = 0.08–0.03), and for placebo p-value (p = 0.09–0.07). Comparatively to patients receiving gel but without phonophoresis, WOMAC stiffness was also significantly decreasing in a dose-dependent manner (Figure 7b): 6% DK gel (p = 0.08–0.03), 4% DK gel (p = 0.09–0.04), 2% DK gel (p = 0.09–0.05), 4% DS gel (p = 0.08–0.04), and for placebo (p = 0.09–0.08).

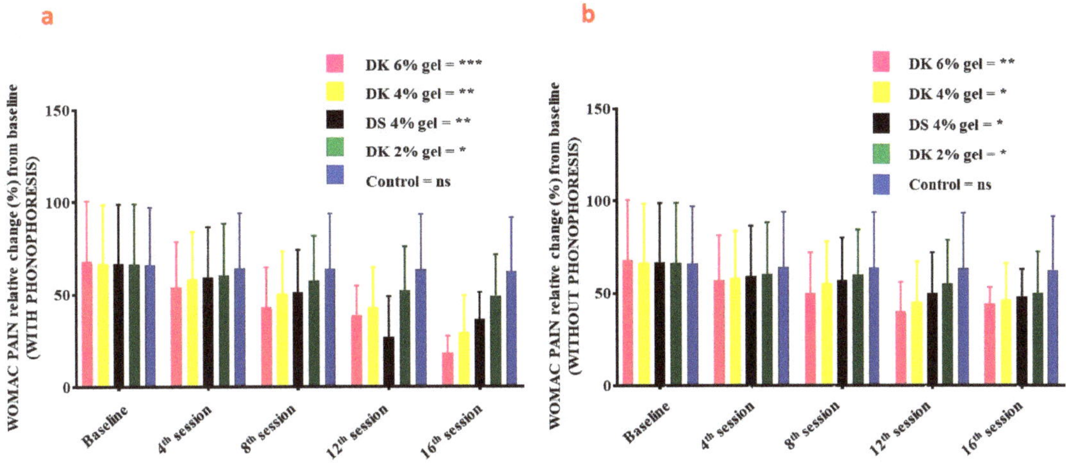

Figure 6. Index of WOMAC pain on the baseline, 4th session, 8th session, 12th session, and 16th session of Diclofenac potassium (DK) treatment (**a**) with phonophoresis (**b**) without phonophoresis, in sports-injured patients. Diclofenac sodium (DS) was used as a standard control. Control group is the placebo. Statistical significance (p-values) is detailed in the main text. ns = no significant; $p < 0.05$ *, $p < 0.01$ **, $p < 0.001$ ***. WOMAC: Western Ontario and McMaster Universities Arthritis Index.

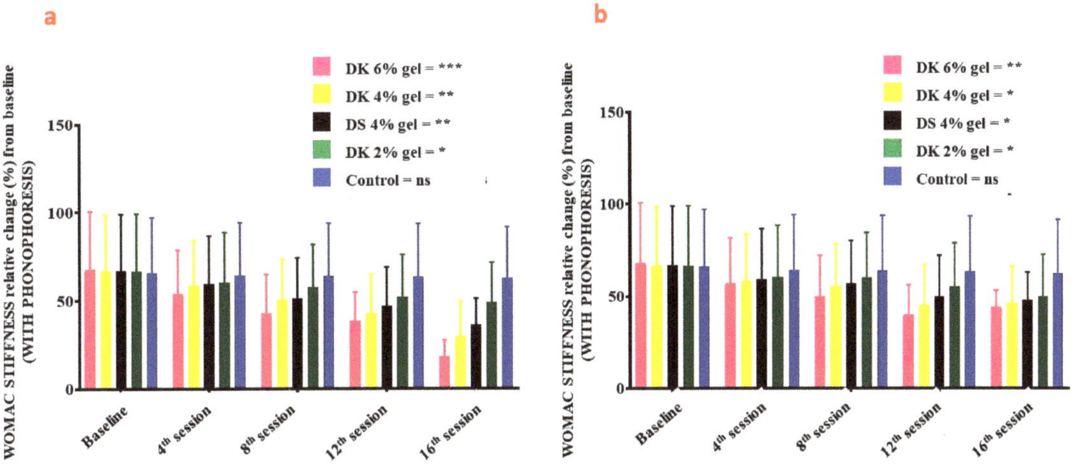

Figure 7. Index of WOMAC stiffness on the baseline, 4th session, 8th session, 12th session, and 16th session of Diclofenac potassium (DK) treatment (**a**) with phonophoresis (**b**) without phonophoresis, in sports-injured patients. Diclofenac sodium (DS) was used as a standard control. Control group is the placebo. Statistical significance (p-values) is detailed in the main text. ns = no significant; $p < 0.05$ *, $p < 0.01$ **, $p < 0.001$ ***. WOMAC: Western Ontario and McMaster Universities Arthritis Index.

Diclofenac is an antipyretic, analgesic, and steroidal anti-inflammatory drug (NSAID) that reduces pain and inflammation by inhibiting the production of prostaglandin (PG) cyclooxygenase-2 [3,10]. In a more detailed way, PGE2 activates the Gq-coupled EP1 receptor, leading to increased activity of the inositol triphosphate/phospholipase C pathway. Activation of this pathway releases intracellular stores of calcium, which directly reduces the action potential threshold and activates protein kinase C (PKC), which contributes to several indirect mechanisms. PGE2 also activates the EP4 receptor, coupled to Gs, which activates the adenylyl cyclase/protein kinase A (AC/PKA) signaling pathway. PKA and PKC both contribute to the potentiation of transient receptor potential action channel subfamily V member 1 (TRPV1) potentiation, which increases sensitivity to heat stimuli. They also activate tetrodotoxin-resistant sodium channels and inhibit inward potassium currents. PKA further contributes to the activation of the P2X3 purine receptor and the sensitization of T-type calcium channels. The activation and sensitization of depolarizing ion channels and inhibition of inward potassium currents serve to reduce the intensity of stimulus necessary to generate action potentials in nociceptive sensory afferents. PGE2 acts via EP3 to increase sensitivity to bradykinin and via EP2 to further increase heat sensitivity. Central sensitization occurs in the dorsal horn of the spinal cord and is mediated by the EP2 receptor, which couples to Gs. Pre-synaptically, this receptor increases the release of pro-nociceptive neurotransmitters such as glutamate, CGRP, and substance P. Post-synaptically, it increases the activity of AMPA and NMDA receptors and produces inhibition of inhibitory glycinergic neurons. Together, these lead to a reduced threshold of activation, allowing low-intensity stimuli to generate pain signals. PGI2 is known to play a role via its Gs-coupled IP receptor, although the magnitude of its contribution varies. It has been proposed to be of greater importance in painful inflammatory conditions such as arthritis. By limiting sensitization, both peripheral and central, via these pathways, NSAIDs can effectively reduce inflammatory pain. PGI2 and PGE2 contribute to acute inflammation via their IP and EP2 receptors. As with β-adrenergic receptors, these are Gs-coupled and mediate vasodilation through the AC/PKA pathway. PGE2 also contributes by increasing leukocyte adhesion to the endothelium and attracting the cells to the site of injury. PGD2 plays a role in the activation of endothelial cell release of cytokines through its DP1 receptor. PGI2 and PGE2 modulate T-helper cell activation and differentiation through IP, EP2, and EP4 receptors, which is believed to be an important activity in the pathology of arthritic conditions. By limiting the production of these PGs at the site of injury, NSAIDs can reduce inflammation.

The goal of this randomized, single-blinded study was to determine in a relatively large cohort population (N = 200 including n = 100 with phonophoresis and n = 100 without phonophoresis) the dose and time (sessions) effects of a newly formulated topical gel product, namely DK gel, with or without phonophoresis, in comparison with DS gel. DS gel, marketed under the name of Voltaren®, is known to reduce pain and inflammation at a usage concentration of 1–3%. While 10% DS is generally used for severe clinical cases, it is avoided in patients with certain pathologies such as gastritis or renal/kidneys failure or undergoing vaccination against diseases such as yellow fever or COVID. DK was then used in this study at reasonable concentrations (2–6%). DS represented the standard product, and placebo was used as the control. Phonophoresis is commonly used with a frequency of about 0.8 MHz, intensity 1.5 W/cm^2, and continuous mode 2:1 [27,37]. Therefore, we applied the same experimental settings in the present study.

To the best of our knowledge and as evoked in the Introduction, this study is the first of its kind. The pertinence of such study resides in the fact that (i) oral DK has shown better analgesic and anti-inflammatory effects compared to DS when they were administered orally; (ii) DS exists in (spray) gels (and patches) for topical application [29] and is commercialized, whereas we failed to find studies related to marketed DK gels; (iii) phonophoresis is commonly used in combination with DS gels for the treatment of musculoskeletal injuries [38,39], and it was proved more effective than DS alone [39,40], but no studies have evaluated phonophoresis in combination with diclofenac derivatives, such as DK.

DK was combined with phonophoresis to increase the pain-relieving effects of the phonophoresis. Indeed, medication particles are pushed deep into the skin tissue by ultrasound waves. A resting membrane potential caused by DK relaxed muscles and phonophoresis increases the penetration of the drug into tissues and circulation, which in turn decreased pain and inflammation [37–40].

The results showed that (i) DK gel alone is more effective than DS gel alone at equivalent concentration, which is a promising result and offers an alternative option in the treatment of pain; (ii) DK gels 6%, 4%, 2%, or 4% DS gel with phonophoresis (as adjuvant treatment) were more effective than the DK gel 6%, 4%, 2%, or 4% DS gel without phonophoresis, confirming the important role of phonophoresis as adjuvant treatment in pain [35,36]; (iii) the pain reduction (likewise other parameters tested) was dose- and time-dependent, highlighting the necessity of several sessions of cycles of gel therapy with phonophoresis to treat effectively and in a personalized manner patients (according to the degree of suffering from sport injuries, the clinical history of the patient but also from a holistic perspective).

At all time points, the percentage of responders (defined as the percentage of subjects achieving a 50% reduction in swelling of the injured soft tissues for 4 weeks after three to four times treatment in a week) was significantly greater in the group treated with DK gel 6% compared to the other groups. DK gel 6% also resulted in a significantly faster decrease in pain, inflammation, and swelling of the injured soft tissues compared to the other groups. Any comparison of the current study's results with other intervention programs for soft tissue injuries to assess the effectiveness of DK gel compared to other established products is hampered by differences in study design, inclusion criteria, duration of treatment, and efficacy assessment methods. Furthermore, in recent years, a variety of biomaterials (e.g., patches-based hydrogels, cryogels, nanofibers) together with their distinct physicochemical features have been extensively investigated and developed in the fields of drug-delivery, tissue engineering, medicine, and public health, including disease diagnosis [21,41–49]. Thereby, a most recent study in rabbits has reported an injectable thermosenstive hydrogel for dual delivery of diclofenac and avastin (an anti-VEGF) to reduce the inflammation of the corneal neovascularization more effectively [49]. Future studies shall compare DK gels + phonophoresis with hydrogel patches and pain killers-loaded nanomaterials not only in animals but also in a cohort population of patients.

3. Conclusions

Group-1 DK gel (6%) with the help of phonophoresis proved highly significant benefits when compared to group-2, group-3, group-4 and group-5 and similar groups without phonophoresis in a good number of patients suffering from sports injuries. Not only did the freshly prepared DK gel 6% quickly help in relieving pain, but it also improved patient mobility because phonophoresis has more penetration of the gel into the skin as compared to direct apply (massage) in acute, uncomplicated soft tissue injuries (e.g., plantar fasciitis, bursitis stress injuries, tendinitis). Strain was also well tolerated. DK gel 6% alone proved more helpful in relieving pain, stiffness, and morbidity than DS gel alone. The promising data open new avenues in the management of musculoskeletal pain due to inflammation or sports injuries because it offers a great alternative to DS in eligible patients. Ongoing clinical studies aim at (i) studying the dose-dependent adverse effects of DK gel at higher concentrations (e.g., 10–15%) in comparison to DS gels, with or without phonophoresis; (ii) developing smart DK hydrogel patches and nanoformulations, and (iii) performing comparisons studies between hydrogel patches, diclofenac nanoformulations and such a present study in different populations.

4. Materials and Methods

4.1. Study, Patients, and Ethics

A single-blinded, randomized controlled trial (IRCT) was started after approval from the Muhammad Institute of Medical and Allied Sciences' ethics committee in Multan, Pakistan (2021/IRB/2/PT/01). The study was conducted from November to April 2022. By using the Formula (1), the sample size was determined.

The sample size was 100 subjects calculated with Borkowf formula [50]:

$$n = \frac{2\sigma^2(Z1 - \alpha + Z1 - \beta)^2}{(\mu o - \mu\alpha)^2} \tag{1}$$

Patients (N = 200) must have met the inclusion requirements to be enrolled in the study. Inclusion criteria considered both genders, adult patients between the ages of 18 and 50, and patients with injuries to the soft tissues (e.g., acute, uncomplicated plantar fasciitis, bursitis, tendinitis, strains, and others mentioned in Figure 2) that occurred between two and eighteen hours before the study enrollment. Exclusion criteria were the following: use of any other medicine, inflammatory or painful disorders as well as fractures and ligament ruptures that were not thought to be amenable to treatment with topical NSAIDS alone, patients in whom NSAIDs treatment may cause serious adverse effects or is not indicated because of other diseases (e.g., kidneys failure, gastritis). The patients were screened by using a numerical pain rating scale (NPRS) and WOMAC index/scale.

All participants signed informed consent. The research was conducted in accordance with the Helsinki Code of Conduct. As shown in Figure 8, out of N = 200 patients, n = 100 were randomly assigned/randomized into each of five groups with and n = 100 without phonophoresis for the treatment of sports injuries. This was completed using the lottery approach. Gel was applied just onto the affected area. Group-1 (n = 20) was given 6% DK gel (1.0–1.4 g containing 40–56 µg, corresponding to daily dose of 96–120 mg DK), Group-2 (n = 20) was given 4% DK gel (0.8–1.0 g containing 32–40 µg, corresponding to daily dose of 96–120 mg DK), Group-3 (n = 20) was given 2% DK gel (0.4–0.8 g containing 16–32 µg, corresponding to daily dose of 48–96 mg DK), Group-4 (n = 20) was given 4% DS gel (0.8–1.0 g containing 50–147 µg, corresponding to daily dose of 50–100 mg diclofenac potassium), and Group-5 (n = 20) was given placebo (water, vehicle only, no active ingredients used here as control group). The treatment was repeated 3–4 times in a week for 4 weeks. Patients were examined on the baseline, day-7, day-14, day-21, and day-28. Phonophoresis (Figure 9A–D), was applied as professionally instructed [51] and set at a frequency of 0.8 MHz, an intensity of about 1.5W/cm^2, and a continuous mode (2:1) [37]. All the patients were blinded to the treatment.

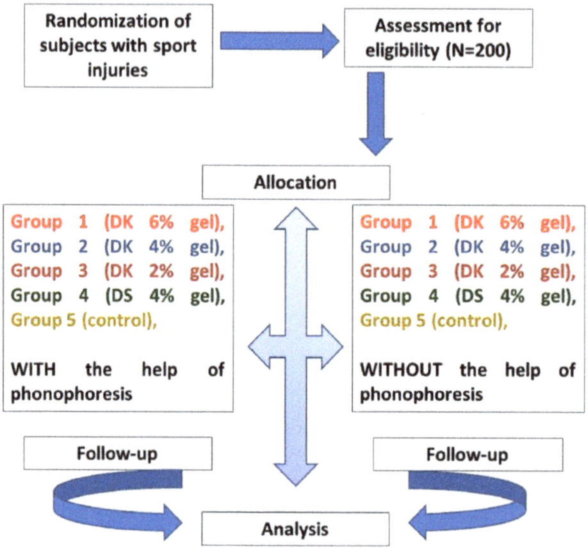

Figure 8. Consolidated Standards of Reporting Trials (CONSORT). DK stands for Diclofenac potassium, and DS means Diclofenac sodium (used as a standard). Control group is the placebo.

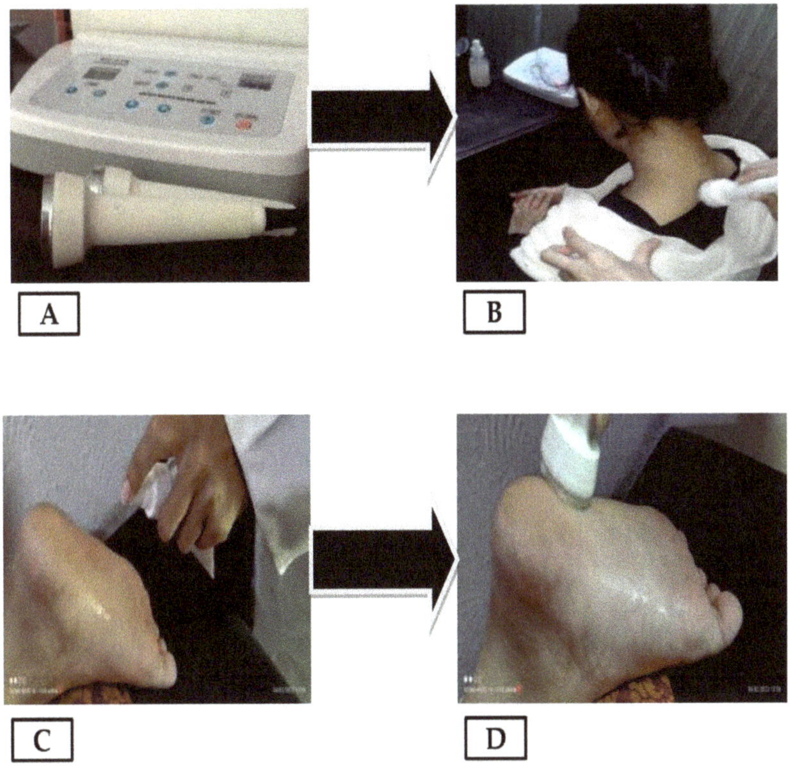

Figure 9. (**A**) Phonophoresis (ultrasound) for the treatment of sport injuries, (**B**) Phonophoresis on a patient with neck pain after applying 4 pumps of DK gel, (**C**) Applying 4 pumps of DK gel on a patient foot, (**D**) Phonophoresis on a patient foot.

4.2. Statistical Analysis

Data were analyzed by using SPSS version-22 (IBM SPSS, Inc., Chicago, IL, USA). The quantitative data were presented for mean ± S.D. Significance was considered if p-value < 0.05. One-way ANOVA was used to determine whether there are any statistically significant differences between the means of two or more independent (unrelated) groups.

4.3. Preparation of Diclofenac Potassium Gel

First, 6% DK of analytical grade was purchased from Bukhari's pharmaceuticals, and the gel was formulated by adding carbazole-940 (6%) into aqueous–methanolic diclofenac potassium (6%). Continuous stirring was used to settle the mixture down until it gained consistency like gel.

4.4. Numerical Pain Rating Scale

NPRS [52] is often made up of a sequence of numbers and oral anchors that indicate the full range of pain intensity (Figure 10). Patients usually rate their pain on a scale of 0 to 10, 0 to 20, or 0 to 100. They are closely related to other levels of pain and exhibit sensitivity to treatment that is expected to alter pain. Zero stands for "no pain", while 10, 20, or 100 denotes the extreme of the pain cycle. NPRS can be performed orally or in writing, is basic, easy to grasp, and can be readily controlled and scored. The NPRS' fundamental flaw is that it lacks the mathematical criteria to be effective. The 11-point numeric scale ranges from 0 to 10, with 0 signifying no pain and 10 reflecting the other extreme degree of pain.

We used NPRS in our study before and after the treatment. Then, we checked the scores of pains to improve the stigma of receiving small changes. NPRS numerically modified NPRS was used for scoring pain. The benefits of NPRS include simplicity, reproduction, ease of understanding, and sensitivity to small changes in pain. Five-year-old children who can count and have a certain sense of number (i.e., that 8 is greater than 4) can use this scale. NPRS (whole and each body region) and the specific outcome measure for each region in all 3 levels of change, using the receiver's operating factor curve. Overall, 64% of NPRS is a valid standard and should be part of a small database of clinical trials, while 14% of NPRS estimates are valid but should only be part of the extended database, 20% of NPRS needs further research to establish credibility and legitimacy before it can be recommended, and 2% of NPRS is not valid or should be used.

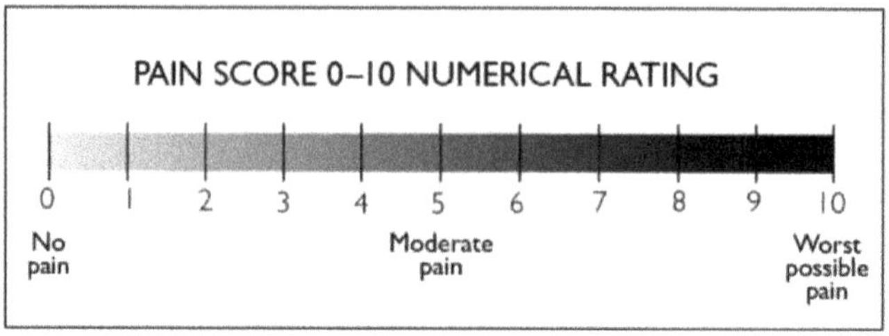

Figure 10. Numerical Pain Rating Scale (NPRS) indicating different intensities of pain: 0 means no pain, 5 reveals moderate pain, and 10 indicates severe pains.

4.5. WOMAC Index

WOMAC [53] is scale estimation for osteoarthritis (OA) of the knee, constituting 24 options in three elements: pain, activities of daily living, and stiffness (Figure 11). However, we assessed only functional disability through this scale. The WOMAC's test–retest reliability depends greatly on the subscale. Although the pain subscale has not been constant across research, it fulfills the basic criteria. The physical function subscale has higher test–retest reliability and is more reliable. Test–retest reliability for the stiffness subscale is minimal.

Severity, on average, during the last 48 hours, of:

Pain

	None	Slight	Moderate	Severe	Extreme
Pain – Walking	☐	☐	☐	☐	☐
Pain – Stair climbing	☐	☐	☐	☐	☐
Pain – Nocturnal	☐	☐	☐	☐	☐
Pain – Rest	☐	☐	☐	☐	☐
Pain – Weightbearing	☐	☐	☐	☐	☐
Stiffness:					
Morning Stiffness	☐	☐	☐	☐	☐
Stiffness occuring during the day	☐	☐	☐	☐	☐

Level of difficulty performing the following functions, on average, during the last 48 hours:

	None	Slight	Moderate	Severe	Extreme
Descending stairs	☐	☐	☐	☐	☐
Ascending stairs	☐	☐	☐	☐	☐
Rising from sitting	☐	☐	☐	☐	☐
Standing	☐	☐	☐	☐	☐
Bending to the floor	☐	☐	☐	☐	☐
Walking on flat	☐	☐	☐	☐	☐
Getting in/out of a car	☐	☐	☐	☐	☐
Going shopping	☐	☐	☐	☐	☐
Putting on socks	☐	☐	☐	☐	☐
Rising from bed	☐	☐	☐	☐	☐
Taking of socks	☐	☐	☐	☐	☐
Lying in bed	☐	☐	☐	☐	☐
Getting in/out of bath	☐	☐	☐	☐	☐
Sitting	☐	☐	☐	☐	☐
Getting on/off toilet	☐	☐	☐	☐	☐
Performing heavy domestic duties	☐	☐	☐	☐	☐
Performing light domestic duties	☐	☐	☐	☐	☐

The WOMAC parameters are:
0 – none, 1 – slight, 2 – moderate, 3 – severe, 4 – extreme.
The index is out of a total of 96 possible points, with 0 being the best and 96 being the worst

Figure 11. Western Ontario McMaster Osteo-Arthritis (WOMAC) scale/index indicating three parameters: pain, stiffness, and functional activities.

Author Contributions: Conceptualization, I.A.K., S.I., F.M. and M.O.I.; methodology, software, validation, formal analysis, investigation, resources, data curation, K.A.B., M.O.I., A.M.A. and F.M.; writing—original draft preparation, visualization, supervision, I.A.K., S.I., F.M. and T.A.; writing—review and editing: F.M.; project administration, funding acquisition, T.A., F.M. and A.M.A.; submission: F.M. All authors have read and agreed to the published version of the manuscript.

Funding: The authors are grateful to the Deanship of Scientific Research at King Khalid University, Saudi Arabia, for funding this study through the Large Research Group Project, under grant number RGP2/3/43.

Institutional Review Board Statement: The single-blinded, randomized controlled trial (IRCT) was conducted in accordance with the Declaration of Helsinki, and approved by the Institutional Review Board of Muhammad Institute of Medical as well as by the Allied Sciences' ethics committee in Multan, Pakistan (protocol code 2021/IRB/2/PT/01 on 23 August 2021). for studies involving humans.

Informed Consent Statement: Informed consent was obtained from all subjects involved in the study.

Data Availability Statement: Data may be available upon request to the corresponding author(s).

Acknowledgments: The authors are grateful to the Deanship of Scientific Research at King Khalid University, Saudi Arabia, for funding this study.

Conflicts of Interest: The authors declare no conflict of interest.

References

1. Ivins, D. Acute ankle sprain: An update. *Am. Fam. Physician* **2006**, *74*, 1714–1720. [PubMed]
2. Kerkhoffs, G.M.; Rowe, B.H.; Assendelft, W.J.; Kelly, P.A.; Struijs, C.N. Immobilisation and functional treatment for acute lateral ankle ligament injuries in adults. *Cochrane Database Syst. Rev.* **2002**, *3*, CD003762. [CrossRef]
3. Zacher, J.; Altman, R.; Bellamy, N.; Brühlmann, J.P.; Da Silva, E.; Huskisson, R.S. Topical diclofenac and its role in pain and inflammation: An evidence-based review. *Curr. Med. Res. Opinion* **2008**, *24*, 925–950. [CrossRef]
4. Osteoarthritis: The care and management of osteoarthritis in adults. *Inst. Clin. Excell. Clin. Guidel.* **2008**. Available online: http://guidance.nice.org.uk/CG59 (accessed on 24 June 2013).
5. Massey, T.; Derry, S.; Moore, R.A.; McQuay, H.J. Topical NSAIDs for acute pain in adults. *Cochrane Database Syst. Rev.* **2010**, *16*, CD007402. [CrossRef]
6. Brune, K. Persistence of NSAIDs at effect sites and rapid disappearance from side-effect compartments contributes to tolerability. *Curr. Med. Res. Opin.* **2007**, *23*, 2985–2995. [CrossRef] [PubMed]
7. Bandolier: Evidence-Based Health Care. Topical Analgesics: A Review Of reviews and a Bit of Perspective. 2005. Available online: https://www.medi-cine.ox.ac.uk/bandolier/Extraforbando/Topextra3.pdf (accessed on 20 January 2013).
8. Davies, N.M.; Anderson, K.E. Clinical pharmacokinetics of diclofenac. Therapeutic insights and pitfalls. *Clin. Pharmacokinet* **1997**, *33*, 184–213. [CrossRef] [PubMed]
9. Riess, W.; Schmid, K.; Bott, L.; Kobayashi, J.K.; Moppert, W.; Schneider, A.; Sioufi, A.; Strusberg, M. The percutaneous absorption of diclofenac. *Arzneim. Schung.* **1986**, *36*, 1092–1096.
10. Bjordal, M.; Ljunggren, A.E.; Klovning, A.; Moppert, J.; Schneider, W.; Sioufi, A.; Strusberg, A.; Tomasi, M. Non-steroidal anti-inflammatory drugs, including cyclo-oxygenase-2 inhibitors, in osteoarthritic knee pain: Meta-analysis of randomised placebo-controlled trials. *BMJ* **2004**, *329*, 1317. [CrossRef]
11. Jordan, K.; Arden, N.; Doherty, M.; Bannwarth, B.; Bijlsma, J.W.J.; Dieppe, P.; Gunther, K.; Hauselmann, H.; Herrero-Beaumont, G.; Kaklamanis, P.; et al. EULAR recommendations 2003: An evidence-based approach to the management of knee osteoarthritis: Report of a task force of the standing committee for international clinical studies including therapeutic trials (ESCISIT). *Ann. Rheum. Dis.* **2003**, *62*, 1145–1155. [CrossRef]
12. Mason, L.; Moore, R.A.; Edwards, J.E.; Moore, A.R. Topical NSAIDs for acute pain: A meta- analysis. *BMC Fam. Pract.* **2004**, *5*, 10. [CrossRef] [PubMed]
13. Moore, R.A.; Tramer, M.R.; Carroll, D.; Wiffen, P.J.; McQuay, H.J. Quantitative systematic review of topically applied non-steroidal anti-inflammatory drugs. *BMJ* **1998**, *316*, 333–338. [CrossRef] [PubMed]
14. van Tulder, M.W.; Scholten, R.J.; Koes, B.W. Non-steroidal anti-inflammatory drugs for low-back pain. *Cochrane Database Syst. Rev.* **2000**, *2*, 31. [CrossRef]
15. Cordero, J.A.; Alarcon, L.; Escribano, E.; Obach, R.; Domenech, J. A comparative study of the transdermal penetration of a series of nonsteroidal antiinflammatory drugs. *J. Pharm. Sci.* **1997**, *86*, 503–508. [CrossRef] [PubMed]
16. Brunner, M.; Dehghanyar, P.; Seigfried, B.; Wolfgang, M.; Georg, M.; Markus, M. Favourable dermal penetration of diclofenac after administration to the skin using a novel spray gel formulation. *Br. J. Clin. Pharmacol.* **2005**, *60*, 573–577. [CrossRef] [PubMed]
17. Lionberger, D.R.; Joussellin, E.; Lanzarotti, A.; Yanchick, J.; Magelli, M. Diclofenac epolamine topical patch relieves pain associated with ankle sprain. *J. Pain Res.* **2011**, *4*, 47–53. [PubMed]
18. van Rijn, R.M.; van Os, A.G.; Bernsen, R.M.; Luijsterburg, P.A.; Koes, B.W.; Bierma-Zeinstra, S.M. What is the clinical course of acute ankle sprains? A systematic literature review. *Am. J. Med.* **2008**, *121*, 324–331.e6. [CrossRef] [PubMed]
19. Purves, R.D. Optimum numerical integration methods for estimation of area-under-the- curve (AUC) and area-under-the-moment-curve (AUMC). *J. Pharmacokinet Biopharm.* **1992**, *20*, 211–226. [CrossRef]
20. Schermer, R. Topical therapy for sprains of the ankle joint. A double-blind study of the efficacy of diclofenac diethylammonium in an emulsion-gel. *Wehrmed Mschr.* **1991**, *9*, 415–419.
21. Yamaguchi, S.; Terahara, T.; Okawa, K.; Inakura, H. A multicenter, randomized, double-blind, placebo-controlled, comparative study to evaluate the efficacy and safety of newly developed diclofenac patches in patients with cancer pain. *Pain* **2021**. [CrossRef]
22. Benaouda, F.; Inacio, R.; Lim, C.H.; Park, H.; Pitcher, T.; Alhnan, M.A.; Aly, M.M.S.; Al-Jamal, K.T.; Chan, K.L.; Gala, R.P.; et al. Needleless administration of advanced therapies into the skin via the appendages using a hypobaric patch. *Proc. Natl. Acad. Sci. USA* **2022**, *119*, e2120340119. [CrossRef] [PubMed]
23. Tabosa, M.A.M.; Cordery, S.F.; Jane White, K.A.; Bunge, A.L.; Guy, R.H.; Delgado-Charro, M.B. Skin pharmacokinetics of diclofenac and co-delivered functional excipients. *Int. J. Pharm.* **2022**, *612*, 121469.
24. Ahmad, M.; Iqbal, M.; Murtaza, G. Comparison of bioavailability and pharmacokinetics of diclofenac sodium and diclofenac potassium in normal and dehydrated rabbits. *Yao Xue Xue Bao* **2009**, *44*, 80–84. [PubMed]
25. Hinz, B.; Chevts, J.; Renner, B.; Wuttke, H.; Rau, T.; Schmidt, A.; Szelenyi, I.; Brune, K.; Werner, U. Bioavailability of diclofenac potassium at low doses. *Br. J. Clin. Pharmacol.* **2005**, *59*, 80–84. [CrossRef]
26. Chuasuwan, B.; Binjesoh, V.; Polli, J.E.; Zhang, H.; Amidon, G.L.; Junginger, H.E.; Midha, K.K.; Shah, V.P.; Stavchansky, S.; Dressman, J.B.; et al. Biowaiver monographs for immediate release solid oral dosage forms: Diclofenac sodium and diclofenac potassium. *J. Pharm. Sci.* **2009**, *98*, 1209–1219. [CrossRef] [PubMed]
27. LEHMANN, J.F. The biophysical basis of biologic ultrasonic reactions with special reference to ultrasonic therapy. *Arch. Phys. Med. Rehabil.* **1953**, *34*, 39–52.

28. Kamel, M.; Eid, H.; Mansour, R.J.T. Ultrasound detection of heel enthesitis: A comparison with magnetic resonance imaging. *J. Rheumatol.* **2003**, *30*, 774–778.
29. Argerakis, N.G.; Positano, R.G.; Positano, R.C.; Boccio, A.K.; Adler, R.S.; Saboeiro, G.R.; Joshua, S.D. Ultrasound diagnosis and evaluation of plantar heel pain. *J. Am. Podiatr. Med. Assoc.* **2015**, *105*, 135–140. [CrossRef]
30. McNally, E.G.; Shetty, S. Plantar fascia: Imaging diagnosis and guided treatment. *Semin. Musculoskelet Radiol.* **2010**, *14*, 334–343. [CrossRef]
31. Cheing, G.; Chang, H.; Lo, S.K.J.S.W. A comparison of the effectiveness of extracorporeal shock wave and ultrasound therapy in the management of heel pain. *Medicine* **2007**, *17*, 195–201. [CrossRef]
32. Babatunde, O.O.; Legha, A.; Littlewood, C.; Chesterton, L.S.; Thomas, M.J.; Menz, H.B.; Danielle, W.; Edward, R. Comparative effectiveness of treatment options for plantar heel pain: A systematic review with network meta-analysis. *Meta-Analyst* **2019**, *53*, 182–194. [CrossRef]
33. Crawford, F.; Snaith, M.J. How effective is therapeutic ultrasound in the treatment of heel pain? *Ann. Rheum. Dis.* **1996**, *55*, 265–267. [CrossRef] [PubMed]
34. Shanks, P.; Curran, M.; Fletcher, P.; Thompson, R.J.T.F. The effectiveness of therapeutic ultrasound for musculoskeletal conditions of the lower limb: A literature review. *Foot* **2010**, *20*, 133–139. [CrossRef] [PubMed]
35. Katzap, Y.; Haidukov, M.; Berland, O.M.; Itzhak, R.B.; Kalichman, L.J. Additive effect of therapeutic ultrasound in the treatment of plantar fasciitis: A randomized controlled trial. *Randomized Control Trial* **2018**, *48*, 847–855. [CrossRef]
36. Dorji, K.; Graham, N.; Macedo, L.; Gravesande, J.; Goldsmith, C.H.; Gelley, G.; Rice, M.; Solomon, P. The effect of ultrasound or phonophoresis as an adjuvant treatment for non-specific neck pain: Systematic review of randomised controlled trials. *Disabil Rehabil.* **2002**, *44*, 2968–2974. [CrossRef] [PubMed]
37. Miller, D.L.; Smith, N.B.; Bailey, M.R.; Czarnota, G.J.; Hynynen, K.; Makin, I.R.; Bioeffects Committee of the American Institute of Ultrasound in Medicine. Overview of therapeutic ultrasound applications and safety considerations. *J. Ultrasound Med.* **2012**, *31*, 623–634. [CrossRef]
38. Ahn, J.H.; Lee, C.W.; Park, C.; Kim, Y.C. Ultrasonographic examination of plantar fasciitis: A comparison of patient positions during examination. *J. Foot Ankle Res.* **2016**, *9*, 38. [CrossRef]
39. Cameron, M.H.; Monro, L.G. Relative transmission of ultrasound by media customarily used for phonophoresis. *Phys. Ther.* **1992**, *72*, 142–148. [CrossRef]
40. D'vaz, A.; Ostor, A.J.; Speed, C.; Jenner, J.; Bradley, M.; Prevost, A.; Hazleman, B.L. Pulsed low-intensity ultrasound therapy for chronic lateral epicondylitis: A randomized controlled trial. *Bmj* **2006**, *45*, 566–570. [CrossRef]
41. Tolu, S.; Köse, M.M.; Korkmaz, M.C.; Üşen, A.; Rezvani, A. Comparison of the Efficacy of Different Concentrations of Diclofenac Sodium Phonophoresis (1.16% vs 2.32%) in Patients with Knee Osteoarthritis: A Randomized Double-Blind Controlled Trial. *Acta Chir. Orthop. Traumatol. Cech.* **2021**, *88*, 117–123.
42. Wang, X.; Dan, W.; Wenzhao, L.; Lei, Y. Emerging biomaterials for reproductive medicine. *Eng. Regen.* **2021**, *2*, 230–245. [CrossRef]
43. Khan, Z.U.; Razzaq, A.; Khan, A.; Rehman, N.U.; Khan, H.; Khan, T.; Khan, A.U.; Althobaiti, N.A.; Menaa, F.; Iqbal, H.; et al. Physicochemical Characterizations and Pharmacokinetic Evaluation of Pentazocine Solid Lipid Nanoparticles against Inflammatory Pain Model. *Pharmaceutics* **2022**, *14*, 409. [CrossRef] [PubMed]
44. Sa'adon, S.; Ansari, M.N.M.; Razak, S.I.A.; Yusof, A.H.M.; Faudzi, A.A.M.; Sagadevan, S.; Nayan, N.H.M.; Anand, J.S.; Amin, K.A.M. Electrospun Nanofiber and Cryogel of Polyvinyl Alcohol Transdermal Patch Containing Diclofenac Sodium: Preparation, Characterization, and In Vitro Release Studies. *Pharmaceutics* **2021**, *13*, 1900. [CrossRef] [PubMed]
45. Branco, A.C.; Oliveira, A.S.; Monteiro, I.; Nolasco, P.; Silva, D.C.; Figueiredo-Pina, C.G.; Colaço, R.; Serro, A.P. PVA-Based Hydrogels Loaded with Diclofenac for Cartilage Replacement. *Gels* **2022**, *8*, 143. [CrossRef] [PubMed]
46. Petersen, B.; Rovati, S. Diclofenac epolamine (Flector®) patch. *Clin. Drug Investig.* **2009**, *29*, 1–9. [CrossRef]
47. Hu, Y.; Hui, Z.; Hao, W.; Hong, C.; Jiaying, C.; Xiaoyan, C.; Lin, X.; Huan, W.; Renjie, C. Scaffolds with Anisotropic Structure for Neural Tissue Engineering. *Eng. Regen.* **2022**, *3*, 154–162. [CrossRef]
48. Yang, L.; Lingyu, S.; Han, Z.; Feika, B.; Yuanjin, Z. Ice-inspired lubricated drug delivery particles from microfluidic electrospray for osteoarthritis treatment. *ACS Nano* **2021**, *12*, 20600–20606. [CrossRef] [PubMed]
49. Shi, H.; Zhu, Y.; Xing, C.; Li, S.; Bao, Z.; Lei, L.; Lin, D.; Wang, Y.; Chen, H.; Xu, X. An injectable thermosensitive hydrogel for dual delivery of diclofenac and Avastin® to effectively suppress inflammatory corneal neovascularization. *Int. J. Pharm.* **2022**, *625*, 122081. [CrossRef]
50. Borkowf, C.B. A New Method for Approximating the Asymptotic Variance of Spearman's Rank Correlation. *Stat. Sin.* **1999**, *9*, 535–558.
51. Lin, J.; Fessell, D.P.; Jacobson, J.A.; Weadock, W.J.; Hayes, C.W.J. An illustrated tutorial of musculoskeletal sonography: Part 3, lower extremity. *AJR Am. J. Roentgenol.* **2000**, *175*, 1313–1321. [CrossRef]
52. Provenzano, A.H.; Åström, E.; Löwing, K. Exploring pain interference and self-perceived health status in children with osteogenesis imperfecta—A cross-sectional study. *BMC Musculoskelet Disord.* **2022**, *23*, 876. [CrossRef]
53. Li, L.; Zeng, Z.; Zhang, H.; Xu, L.; Lin, Y.; Zhang, Y.; Deng, M.; Fan, P. Different Prevalence of Neuropathic Pain, and Risk Factors in Patients with Knee Osteoarthritis at Stages of Outpatient, Awaiting and after Total Knee Arthroplasty. *Orthop Surg.* **2020**. Epub ahead of print. [CrossRef]

gels

MDPI

Article

Development of Carvedilol Nanoformulation-Loaded Poloxamer-Based In Situ Gel for the Management of Glaucoma

Bjad K. Almutairy [1], El-Sayed Khafagy [1,2,*] and Amr Selim Abu Lila [3,4,5]

[1] Department of Pharmaceutics, College of Pharmacy, Prince Sattam Bin Abdulaziz University, Al-Kharj 11942, Saudi Arabia; b.almutairy@psau.edu.sa
[2] Department of Pharmaceutics and Industrial Pharmacy, Faculty of Pharmacy, Suez Canal University, Ismailia 41522, Egypt
[3] Department of Pharmaceutics and Industrial Pharmacy, Faculty of Pharmacy, Zagazig University, Zagazig 44519, Egypt; a.abulila@uoh.edu.sa
[4] Department of Pharmaceutics, College of Pharmacy, University of Hail, Hail 81442, Saudi Arabia
[5] Medical and Diagnostic Research Center, University of Hail, Hail 81442, Saudi Arabia
* Correspondence: e.khafagy@psau.edu.sa; Tel.: +966-533-564-286

Abstract: The objective of the current study was to fabricate a thermosensitive in situ gelling system for the ocular delivery of carvedilol-loaded spanlastics (CRV-SPLs). In situ gel formulations were prepared using poloxamer analogs by a cold method and was further laden with carvedilol-loaded spanlastics to boost the precorneal retention of the drug. The gelation capacity, rheological characteristics, muco-adhesion force and in vitro release of various in situ gel formulations (CS-ISGs) were studied. The optimized formula (F2) obtained at 22% *w*/*v* poloxamer 407 and 5% *w*/*v* poloxamer 188 was found to have good gelation capacity at body temperature with acceptable muco-adhesion properties, appropriate viscosity at 25 °C that would ease its ocular application, and relatively higher viscosity at 37 °C that promoted prolonged ocular residence of the formulation post eye instillation and displayed a sustained in vitro drug release pattern. Ex vivo transcorneal penetration studies through excised rabbit cornea revealed that F2 elicited a remarkable ($p < 0.05$) improvement in CRV apparent permeation coefficient ($P_{app} = 6.39 \times 10^{-6}$ cm/s) compared to plain carvedilol-loaded in situ gel (CRV-ISG; $P_{app} = 2.67 \times 10^{-6}$ cm/s). Most importantly, in normal rabbits, the optimized formula (F2) resulted in a sustained intraocular pressure reduction and a significant enhancement in the ocular bioavailability of carvedilol, as manifested by a 2-fold increase in the $AUC_{0–6h}$ of CRV in the aqueous humor, compared to plain CRV-ISG formulation. To sum up, the developed thermosensitive in situ gelling system might represent a plausible carrier for ophthalmic drug delivery for better management of glaucoma.

Keywords: carvedilol; glaucoma; in situ gel; poloxamer; spanlastics

check for updates

Citation: Almutairy, B.K.; Khafagy, E.-S.; Abu Lila, A.S. Development of Carvedilol Nanoformulation-Loaded Poloxamer-Based In Situ Gel for the Management of Glaucoma. *Gels* **2023**, *9*, 952. https://doi.org/10.3390/gels9120952

Academic Editor: Ana Paula Serro

Received: 6 November 2023
Revised: 24 November 2023
Accepted: 30 November 2023
Published: 4 December 2023

1. Introduction

Glaucoma is a neurodegenerative disorder that is characterized by progressive optic nerve degeneration, leading to permanent blindness [1]. Elevated intraocular pressure (IOP) stands as a major risk factor for developing glaucoma. According to the World Health Organization (WHO), glaucoma is considered the second leading cause of visual impairment and blindness in the world [2]. Medical treatment of glaucoma includes topical beta-adrenergic blockers (timolol, betaxolol and metipranolol), alpha agonists (Brimonidine), carbonic anhydrase inhibitors (methazolamide and acetazolamide), prostaglandin analogs (latanoprost, bimatoprost and travoprost), and rho kinase inhibitors (netarsudil and ripasudil) [3]. Carvedilol is a beta-adrenergic blocker that is used to treat hypertension and ischemic heart diseases. Carvedilol is a class II (high permeability/low solubility) drug according to Biopharmaceutical Classification System. It shows poor bioavailability (~25%);

owing to extensive first-pass metabolism, following oral administration [4]. Nevertheless, recent research has underscored the plausible use of topical carvedilol to treat high intraocular pressure [5,6]. For instance, Hassan et al. have affirmed the efficacy of ocularly applied carvedilol-loaded cationic nanoparticle (leciplex) in reducing the intraocular pressure to a normal range in ocular hypertensive rabbits [6].

A major challenge in ocular drug delivery is to tailor a delivery system that can grant adequate drug concentrations at the target region for sufficient time interval. Conventional ocular drug delivery systems, such as eye drops and ointments, usually suffer from poor drug bioavailability, presumably, due to the complex anatomy and highly selective physiological corneal barriers, which hinder the entry of exogenous materials to the ocular tissues [7]. In addition, excessive and rapid loss of drug due to high tear fluid turnover and the nasolacrimal drainage could reduce ocular absorption by reducing the contact time of instilled drug at the site of action [8].

Colloidal drug delivery systems such as nanoparticles, nanoemulsions, niosomes, and liposomes have been extensively explored in ocular drug delivery [9–11]. The benefits of colloidal carriers include controlled and/or sustained drug release at the targeted site, the ability to deliver both hydrophilic and hydrophobic drugs to eye tissue, and the potential to circumvent blood-ocular barriers/efflux-related problems encountered with the parent drug [10,11]. Among a wide range of colloidal carriers in the realm of ocular drug delivery, spanlastics (SLs), an elastic non-ionic surfactant based vesicular system, have emerged as a viable delivery vehicle that could efficiently circumvent the limitations of conventional ophthalmic drug delivery systems [12,13]. Spanlastics are composed of a non-ionic surfactant as a main vesicle forming component, and an edge activator (EA). The inclusion of EA to vesicles provides SLs with a great elasticity, compared to conventional niosomes [14]. Such elasticity of the vesicles enhances the corneal permeability of the entrapped drugs, underscoring the potential use of SLs as efficient drug delivery vehicles for ocular administration [15].

In situ gelling systems are stimuli-sensitive polymeric viscous liquids that undergo sol to gel transformation upon application to the human body in response to small changes in specific conditions like pH, temperature and/or ionic strength [16]. Recently, ophthalmic in situ gels have emerged as an ideal choice for ocular delivery. Compared to conventional gels, in situ gels are applied as solutions or suspensions, which provides ease of administration while maintaining dose accuracy. Afterwards, they are converted into a gel state upon contact with tear fluids in response to definite stimuli such as temperature, pH, etc. [17]. This would extend ocular residence time, decrease pre-corneal elimination, and consequently, enhance the ocular availability of the administered drug along with improving patient compliance via reducing dosing frequency [18]. Rawat et al. [8] have recently emphasized the efficacy of dual-responsive in situ gel, composed of a combination of the thermo-sensitive poloxamers (poloxamer 407/poloxamer 188) and the ion-sensitive polymer kappa-carrageenan, for enhancing the anti-glaucoma potential of the β-adrenergic antagonist, nebivolol. Intriguingly, the ocular applicability of in situ gels can be augmented by the incorporation of nanoparticulate systems within in situ gel with the goal of extending drug release and enhancing the therapeutic outcomes [19,20].

This study, therefore, aimed at formulating carvedilol in a dosage form, spanlastic-laden in situ gel system, that combine the advantages of both nano-systems (spanlastics) and in situ gels as a plausible tool to sustain drug delivery, enhance drug transcorneal permeation and eventually ameliorate its ocular bioavailability for the management of glaucoma. For such purpose, thermosensitive polymers; poloxamer 407/poloxamer 188, were utilized for the formulation of in situ gel system. Formulation parameters such as gelation temperature, muco-adhesion force and in vitro release behavior of in situ gel systems were optimized. Finally, the ex-vivo permeation, the in vivo fate, and the in vivo efficacy of optimized carvedilol-loaded in situ gel system in lowering IOP were investigated.

2. Results and Discussion

2.1. Formulation of Carvedilol-Loaded Spanlastic In Situ Gel (CS-ISG)

In our previous study, we succeeded to formulate carvedilol-loaded spanlastics (CRV-SPLs) for augmenting the therapeutic efficacy of carvedilol in a hypertensive rat model [21]. In that study, a combination of Span 60 as main vesicle component and different edge activators (EA), namely Tween 80 and Brij 97, were adopted for the fabrication of CRV-loaded spanlastics at two Span:EA ratios (90:10 and 80:20). It was evident that spanlastics prepared with Brij 97 as an edge activator at a Span:EA ratio of 80:20 showed optimal formulation attributes such as minimum vesicle diameter, high entrapment efficiency, and efficient drug permeability. In this study, we tried to extend our work via challenging the efficacy of CRV-loaded spanlastics (CRV-SPLs) in the management of glaucoma. For such purpose, the optimized CRV-loaded spanlastic formulation (CRV-SPLs) was incorporated into an in situ gel formulation (CS-ISG) to facilitate its application, extend the residence time onto the corneal surface after application, and subsequently, augmenting drug penetration through the cornea. Herein, poloxamer-based thermo-sensitive polymers (poloxamer 407/poloxamer 188) were utilized for the fabrication of carvedilol-loaded spanlastic in situ gel (CS-ISG) formulations. Poloxamer-based polymers were selected because of their advantages of high-water solubility, thermo-reversible gelation properties, and their ability to produce translucent gels that do not interfere with normal vision. The composition of different poloxamer-based in situ gel formulation is summarized in Table 1.

Table 1. Composition of various CRV-SPLs-loaded in situ gels (CS-ISGs) formulations.

Formula	Poloxamer 407	Poloxamer 188
F1	20	5
F2	22	5
F3	25	5
F4	20	7.5
F5	22	7.5
F6	25	7.5
F7	20	10
F8	22	10
F9	25	10

2.2. Evaluation of CS-ISGs

2.2.1. Clarity

Clarity and homogeneity of various CS-ISG formulations were assessed visually before and after loading with CRV-SPLs. Plain ISGs were transparent and homogeneous; however, following CRV-SPLs loading, CS-ISGs became less transparent, but with no suspended particles.

2.2.2. pH

Generally, for ophthalmic products to be well tolerated by eyes, the pH of the fabricated products should fall within the normal ocular comfort range (pH range of 6.5 to 7.5) [22]. As illustrated in Table 2, the pH of all CS-ISG formulations was determined to be between 6.48 ± 0.04 and 7.32 ± 0.05, which is considered within the acceptable limit for ophthalmic treatments.

Table 2. Physicochemical properties of different CS-ISG formulations.

Formula	pH	Drug Content (%)	Gelation Temperature (T_G; °C)	Muco-Adhesion Force (mN)	Viscosity (cp)	
					At 25 °C	At 37 °C
F1	6.48 ± 0.04	99.3 ± 0.4	36.8 ± 0.3	58.7 ± 4.9	58.6 ± 4.2	276.3 ± 13.8
F2	6.94 ± 0.03	99.1 ± 0.5	32.4 ± 0.5	87.2 ± 5.2	83.6 ± 5.9	369.3 ± 10.9
F3	7.13 ± 0.07	98.6 ± 0.7	28.8 ± 0.6	112.3 ± 6.1	112.8 ± 9.3	424.3 ± 13.7
F4	6.71 ± 0.06	98.4 ± 0.4	37.9 ± 0.5	78.1 ± 7.6	76.3 ± 4.4	303.5 ± 10.8
F5	7.03 ± 0.05	98.3 ± 0.6	34.9 ± 0.7	96.4 ± 8.3	103.3 ± 6.1	363.7 ± 18.8
F6	7.29 ± 0.08	97.9 ± 0.9	31.5 ± 0.4	134.5 ± 9.7	146.4 ± 11.5	485.3 ± 43.9
F7	6.87 ± 0.08	98.2 ± 0.7	40.5 ± 0.8	92.3 ± 8.2	96.1 ± 6.5	362.7 ± 11.7
F8	7.18 ± 0.06	97.8 ± 1.1	36.9 ± 0.9	161.5 ± 7.3	114.7 ± 9.6	454.9 ± 11.4
F9	7.32 ± 0.05	97.3 ± 0.9	33.6 ± 0.7	174.8 ± 11.2	164.3 ± 12.0	549.8 ± 15.2

All data represents mean ± SD. (n = 3).

2.2.3. Drug Content

All the prepared CS-ISG formulations showed high drug content fluctuating from 97.3 ± 0.9% to 99.3 ± 0.4% (Table 2). These results suggest that CRV-SPLs were uniformly distributed within the prepared ISGs, and the preparation method was reproducible.

2.2.4. Gelation Temperature

Gelation temperature (T_G) is an important metric in determining the ability of the applied ISG formulation to be transformed into gel state at ocular temperature upon eye installation. To be readily instilled into the eye, an ophthalmic in situ gel should have a T_G greater than room temperature (25 °C) and be transformed into a gel at pre-corneal temperature (35 °C). Formulation with T_G greater than 37 °C is not desirable for ophthalmic use because these formulations would stay in the sol state after ocular administration and could suffer from nasolacrimal drainage prior exhibiting its pharmacological effect. In this study, T_G of various CS-ISG formulations ranged from 28.8 ± 0.6 °C (F6) to 40.5 ± 0.8 °C (F7) (Table 2). It was obvious that, at fixed P188 concentration, increasing P407 concentrations from 20% *w*/*v* to 25% *w*/*v* results in a significant decrease in gelation temperature. For instant, in situ gel formula prepared at 25% *w*/*v* P407 (F3) showed a gelation temperature of 28.8 ± 0.6 °C, which was significantly lower than that prepared at 20% *w*/*v* P407 (F1; 36.8 ± 0.3 °C). On the other hand, at fixed P407 concentration, increasing P188 concentration from 5% *w*/*v* to 10% *w*/*v* tended to elevate gelation temperature, as summarized in Table 2. The gelation temperature of F7 (40.5 ± 0.8 °C), prepared at 10% *w*/*v* P188 was remarkably higher than that of F1 (36.8 ± 0.3 °C), prepared at 5% *w*/*v* P188. Poloxamers are triblock copolymers composed of a core poly (propylene oxide) hydrophobic portion (PPO) and two poly (ethylene oxide) hydrophilic parts (PEO) [23]. Many reports have revealed that gelation properties of poloxamers relies on the ratio between the hydrophobic and hydrophilic sub-units (PPO/PEO) in the polymer chain [23–25]. Reducing PPO/PEO ratio, by either decreasing P407 concentration or increasing P188 concentration, resulted in a significant rise ($p < 0.05$) in the gelation temperature of in situ gel formulation. This might be accounted for the abundant hydrogen bonds between the comparatively hydrophilic PEO blocks and water, which raises the energy necessary to break down the hydrogen bonds between water and PEO blocks and, as a result, raises the sol-to-gel transition temperature [26]. Similar results were reported by Cao et al. who highlighted the impact of increasing poloxamer 407 concentrations in promoting gel formation at lower temperatures for poloxamer-based azithromycin in situ gel formulations [27].

2.2.5. In Vitro Muco-Adhesion Force

Muco-adhesion force is the force with which the formulation binds to the corneal surface. It is one of the essential parameters for ocular ISGs formulation since the ability of ISGs to increase pre-corneal residence does not depend only on the ability to be converted into gel after instillation into the eye but on the muco-adhesion power of the formed ISGs as well. Generally, polymers with good muco-adhesion properties will increase the muco-adhesion force, prolong the pre-corneal residence time of the formulation, and thereby, enhance the overall ocular bioavailability [28,29]. As summarized in Table 2 and Figure 1, the muco-adhesion force of all ISGs formulations fluctuates from 58.7 ± 4.9 mN (F1) to 174.8 ± 11.2 mN (F9). It was evident that increasing either P407 or P188 concentration could result in a proportional enhancement in the muco-adhesion force between the prepared ISGs formulations and corneal surface, presumably, owing to the elevated number of co-polymer chains penetrating glycoprotein chains per unit volume of mucin [30].

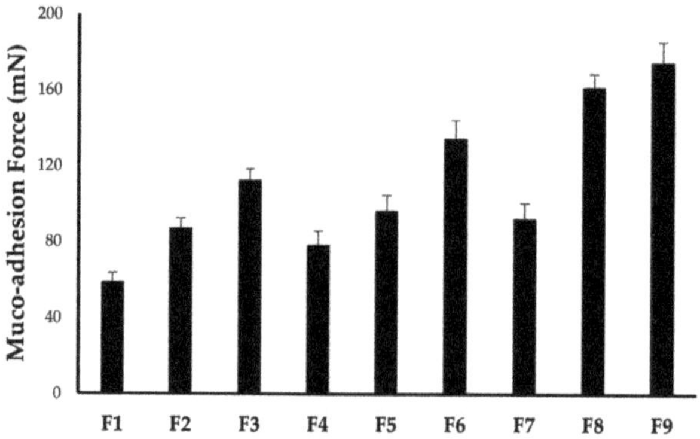

Figure 1. In vitro muco-adhesion force of various CS-ISG formulations. The data represent the mean ± SD of three independent experiments.

2.2.6. Rheological Studies

Rheological properties play a key role in the formulation of in situ gelling solutions. In general, ocular in situ gels should have a viscosity that permits easy instillation into the eye and a rapid transition from sol to gel upon instillation [31]. Herein, the rheological properties of all CS-ISG formulations were examined as a function of temperature. The average results of the viscosity of the prepared ISGs before (at 25 °C) and after (35 °C) gelling are tabulated in Table 2. As presented in Table 2, all CS-ISG formulations had low viscosities at ambient temperature (25 °C), but when the temperature was raised to 37 °C, a significant rise in the viscosity of all formulations was observed, presumably, because of the thermosensitive in situ gelling property of these polymers systems. In addition, it was obvious that, for all tested formulations, there was a remarkable increase in the viscosity with increments in the concentrations of either P407 or P188. For instance, at fixed P188 concentration, the viscosity of CS-ISG formulation prepared at 25% *w*/*v* P407 (F3; 112.8 ± 9.3 cp) was significantly higher than that prepared at 20% *w*/*v* P407 (F1; 58.6 ± 4.2 cp). In the same context, increasing co-polymer (P188) concentration was associated with a pronounced increase in CS-ISGs viscosities. For example, at fixed P407 concentration, the viscosity of CS-ISG formulation prepared at 10% *w*/*v* P188 (F9; 164.3 ± 12.0 cp) was obviously higher than that prepared at 5% *w*/*v* P188 (F3; 112.8 ± 9.3 cp). The same trend of a mutual increase in CS-ISGs viscosities with increasing either P407 or P188 concentrations was observed at 35 °C. This increase in viscosity might be explained by the interaction of the co-polymer (P188) with the micellar entanglement of P407, which could cause the creation of stronger bonds and hence an increase in formulation viscosity. Of note, viscosity results revealed that F1, F2, F4 and F7 met the requirement of ophthalmic in situ gel viscosity 5–100 cPs at room temperature [32]. Nevertheless, F1, F4 and F7 did not fulfill the criterion of having appropriate gelation temperature for ophthalmic application; all these formulations showed TG values greater than pre-corneal temperature (35 °C).

Based on various characterization parameters, particularly, gelation temperature and viscosity measurements, F2 was selected for further studies since formula F1, F4, F5, F7 and F8 had high gelation temperature (>35 °C), which would hinder their transformation into gel state at corneal temperature. Whilst F3, F6, and F9 had relatively higher viscosity values at room temperature (>100 cPs) that might hinder proper in situ gel application. By contrary, the selected formula (F2) has good gelation capacity at body temperature, appropriate viscosity at 25 °C that would ease its ocular application, and relatively higher viscosity at 37 °C, which would promote prolonged ocular residence of the formulation post its instillation into eyes.

2.3. In Vitro Release Studies

The in vitro release profiles of CRV from different formulations was investigated using STF (pH 7.4) as a dissolution medium. As depicted in Figure 2A, it was clear that entrapping CRV within spanlastic (SPLs) system greatly slowed drug release. For CRV suspension, ~30% of CRV was released after 2 h and ~95% after 6 h. On the other hand, CRV-SPLs demonstrated biphasic release, with ~40% of CRV was rapidly released in the first 4 h, presumably owing to surface-adsorbed free CRV, followed by continuous release from the vesicle core for up to 24 h. Such slower release pattern of CRV from spanlastic formulation was ascribed to the entrapment of CRV within vesicular system (spanlastics), which is known to operate as a reservoir that slows down drug release, resulting in a prolonged release profile.

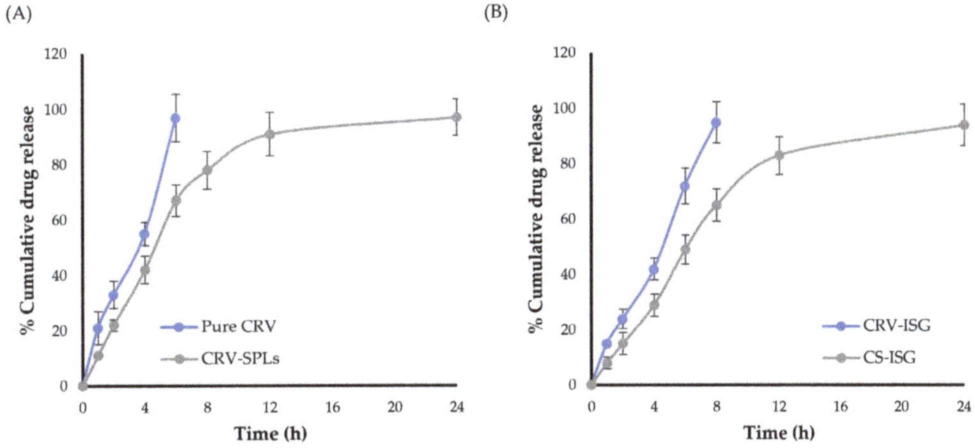

Figure 2. In vitro release profile of CRV from (**A**) SPLs in comparison with pure CRV and (**B**) Spanlastic laden in situ gel (F2) in comparison with CRV-loaded in situ gel (CRV-ISG). The data represent the mean ± SD of three independent experiments.

The in vitro release profiles of selected CS-ISG formulation (F2), compared to CRV-loaded in situ gel (CRV-ISG) were graphically illustrated in Figure 2B. In vitro release results inferred that incorporating CRV-SPLs into ISGs had significantly sustained drug release, compared to plain CRV-ISG. The percentage cumulative CRV released from CS-ISG formulation (F2) in 8 h was ~70%, compared to 95% for plain CRV-ISG. This slower drug release from CS-ISG formulation (2) might be related to the dual action of including the vesicular system (CRV-SPLs) within in situ gelling system.

2.4. Ex Vivo Corneal Permeability

Ex-vivo drug permeation of plain CRV-ISG and CS-ISG formulation (F2) was conducted using goat corneal membrane since it simulates the condition of the human corneal membrane [33]. Figure 3 depicts the cumulative amount of CRV permeated through the cornea membrane from both CRV-ISG and CS-ISG formulation. Figure 3 inferred that CRV release from either CRV-ISG and CS-ISG formulation (F2) was comparable in the first hour. Following that, an increase in CRV permeation was observed with CS-ISG formulation (F2) when compared to plain CRV-ISG formulation. The cumulative amount of CRV permeated after 6 h (Q_{6h}) from CS-ISG formulation (F2) was 110.4 ± 9.8 μg, which was ~3 times higher than that of plain CRV-ISG (Q_{6h} 46.2 ± 3.9 μg). The remarkable increase in drug permeation from CS-ISG formulation might be ascribed primarily to the encapsulation of the drug with nano-size vesicles (spanlastics) with superior corneal penetration properties. Similar results were reported by Hassan et al. [6] who underscored the potential of entrapping carvedilol within cationic nanoparticles (leciplex) for enhancing the transcorneal penetration of CRV compared to plain drug solution.

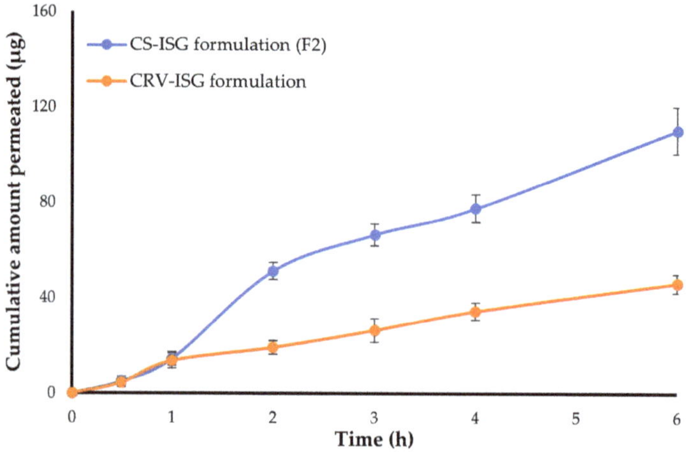

Figure 3. Transcorneal permeability of CRV across rabbit cornea from CRV-loaded in situ gel (CRV-ISG) and optimized CS-ISG (F2) (mean ± SD, n = 3).

In addition, transcorneal permeation parameters for both plain CRV-ISG and CS-ISG formulation (F2) were calculated (Table 3). As summarized in Table 3, CS-ISG formulation (F2) showed greater steady state transcorneal flux (J_{ss} 22.37 ± 2.1 μg/h) compared to plain CRV-ISG (J_{ss} 8.38 ± 1.1 μg/h). Interestingly, F2 triggered a considerable enhancement in CRV apparent permeation coefficient (P_{app} 6.39 × 10^{-6} cm/s) compared to plain CRV-ISG (P_{app} 2.67 × 10^{-6} cm/s). This might be ascribed to the presence of span 60 and EA in the spanlastics, which would promote higher drug permeation through corneal membrane [34]. Collectively, it is possible to infer that entrapment of CRV within SPLs greatly improved CRV corneal penetration over the plain ISG formulation.

Table 3. Ex vivo permeation parameters of different CRV formulations.

Formula	J_{ss} (μg/h)	P_{app} (cm/s) × 10^{-6}	Q_{6h} (μg)
Plain CRV-ISG	8.38 ± 0.95	2.67	46.2 ± 3.9
CS-ISG formulation (F2)	22.37 ± 2.24	6.39	110.4 ± 9.8

2.5. *In Vitro Stability Study*

The optimized CS-ISG formulation (F2) was subjected to stability study following storage at 4 °C for 8 weeks. Visual appearance, pH, drug content and gelling capability were examined. Based on a 2-month investigation, there is little difference in visual appearance, pH, drug content and gelling capability, inferring that the formulation is stable (Table 4).

Table 4. Stability study of optimized CS-ISG formulation (F2).

Time	Visual Appearance	pH	Drug Content	Gelling Capacity
0	Clear	6.94 ± 0.03	99.1 ± 0.5	+++
4th week	Clear	7.05 ± 0.10	98.3 ± 1.0	+++
8th week	Clear	7.21 ± 0.09	97.5 ± 1.3	+++

Data represent mean ± SD. (n = 3). +++ Gelation immediately, remains for extended period.

2.6. *In Vivo Pharmacokinetic Study*

In vivo pharmacokinetic study was performed to estimate the ocular bioavailability, based on calculating the amount of CRV penetrated to the aqueous humor of rabbit eyes, following a single instillation of either plain CRV-ISG and CS-ISG formulation (F2). Non-compartment model analysis was implemented to calculate several pharmacokinetic parameters such as C_{max}, t_{max}, and AUC from a graph drawn between CRV concentrations (ng/mL) in aqueous humor and time [35]. As shown in Figure 4, CRV levels in aqueous humor were elevated rapidly within 1 h post instillation of either plain CRV-ISG and CS-ISG formulation (F2), indicating relatively rapid onset of action. Nevertheless, for plain CRV-ISG, drug levels in aqueous humor declined rapidly, where very low concentrations of CRV were detected in aqueous humor at 4 h post instillation. On the other hand, F2 showed higher drug levels in the aqueous humor for an extended period of time, suggesting a dramatic increase in drug penetration through corneal membrane.

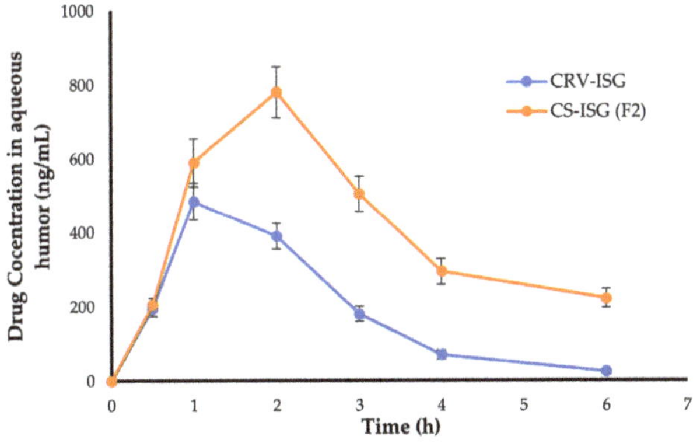

Figure 4. Carvedilol concentration in the aqueous humor following single ocular instillation of various formulations. The data represent mean ± SD (n = 6).

The computed pharmacokinetic parameters for both plain CRV-ISG and CS-ISG formulation (F2) were tabulated in Table 5. As depicted in Table 5, plain CRV-ISG showed t_{max} of 1 h, which was remarkably shorter than that observed with CS-ISG formulation (F2; t_{max} = 2 h). This delayed t_{max} of F2 compared to plain CRV-ISG might be accounted to the entrapment of CRV into spanlastic vesicles, which might pose an additional diffusion barrier for drug release into aqueous humor. Nevertheless, it was obvious that F2 had a significantly higher peak concentration (C_{max} 781.4 ± 69.4 ng/mL) and greater $AUC_{0–6h}$ (2494.5 ± 113.7 ng·h/mL) compared to plain CRV-ISG (C_{max}; 485.7 ± 52.9 ng/mL and $AUC_{0–6h}$ 1161.3 ± 98.6 ng·h/mL). These results are consistent with ex vivo permeation data, in which, the flux (J_{ss}) and apparent permeability coefficient (P_{app}) were considerably higher in F2 (Figure 3 and Table 4). Furthermore, the mean residence time (MRT) of F2 was 4.11 ± 0.5 h, which was longer than that of CRV-ISG (MRT = 2.15 ± 0.3 h). This increase in MRT for F2 could be ascribed to the gradual and prolonged release of CRV from spanlastic vesicles. Collectively, our results underscored the potential of CRV-loaded spanlastic vesicles to augment the ocular bioavailability of CRV. These findings are consistent with previous research on the effect of drug encapsulation into nanoparticulate systems on drug ocular pharmacokinetics. For instance, Huang et al. discovered that, as compared to commercial timolol eye drops, cubosomes had the ability to sustain timolol release, and thereby, foster its retention in the aqueous humor and the anterior segment of eye [36]. Similarly, Ban et al. accentuated the efficacy of charged lipid nanoparticle to extend dexamethasone retention time and to boost its permeation through the cornea, resulting in higher ocular bioavailability when compared to dexamethasone solution [37].

Table 5. Pharmacokinetic parameters of different carvedilol formulations in aqueous humor.

Pharmacokinetic Parameter	CRV-ISG	CS-ISG (F2)
C_{max} (ng/mL)	485.7 ± 52.9	781.4 ± 69.4
t_{max} (h)	1	2
$t_{1/2(h)}$	1.01 ± 0.2	2.21 ± 0.4
$AUC_{0–6h}$ (ng·h/mL)	1161.3 ± 98.6	2494.5 ± 113.7
MRT (h)	2.15 ± 0.3	4.11 ± 0.5

2.7. *In Vivo Pharmacodynamic Study*

The in vivo efficacy of the optimized CS-ISG formulation (F2) on reducing IOP was investigated and compared to that of plain CRV-ISG. The change in IOP from baseline with time following ocular instillation of either optimized F2 formula or plain CRV-ISG was plot in Figure 5. As shown in Figure 5, plain CRV-ISG succeeded to elicit a rapid drop in IOP (16.1 ± 0.5 mmHg) after one hour following ocular administration that lasted for three hours, following which IOP progressively increased to its initial value (20.6 ± 0.7 mmHg) at eight hours. On the other hand, in comparison to plain CRV-ISG, the optimized F2 formula triggered a substantial decrease in IOP readings after 2 h, with a maximum reduction of 15.1 ± 0.4 mmHg. This IOP lowering activity was maintained for up to 8 h post F2 instillation, indicating a sustained action of F2. The average IOP at 8 h post F2 instillation was 17.9 ± 0.9 mmHg, compared to 21.6 ± 0.6 mmHg for control eye. These results suggest that the inclusion of CRV-SPLs into ISGs formulation would sustain drug release for more prolonged time than plain CRV-ISG, and subsequently, the instilled dose could be decreased. The double-layered structure of spanlastic, gel viscosity, and gel matrix structure all contributed to the prolonged and sustained impact [38]. Similar results were stated by Leonardi et al. who investigated the IOP lowering activity of cationic solid lipid nanoparticle encapsulating melatonin. They revealed that solid lipid nanoparticle entrapping melatonin had a superior IOP lowering activity that lasted for 24 h after instillation, compared to that of free drug, which exerted its effect for only 4 h post ocular application [39].

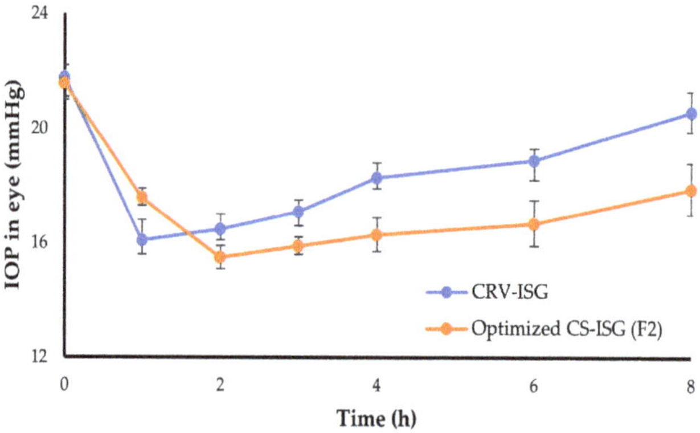

Figure 5. IOP lowering effect CRV-ISG and CS-ISG (F2). The data represent mean ± SD (n = 6).

2.8. Ocular Irritation

The Draize rabbit eye test was adopted to scrutinize the possible irritation potential of optimized F2 formula. In this study, ocular irritation was studied after ocular instillation of CRV-ISG and optimized CS-ISG formulations. Both treated groups showed no symptoms of ocular irritation such as tears, redness, or edema during the test (Table S1). These findings ruled out the irritating potential of the test formulations.

3. Conclusions

In this study, we explored the influence of incorporating carvedilol-loaded spanlastics into in situ gelling system on the anti-glaucoma action of carvedilol. Thermosensitive in situ gel was prepared with a blend of two poloxamer analogs; 22% *w*/*v* poloxamer 407 and 5% *w*/*v* poloxamer 188. Incorporating CRV-loaded spanlastics within poloxamer-based in situ gel provided a dual action on sustaining drug release and prolonging the corneal retention time. In addition, ex vivo permeation studies demonstrated that the optimized spanlastic-laden in situ gel formulation (F2) significantly enhanced CRV permeation across the rabbit cornea by a 2.4-fold compared to plain CRV-ISG formulation. Most importantly, in vivo studies verified that incorporating CRV-loaded spanlastics within poloxamer-based in situ gel triggered a 2-fold increase in the AUC of optimized formula (F2), compared to plain CRV-ISG formulation. This enhancement in CRV ocular bioavailability was synchronized with a superior IOP lowering potential of optimized formula (F2). Collectively, spanlastic laden in situ gel might represent a promising alternative to conventional dosage forms for promoting efficient corneal delivery of anti-glaucoma drugs.

4. Materials and Methods

4.1. Materials

Carvedilol (CRV) was generously obtained from SAGA Pharmaceutical Company (Cairo, Egypt). Brij 97, poloxamer 407, poloxamer 188 and Span 60 were provided by Sigma Aldrich (St. Louis, MO, USA). All other used chemicals were of analytical grade.

4.2. Formulation of Carvedilol-Loaded Spanlastics (CRV-SPLs)

Carvedilol (CRV)-loaded spanlastics (CRV-SPLs) were fabricated by the ethanol injection method, as described previously [21]. Span 60 was adopted as the main vesicle forming component, while Brij 97 was used as an edge activator (EA) at Span 60:EA weight ratio of 90:10. To prepare CRV-loaded SPLs, CRV (62.5 mg) and a definite weight of Span 60 were dissolved in 5 mL ethanol. The ethanolic solution was then added dropwise to 15 mL of a preheated Brij 97 aqueous solution (70 °C). The dispersion was continuously stirred on a magnetic agitator (Jenway 1000, Jenway, UK) till the formation of milky spanlastic dispersion. The resultant dispersion was sonicated for 5 min to minimize the particle size. Finally, the obtained dispersion was stored at 4 °C until being used in further experiments.

4.3. Incorporation of CRV-SPLs into In Situ Gels (CS-ISGs)

Poloxamer-based hydrogel containing CRV-SPLs equivalent to 0.5% *w*/*w* of the drug were fabricated by the cold method [40] using definite concentrations of Poloxamer 407 (P407) and Poloxamer 188 (P 188) as summarized in Table 1. Briefly, definite concentrations of both P407 and P188 were mixed together and were dissolved in specific volumes of deionized water at 4 ± 1 °C. The mixture was stirred continuously overnight until a clear homogenous solution without lumps was obtained. Finally, an accurately weighed amount of CRV-SPLs was uniformly dispersed in the preformed poloxamer-based hydrogel and stored overnight in a refrigerator to exclude any entrapped air bubbles for further examinations.

4.4. Characterization of CRV-SPLs-Loaded In Situ Gels (CS-ISGs)

4.4.1. Visual Appearance

Clarity and homogeneity of the prepared CS-ISGs were observed by visual inspection of different formulations against a black and white background.

4.4.2. pH

pH meter (CG820 Schott Geräte,, Gerbershausen, Germany) was adopted for measuring the pH values of various formulation.

4.4.3. Drug Content

0.5 g of CS-ISGs were dissolved in ethanol using sonication to thoroughly lyse the vesicles then filtered through a 0.45 m milipore filter. The filtrate was suitably diluted, and the drug content was finally quantified spectrophotometrically at 242 nm using Ultraviolet–visible (UV–Vis) spectrophotometer (Shimadzu, Kyoto, Japan). The drug content (%) was calculated using the following formula:

$$Drug\ content\ (\%) = \frac{Actual\ amount\ of\ CRV}{Theoretical\ amount\ of\ CRV} \times 100$$

4.4.4. Determination of Gelation Temperature

Gelation temperature (T_G) was assessed by an inversion method [41]. Briefly, 1 mL of each CS-ISG formulation was placed into a 2 mL Eppendorf tube and allowed to equilibrate for 5 min at room temperature. The tubes were then placed in a thermomixer (Eppendorf ThermoMixer® C, Enfield, CT, USA) that had previously adjusted at 20 °C and subjected to a temperature rise of 1 °C every 2 min. The temperature at which no movement into the liquid was detected upon tilting up the tubes at 90° is referred to as the sol-gel transition temperature.

4.4.5. Rheological Studies

Brookfield viscometer model DVII (Haake Inc., Osterode am Harz, Germany) was used for viscosity measurements of different CS-ISG formulations before and after gelation (at 25 °C and 37 °C). Briefly, one gram of gel under investigation was put in the sample holder, and spindle no. 4 was lowered perpendicularly into it. The spindle was rotated at a constant speed of 100 rpm, and all measures were done in triplicates.

4.4.6. Measurement of Muco-Adhesion Force

The muco-adhesion force was determined using the modified analytical two-pan balance [28]. Freshly excised goat cornea was obtained from a local slaughterhouse. The cornea was excised from the ocular tissue and rinsed several times with cold PBS (pH 7.4) to get off any protein debris. Two same glass slides were used; one was fitted on the lower pan using double-sided adhesive tape, and the other was fitted on a table bench. Two same pieces of the cornea (2.5 cm^2) were adhered to each slide using glue. A very thin layer of a specific weight (0.5 gm) of each CS-ISG formulation was applied between corneal tissues. A preload of 5 gm was applied over the balance pan above glass slides for 30 s and then removed to ensure intimate contact between the excised cornea and the ISGs formulation. Increasing amounts of water were added in the second pan until the slides detached from each other and the water weight that cause complete detachment was recorded. The force of adhesion (N), defined as the minimum weight needed to detach the cornea from the formulation, was calculated using the following equation [42]:

$$N = \frac{m \cdot g}{1000}$$

where (m) is the weight in grams of water needed to detach the CS-ISGs formulation away from the cornea; (g) is the gravitational acceleration (9.81 m/s^2).

4.5. In Vitro Release Study

In vitro release of CRV from different CS-ISGs formulations was performed using a modified Franz diffusion cell, employing freshly prepared simulated tear fluid (STF; pH 7.4) as the release medium. In brief, a specific volume of each formula corresponding to 5 mg of CRV was transferred to the donor chamber. The donor chamber was then suspended in 250 mL of the release medium placed in the receptor compartment and was kept at 37 ± 0.5 °C and constantly agitated at 100 rpm. At predetermined time intervals, 1 mL samples were collected and replaced with an equal volume of fresh medium. The collected aliquot samples were diluted and spectrophotometrically analyzed at 242 nm to quantify the amount of drug released.

4.6. Ex Vivo Corneal Permeability Study

Ex vivo corneal permeability across freshly excised rabbit cornea was investigated using the membrane diffusion method [43]. Briefly, the excised cornea was sandwiched between the donor and receptor compartments of Franz diffusion cell. Both free CRV solution and selected CS-ISG formulation (equal to 1 mg CRV) were applied to the corneal epithelium in the donor compartment. 25 mL of fresh STF (pH 7.4) was used as the receptor medium. At predetermined time intervals, 1 mL aliquots were removed and replaced with fresh medium. The samples were spectrophotometrically analyzed at 242 nm to quantify drug content in each sample. The cumulative amount of CRV permeated through corneal membrane per unit area was plotted versus time (h). Drug flux across cornea was determined from the slope of the linear part of the curve, while the apparent permeability coefficient was calculated using the following formula:

$$P_{app} = \frac{\Delta Q}{\Delta t} \cdot \frac{1}{3600 \times A \times C_o}$$

where, $\Delta Q/\Delta t$ is the cumulative amount of drug permeated across the cornea over time t, A is the exposed corneal surface area (0.8 cm^2), and C_o is the initial drug concentration in donor chamber.

4.7. Stability Studies

The stability studies for the optimized CS-ISG formulation were carried out by storing the optimized formula at 4 °C for 8 weeks and then the formula was assessed for visual appearance, pH, drug content and gelling capability.

4.8. In Vivo Experiments

4.8.1. In Vivo Pharmacokinetics

In vivo pharmacokinetic study was conducted on male albino rabbits (2–2.5 kg). The study protocol was reviewed by the Animal Ethics Committee, Prince Sattam Bin Abdulaziz University, Al-Kharj, KSA (approval number: 048/2022). In this study, the animals were categorized into two groups; the first group was treated with 50 µL of CS-ISG formulation (5 mg CRV/mL), while the other group was treated with 50 µL of CRV ophthalmic suspension (0.5% *w*/*v*). Animal eyelids were lightly closed for 1 min to permit better contact of drug with the corneal membrane. Prior to aqueous humor withdrawal, rabbits were anaesthetized with sodium phenobarbital (30 mg/kg), and 100 µL samples of aqueous humor were obtained at 0.5, 1, 2, 4, and 6 h using a 29-gauge insulin syringe needle. The samples were mixed with 500 µL of methanol to precipitate protein, followed by centrifugation at 5000 rpm for 15 min. The concentration of CRV in the supernatant was quantified using an HPLC system equipped with a UV detector (Shimadzu, Tokyo, Japan) at 240 nm and a Hypersil® C-18 column (150 mm × 4.6 mm, 5 µm). The column was eluted with a mobile phase consisting of KH_2PSO_4:acetonitrile (50:50 *v*/*v*), adjusted to pH 3.0 with dilute orthophosphoric acid solution. The flow rate was 1 mL/min and the injection volume was 20 µL. The pharmacokinetic parameters were determined using a PKSolver 2.0 software.

4.8.2. Pharmacodynamic Study

In vivo pharmacodynamic study was conducted on male albino rabbits, with an average IOP value of 21.6 mmHg. Twelve rabbits were divided into two groups: Group I was instilled with 50 µL of CRV suspension (0.5% *w/v*), while Group II was instilled with 50 µL of CS-ISG formulation (5 mg CRV/mL), into the left eye. The right eye received 50 µL physiological saline and served as control. At predetermined time points post treatments (0, 0.5, 1, 2, 3, 4, 5, and 6 h), the IOP was measured under surface anesthesia with 0.2% lidocaine using a tonometer (Riester, Jungingen, Germany).

4.8.3. Assessment of Ocular Irritancy of CS-ISG Formulation

Male albino rabbits (2 groups of 3 animals in each) were employed to examine the ocular tolerability of the formulated CS-ISG formulation. The animals were inspected for any signs of irritation (redness, inflammation, or increased tear production) upon ocular application. A 50 µL aliquot of CS-ISG formulation or an equivalent concentration of CRV-ISG were ocularly applied into the left eye's conjunctival sac, while the contralateral eye served as a control and received no treatment. Direct visual inspection using a slit lamp was used to examine both eyes of the rabbits for any signs of irritation.

Supplementary Materials: The following supporting information can be downloaded at: https://www.mdpi.com/article/10.3390/gels9120952/s1, Table S1: Grading of ocular irritation by Draize irritation test in rabbits.

Author Contributions: Conceptualization, A.S.A.L. and B.K.A.; methodology, B.K.A. and E.-S.K.; validation, E.-S.K.; formal analysis, A.S.A.L.; investigation, B.K.A. and E.-S.K.; resources, B.K.A.; writing—original draft preparation, B.K.A. and E.-S.K.; writing—review and editing, A.S.A.L.; project administration, B.K.A.; funding acquisition, E.-S.K. All authors have read and agreed to the published version of the manuscript.

Funding: This research was supported by Prince Sattam Bin Abdulaziz University project.

Institutional Review Board Statement: The animal study protocol was approved by Ethical Committee, Prince Sattam Bin Abdulaziz University, Al-Kharj, KSA (approval number: 048/2022).

Informed Consent Statement: Not applicable.

Data Availability Statement: Data are contained within the article or Supplementary Materials.

Acknowledgments: This study is supported via funding from Prince Sattam Bin Abdulaziz University project number (PSAU/2023/R/1444).

Conflicts of Interest: The authors declare no conflict of interest.

References

1. Weinreb, R.N.; Aung, T.; Medeiros, F.A. The pathophysiology and treatment of glaucoma: A review. *JAMA* **2014**, *311*, 1901–1911. [CrossRef] [PubMed]
2. Alamri, A.; Bakri, K.A.; Alqarni, S.M.; Alosaimi, M.N.; Alshehri, E.Y.; Alshahrani, M.S. Knowledge and awareness of glaucoma in a population of Abha, Southern Saudi Arabia. *J. Fam. Med. Prim. Care* **2022**, *11*, 6165–6169. [CrossRef] [PubMed]
3. Sharif, N.A. Therapeutic Drugs and Devices for Tackling Ocular Hypertension and Glaucoma, and Need for Neuroprotection and Cytoprotective Therapies. *Front. Pharmacol.* **2021**, *12*, 729249. [CrossRef]
4. Öztürk, K.; Arslan, F.B.; Öztürk, S.C.; Çalış, S. Mixed micelles formulation for carvedilol delivery: In-vitro characterization and in-vivo evaluation. *Int. J. Pharm.* **2022**, *611*, 121294. [CrossRef] [PubMed]
5. Abdelmonem, R.; Elhabal, S.F.; Abdelmalak, N.S.; El-Nabarawi, M.A.; Teaima, M.H. Formulation and Characterization of Acetazolamide/Carvedilol Niosomal Gel for Glaucoma Treatment: In Vitro, and In Vivo Study. *Pharmaceutics* **2021**, *13*, 221. [CrossRef] [PubMed]
6. Hassan, D.H.; Abdelmonem, R.; Abdellatif, M.M. Formulation and Characterization of Carvedilol Leciplex for Glaucoma Treatment: In-Vitro, Ex-Vivo and In-Vivo Study. *Pharmaceutics* **2018**, *10*, 197. [CrossRef] [PubMed]
7. Irimia, T.; Ghica, M.V.; Popa, L.; Anuţa, V.; Arsene, A.L.; Dinu-Pîrvu, C.E. Strategies for Improving Ocular Drug Bioavailability and Corneal Wound Healing with Chitosan-Based Delivery Systems. *Polymers* **2018**, *10*, 1221. [CrossRef]

8. Rawat, P.S.; Ravi, P.R.; Mir, S.I.; Khan, M.S.; Kathuria, H.; Katnapally, P.; Bhatnagar, U. Design, Characterization and Pharmacokinetic–Pharmacodynamic Evaluation of Poloxamer and Kappa-Carrageenan-Based Dual-Responsive In Situ Gel of Nebivolol for Treatment of Open-Angle Glaucoma. *Pharmaceutics* **2023**, *15*, 405. [CrossRef]
9. null, A.; Ali, J.; Fazil, M.; Qumbar, M.; Khan, N.; Ali, A. Colloidal drug delivery system: Amplify the ocular delivery. *Drug Deliv.* **2016**, *23*, 700–716. [CrossRef]
10. Ahmed, S.; Amin, M.M.; Sayed, S. Ocular Drug Delivery: A Comprehensive Review. *AAPS PharmSciTech* **2023**, *24*, 66. [CrossRef]
11. Silva, B.; São Braz, B.; Delgado, E.; Gonçalves, L. Colloidal nanosystems with mucoadhesive properties designed for ocular topical delivery. *Int. J. Pharm.* **2021**, *606*, 120873. [CrossRef] [PubMed]
12. Shukr, M.H.; Ismail, S.; El-Hossary, G.G.; El-Shazly, A.H. Spanlastics nanovesicular ocular insert as a novel ocular delivery of travoprost: Optimization using Box–Behnken design and in vivo evaluation. *J. Liposome Res.* **2022**, *32*, 354–364. [CrossRef] [PubMed]
13. Kakkar, S.; Kaur, I.P. Spanlastics—A novel nanovesicular carrier system for ocular delivery. *Int. J. Pharm.* **2011**, *413*, 202–210. [CrossRef] [PubMed]
14. Rathod, S.; Arya, S.; Shukla, R.; Ray, D.; Aswal, V.K.; Bahadur, P.; Tiwari, S. Investigations on the role of edge activator upon structural transitions in Span vesicles. *Colloids Surf. A Physicochem. Eng. Asp.* **2021**, *627*, 127246. [CrossRef]
15. Abdelbari, M.A.; El-Mancy, S.S.; Elshafeey, A.H.; Abdelbary, A.A. Implementing Spanlastics for Improving the Ocular Delivery of Clotrimazole: In vitro Characterization, Ex vivo Permeability, Microbiological Assessment and In vivo Safety Study. *Int. J. Nanomed.* **2021**, *16*, 6249–6261. [CrossRef]
16. Shymborska, Y.; Budkowski, A.; Raczkowska, J.; Donchak, V.; Melnyk, Y.; Vasiichuk, V.; Stetsyshyn, Y. Switching it Up: The Promise of Stimuli-Responsive Polymer Systems in Biomedical Science. *Chem. Rec.* **2023**, e202300217. [CrossRef]
17. Nagai, N.; Minami, M.; Deguchi, S.; Otake, H.; Sasaki, H.; Yamamoto, N. An in situ Gelling System Based on Methylcellulose and Tranilast Solid Nanoparticles Enhances Ocular Residence Time and Drug Absorption Into the Cornea and Conjunctiva. *Front. Bioeng. Biotechnol.* **2020**, *8*, 764. [CrossRef]
18. Wu, Y.; Liu, Y.; Li, X.; Kebebe, D.; Zhang, B.; Ren, J.; Lu, J.; Li, J.; Du, S.; Liu, Z. Research progress of in-situ gelling ophthalmic drug delivery system. *Asian J. Pharm. Sci.* **2019**, *14*, 1–15. [CrossRef]
19. Abbas, M.N.; Khan, S.A.; Sadozai, S.K.; Khalil, I.A.; Anter, A.; Fouly, M.E.; Osman, A.H.; Kazi, M. Nanoparticles Loaded Thermoresponsive In Situ Gel for Ocular Antibiotic Delivery against Bacterial Keratitis. *Polymers* **2022**, *14*, 1135. [CrossRef]
20. Aldawsari, M.F.; Moglad, E.H.; Alotaibi, H.F.; Alkahtani, H.M.; Khafagy, E.-S. Ophthalmic Bimatoprost-Loaded Niosomal In Situ Gel: Preparation, Optimization, and In Vivo Pharmacodynamics Study. *Polymers* **2023**, *15*, 4336. [CrossRef]
21. Sallam, N.M.; Sanad, R.A.B.; Ahmed, M.M.; Khafagy, E.L.S.; Ghorab, M.; Gad, S. Impact of the mucoadhesive lyophilized wafer loaded with novel carvedilol nano-spanlastics on biochemical markers in the heart of spontaneously hypertensive rat models. *Drug Deliv. Transl. Res.* **2021**, *11*, 1009–1036. [CrossRef]
22. Abelson, M.B.; Udell, I.J.; Weston, J.H. Normal human tear pH by direct measurement. *Arch. Ophthalmol.* **1981**, *99*, 301. [CrossRef]
23. Zarrintaj, P.; Ramsey, J.D.; Samadi, A.; Atoufi, Z.; Yazdi, M.K.; Ganjali, M.R.; Amirabad, L.M.; Zangene, E.; Farokhi, M.; Formela, K.; et al. Poloxamer: A versatile tri-block copolymer for biomedical applications. *Acta Biomater.* **2020**, *110*, 37–67. [CrossRef]
24. Liu, S.; Bao, H.; Li, L. Role of PPO–PEO–PPO triblock copolymers in phase transitions of a PEO–PPO–PEO triblock copolymer in aqueous solution. *Eur. Polym. J.* **2015**, *71*, 423–439. [CrossRef]
25. Russo, E.; Villa, C. Poloxamer Hydrogels for Biomedical Applications. *Pharmaceutics* **2019**, *11*, 671. [CrossRef]
26. Abdeltawab, H.; Svirskis, D.; Hill, A.G.; Sharma, M. Increasing the Hydrophobic Component of Poloxamers and the Inclusion of Salt Extend the Release of Bupivacaine from Injectable In Situ Gels, While Common Polymer Additives Have Little Effect. *Gels* **2022**, *8*, 484. [CrossRef]
27. Cao, F.; Zhang, X.; Ping, Q. New method for ophthalmic delivery of azithromycin by poloxamer/carbopol-based in situ gelling system. *Drug Deliv.* **2010**, *17*, 500–507. [CrossRef]
28. Al-Kassas, R.S.; El-Khatib, M.M. Ophthalmic controlled release in situ gelling systems for ciprofloxacin based on polymeric carriers. *Drug Deliv.* **2009**, *16*, 145–152. [CrossRef]
29. Al Khateb, K.; Ozhmukhametova, E.K.; Mussin, M.N.; Seilkhanov, S.K.; Rakhypbekov, T.K.; Lau, W.M.; Khutoryanskiy, V.V. In situ gelling systems based on Pluronic F127/Pluronic F68 formulations for ocular drug delivery. *Int. J. Pharm.* **2016**, *502*, 70–79. [CrossRef]
30. Jaipal, A.; Pandey, M.M.; Charde, S.Y.; Raut, P.P.; Prasanth, K.V.; Prasad, R.G. Effect of HPMC and mannitol on drug release and bioadhesion behavior of buccal discs of buspirone hydrochloride: In-vitro and in-vivo pharmacokinetic studies. *Saudi Pharm. J.* **2015**, *23*, 315–326. [CrossRef]
31. Gupta, H.; Jain, S.; Mathur, R.; Mishra, P.; Mishra, A.K.; Velpandian, T. Sustained ocular drug delivery from a temperature and pH triggered novel in situ gel system. *Drug Deliv.* **2007**, *14*, 507–515. [CrossRef]
32. Kurniawansyah, I.S.; Rusdiana, T.; Sopyan, I.; Ramoko, H.; Wahab, H.A.; Subarnas, A. In situ ophthalmic gel forming systems of poloxamer 407 and hydroxypropyl methyl cellulose mixtures for sustained ocular delivery of chloramphenicole: Optimization study by factorial design. *Heliyon* **2020**, *6*, e05365. [CrossRef]
33. Ranch, K.M.; Maulvi, F.A.; Naik, M.J.; Koli, A.R.; Parikh, R.K.; Shah, D.O. Optimization of a novel in situ gel for sustained ocular drug delivery using Box-Behnken design: In vitro, ex vivo, in vivo and human studies. *Int. J. Pharm.* **2019**, *554*, 264–275. [CrossRef]

34. Gilani, S.J.; Jumah, M.N.B.; Zafar, A.; Imam, S.S.; Yasir, M.; Khalid, M.; Alshehri, S.; Ghuneim, M.M.; Albohairy, F.M. Formulation and Evaluation of Nano Lipid Carrier-Based Ocular Gel System: Optimization to Antibacterial Activity. *Gels* **2022**, *8*, 255. [CrossRef]
35. Nair, A.B.; Shah, J.; Jacob, S.; Al-Dhubiab, B.E.; Sreeharsha, N.; Morsy, M.A.; Gupta, S.; Attimarad, M.; Shinu, P.; Venugopala, K.N. Experimental design, formulation and in vivo evaluation of a novel topical in situ gel system to treat ocular infections. *PLoS ONE* **2021**, *16*, e0248857. [CrossRef]
36. Huang, J.; Peng, T.; Li, Y.; Zhan, Z.; Zeng, Y.; Huang, Y.; Pan, X.; Wu, C.Y.; Wu, C. Ocular Cubosome Drug Delivery System for Timolol Maleate: Preparation, Characterization, Cytotoxicity, Ex Vivo, and In Vivo Evaluation. *AAPS PharmSciTech* **2017**, *18*, 2919–2926. [CrossRef]
37. Ban, J.; Zhang, Y.; Huang, X.; Deng, G.; Hou, D.; Chen, Y.; Lu, Z. Corneal permeation properties of a charged lipid nanoparticle carrier containing dexamethasone. *Int. J. Nanomed.* **2017**, *12*, 1329–1339. [CrossRef]
38. Natarajan, J.V.; Ang, M.; Darwitan, A.; Chattopadhyay, S.; Wong, T.T.; Venkatraman, S.S. Nanomedicine for glaucoma: Liposomes provide sustained release of latanoprost in the eye. *Int. J. Nanomed.* **2012**, *7*, 123–131. [CrossRef]
39. Leonardi, A.; Bucolo, C.; Drago, F.; Salomone, S.; Pignatello, R. Cationic solid lipid nanoparticles enhance ocular hypotensive effect of melatonin in rabbit. *Int. J. Pharm.* **2015**, *478*, 180–186. [CrossRef]
40. Brambilla, E.; Locarno, S.; Gallo, S.; Orsini, F.; Pini, C.; Farronato, M.; Thomaz, D.V.; Lenardi, C.; Piazzoni, M.; Tartaglia, G. Poloxamer-Based Hydrogel as Drug Delivery System: How Polymeric Excipients Influence the Chemical-Physical Properties. *Polymers* **2022**, *14*, 3624. [CrossRef]
41. Balasubramaniam, J.; Pandit, J.K. Ion-activated in situ gelling systems for sustained ophthalmic delivery of ciprofloxacin hydrochloride. *Drug Deliv.* **2003**, *10*, 185–191. [CrossRef] [PubMed]
42. Ali, J.; Khar, R.; Ahuja, A.; Kalra, R. Buccoadhesive erodible disk for treatment of oro-dental infections: Design and characterisation. *Int. J. Pharm.* **2002**, *238*, 93–103. [CrossRef] [PubMed]
43. Bíró, T.; Bocsik, A.; Jurišić Dukovski, B.; Gróf, I.; Lovrić, J.; Csóka, I.; Deli, M.A.; Aigner, Z. New Approach in Ocular Drug Delivery: In vitro and ex vivo Investigation of Cyclodextrin-Containing, Mucoadhesive Eye Drop Formulations. *Drug Des. Devel Ther.* **2021**, *15*, 351–360. [CrossRef] [PubMed]

gels

Article

Exploring Functionalized Magnetic Hydrogel Polyvinyl Alcohol and Chitosan Electrospun Nanofibers

Mónica Guerra [1], Fábio F. F. Garrudo [2], Célia Faustino [1,3,4], Maria Emilia Rosa [5] and Maria H. L. Ribeiro [1,3,4,*]

1 Faculty of Pharmacy, Universidade de Lisboa, Av. Prof. Gama Pinto, 1649-003 Lisboa, Portugal; guerramsc@gmail.com (M.G.); cfaustino@ff.ulisboa.pt (C.F.)
2 Department of Bioengineering, Institute of Telecomunications, Instituto Superior Técnico, Universidade de Lisboa, Av. Rovisco Pais, 1049-001 Lisboa, Portugal; fabio.garrudo@tecnico.ulisboa.pt
3 Research Institute for Medicines (iMed.ULisboa), Faculty of Pharmacy, Universidade de Lisboa, Av. Prof. Gama Pinto, 1649-003 Lisboa, Portugal
4 Department of Pharmaceutical Sciences and Medicines, Faculty of Pharmacy, Universidade de Lisboa, Av. Prof. Gama Pinto, 1649-003 Lisbon, Portugal
5 Instituto de Engenharia Mecânica (IDMEC), Instituto Superior Técnico, Universidade de Lisboa, 1049-001 Lisboa, Portugal; emilia.rosa@ist.utl.pt
* Correspondence: mhribeiro@ff.ul.pt; Tel.: +351-21-7946400; Fax: +351-21-7946470

Abstract: Nanofibrous materials present interesting characteristics, such as higher area/mass ratio and reactivity. These properties have been exploited in different applications, such as drug-controlled release and site-specific targeting of biomolecules for several disease treatments, including cancer. The main goal of this study was to develop magnetized nanofiber systems of lysozyme (Lys) for biological applications. The system envisaged electrospun polyvinyl alcohol (PVA) and PVA/chitosan (CS) nanofibers, loaded with Lys, crosslinked with boronic acids [phenylboronic acid (PBA), including 2-acetylphenylboronic acid (aPBA), 2-formylphenylboronic (fPBA), or bortezomib (BTZ)] and functionalized with magnetic nanobeads (IONPs), which was successfully built and tested using a microscale approach. Evaluation of the morphology of nanofibers, obtained by electrospinning, was carried out using SEM. The biological activities of the Lys-loaded PVA/CS (90:10 and 70:30) nanofibers were evaluated using the *Micrococcus lysodeikticus* method. To evaluate the success of the encapsulation process, the ratio of adsorbed Lys on the nanofibers, Lys activity, and in vitro Lys release were determined in buffer solution at pH values mimicking the environment of cancer cells. The viability of Caco-2 cancer cells was evaluated after being in contact with electrospun PVA + Lys and PVA/CS + Lys nanofibers, with or without boronic acid functionalation, and all were magnetized with IONPs.

Keywords: lysozyme; 3D-electrospun; hydrogels; chitosan; PVA; magnetic nanoparticles; tunable properties

Citation: Guerra, M.; Garrudo, F.F.F.; Faustino, C.; Rosa, M.E.; Ribeiro, M.H.L. Exploring Functionalized Magnetic Hydrogel Polyvinyl Alcohol and Chitosan Electrospun Nanofibers. *Gels* **2023**, *9*, 968. https://doi.org/10.3390/gels9120968

Academic Editor: Tetsu Mitsumata

Received: 6 November 2023
Revised: 8 December 2023
Accepted: 10 December 2023
Published: 11 December 2023

1. Introduction

Nanofibers can be produced from a wide range of natural and synthetic polymers. Some of the natural polymers include hydrogels such as alginate (AL), chitosan (CS), collagen (CO), gelatin (GE), fibrin (FI), and hyaluronic acid (HA) [1,2]. CS, a natural polysaccharide obtained from the deacetylation of chitin, has found widespread use in the pharmaceutical, food, and biotechnology industries due to its biodegradability, biocompatibility, and biological properties, such as antioxidant, and antibacterial properties [3]. In the field of biomedicine, chitosan gel has been employed as a polymeric component in drug delivery, bone tissue regeneration, and the healing of skin lesions [4]. Additionally, it is also appropriate to remark on the possibility of combining chitosan with other types of hydrogels to create suitable materials for improved tissue engineering applications [5].

Synthetic polymers such as poly(vinyl alcohol) (PVA), poly(hydroxyethyl methacrylate), poly(ethylene glycol dimethacrylate), poly(ethylene oxide), poly(propylene-co-ethylene glycol fumarate), polypeptides, and poly(acrylic acid) and its derivatives have also been used to produce nanofibers. PVA is an electrospinnable hydrophilic gel with excellent mechanical properties, biodegradability, non-toxicity, and biocompatibility. It has an acceptable toxicological profile, with LD50 above 15–20 g/Kg, NOAEL of 5 g/Kg, low gastrointestinal absorption, lack of accumulation in the body, no sub-chronic or chronic toxicity events reported, and no mutagenic or carcinogenic effects observed on in vitro assays [6]. PVA is considered a safe compound by the European Food Safety Agency (EFSA) and the Food and Drug Administration (FDA). Owing to its properties, PVA is one of the most extensively studied and widely used polymers in biomedicine, especially in contact lenses, implants, drug delivery systems, tissue engineering, artificial organ development, and immobilization [2].

Polymer blending has emerged as a method of enhancing the chemical and mechanical properties of polymeric materials for practical applications [7–9]. The existence of electrospinning emulsions of poly (L-lactic acid)/poly (vinyl alcohol) with chitosan were used for wound dressing with antibacterial properties [10]. Electrospinning is one of the most popular processes for producing nanofibers, and it has been of interest since the 1980s, inspired by the development of nanotechnology. Electrospinning allows for the conversion of a polymeric solution into solid nanofibers via the application of electrical force [11,12]. The nanofibers obtained using electrospinning are collected in the form of a porous matrix with a high surface area, which is structurally similar to the extracellular matrix [11,12]. These fibers are associated with a low production cost and simplicity of manufacturing, making them a promising substrate with vast applicability in numerous areas of chemistry, biology, medicine, and engineering, such as wound healing, wound care, biosensors, drug delivery systems, medical implants, tissue engineering, dental materials, filtration membranes, protective clothing, and other industrial applications [11,12].

Lysozyme is a widely distributed enzyme found in several organisms, such as bacteria, bacteriophages, fungi, plants, and mammals [13,14]. Chicken egg white lysozyme, which has 129 amino acids and reactivity 3–4 times lower than human lysozyme, has been widely used as an experimental model due to its structural similarity, availability, and low cost, making it one of the most studied enzymes [15].

The antimicrobial activity of lysozyme relies on its ability to catalyze the hydrolysis of the β1-4 glycosidic bond between *N*-acetylmuramic acid and *N*-acetylglucosamine, which are components of the peptidoglycan in bacterial cell walls [16].

Lysozyme's bactericidal properties have been applied in the food and pharmaceutical industries [17]. In addition to its bactericidal properties, antifungal, antiviral, antitumor, and immunomodulatory properties have also been described [13]. Lysozyme's association with cancer began in the 1960s due to its role as a tumor biomarker in hematological cancers [6,18].

In recent years, the immobilization of enzymes and magnetic nanoparticles on polymeric nanofibers has allowed their use in sensors, tissue regeneration structures, drug delivery systems, and other applications [19]. Iron oxide nanoparticles (IONPs) have paramagnetic properties, are biocompatible and non-toxic, and are suitable for biomedical applications.

A different approach was the development of co-immobilized cellulase and lysozyme on the surface of amino-functionalized magnetic nanoparticles using glutaraldehyde [20]. Another approach [21] was the fabrication of magnetic lysozyme@Fe_3O_4 composites via amyloid-like assembly for uranium extraction with magnetism for easy recovery and good binding affinity towards uranium, respectively. These composite adsorbents also showed excellent photothermal properties derived from the Fe_3O_4 nanoparticles.

A 2D protein self-assembly film was reported to capture functional enzymes without any further chemical modification, with enzymes immobilized between Fe_3O_4 nanoparticles and a lysozyme film, preventing enzyme leaching and ensuring contact with substrates [22,23].

To extend the lifespan and bioactivity of lysozyme for use in food packaging, medicine, medical devices, and cosmetics, immobilization of lysozyme on solid supports showed positive results, as demonstrated by the increased stability and extended half-life of the enzyme [24].

Magnetic nanoparticles have gained widespread research interest due to their additional use in hyperthermia-based cancer therapy. The process is based on the increased sensitivity of various types of cancer cells to temperatures above 41 °C. Iron oxide nanoparticles (IONPs), such as magnetite (Fe_3O_4), can heat the surrounding environment to 45 °C when an alternating magnetic field is applied to them (hyperthermia). The damage caused by the application of temperatures in the range of 41–45 °C in normal tissue is reversible, while tumor cells are irreversibly damaged, and cell death occurs [19,25]. Therefore hydrogel nanofibers magnetized with IONPs can be one strategy for tumor treatment based on localized hyperthermia [19]. The use of magnetized scaffolds as a therapeutic system and as a drug delivery system is an interesting approach to colon cancer therapy. The magnetic nanoparticle properties directly rely on their morphology and size. Thus, as the nanoparticle size decreases, the magnetic behavior of the particle enormously decreases [26], which directly impacts their applications as drug carriers or in hyperthermia treatments. Colon cancer is the third most common cancer worldwide and the fourth most common cause of death [27]. Due to its invasive nature, there is a need for alternative colon cancer therapies, especially ones that allow growth control, enclosure of metastatic cells, and recurrence. The human epithelial cell line Caco-2, derived from colon carcinoma, has been widely used as a model of the intestinal epithelial barrier [28–30] and was used in this study to test the cytotoxicity of the developed lysozyme magnetized nanofibers. Therefore, the main goal of this study was to develop a hydrogel nanofiber system using the electrospinning technique, with the encapsulation of lysozyme crosslinked with boronic acids and magnetized with IONPs.

2. Results and Discussion

The morphology of nanofibers depends on several factors, namely, properties of the polymeric solution (concentration, molecular weight, viscosity, conductivity, surface tension), process parameters (voltage, flow rate, collectors, distance between collector and syringe needle), and environmental conditions (humidity and temperature) [31].

2.1. Characterization of Electrospinning Solutions

The specific conductivity (κ) was evaluated for the hydrogel solutions of PVA, PVA + Lys, PVA/CS 90:10, PVA/CS 90:10 + Lys, PVA/CS 70:30, PVA/CS 70:30 + Lys, and CS. Figure 1 presents the specific conductivity profile as a function of polymer concentration. For CS, no conductivity profile was observed.

The specific conductivity increased linearly with hydrogel polymer concentration up to the point of discontinuity (Figure 1). This profile was divided into two linear series that were adjusted for each one and allowed the critical aggregation concentration (c.a.c.) calculation for each solution based on the intersection of the two lines (Table 1). These results suggest that both Lys and CS positively influence the conductivity values of the tested solutions. Similarly, the contribution of the CS to the c.a.c. represents an increase of about 1.5% in the concentration required for the 10% PVA solution to reach the c.a.c. (Table 1). Moreover, Lys in PVA/CS solutions increased the c.a.c. values. On the contrary, for the PVA + Lys solution, the conductivity values increased proportionally with the concentration, with a cutoff point at 0.03% PVA concentration (Figure 1).

The surface tension (γ) was another parameter used in the characterization of the electrospinning solutions. Figure 2 represents the surface tension profiles as a function of the logarithm of polymer concentration. The surface tension of the solutions decreased with increasing concentration of the polymers, attaining an equilibrium at the point where the c.a.c. was achieved for PVA solutions: PVA/CS 70:30 and PVA/CS 70:30 + Lys. The presence of CS did not significantly alter the surface tension values; however, the PVA/CS

70:30 solutions originated higher values. Moreover, the presence of Lys decreased the surface tension values at c.a.c. of all analyzed solutions (Figure 2).

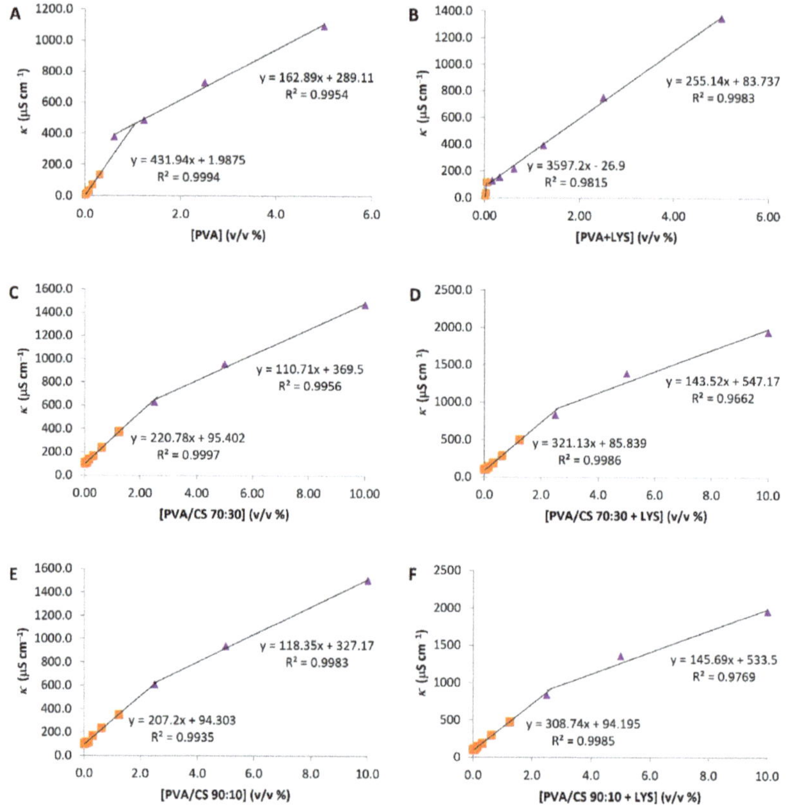

Figure 1. Specific conductivity (average temperature of 18.5 °C). (**A**) PVA; (**B**) PVA + Lys; (**C**) PVA/CS 70:30; (**D**) PVA/CS 70:30 + Lys; (**E**) PVA/CS 90:10; (**F**) PVA/CS 90:10 + Lys. SD (Standard Deviation) was ±0.01, and each point of the graphic was carried out in triplicates.

Table 1. Critical aggregation concentration (c.a.c.), specific conductivity, and surface tension at c.a.c. for each polymer solution.

Polymeric Solution	κ [1] (S cm^{-1})	c.a.c.$_κ$ [2] (%)	γ [3] (mN m^{-1})	c.a.c.$_γ$ [4] (%)
PVA 10% (*m*/*v*)	985	1.07	38.86 ± 0.40	0.68
PVA 10% (*m*/*v*) + Lys	1958	0.03	36.70 ± 0.36	2.27
CS 2% (*m*/*v*)	2324	----	----	----
PVA/CS 90:10 (*v*/*v*)	1503	2.62	37.92 ± 0.34	2.54
PVA/CS 90:10 (*v*/*v*) + Lys	1958	2.69	35.25 ± 0.30	2.48
PVA/CS 70:30 (*v*/*v*)	1466	2.49	42.35 ± 0.37	2.43
PVA/CS 70:30 (*v*/*v*) + Lys	1943	2.60	41.31 ± 0.31	2.62

[1] Specific conductivity of the solution at the concentration that was used in electrospinning. [2] Critical aggregation concentration from conductivity measurements. [3] Surface tension of polymer solutions diluted 1:2 in distilled water. [4] Critical aggregation concentration from surface tension measurements.

The c.a.c. was calculated based on surface tension data and fitted for conductivity data (Table 1). In this case, the values of c.a.c. are similar for all solutions except for the 10%

PVA + Lys solution, which showed some variability between the results calculated from the conductivity and surface tension.

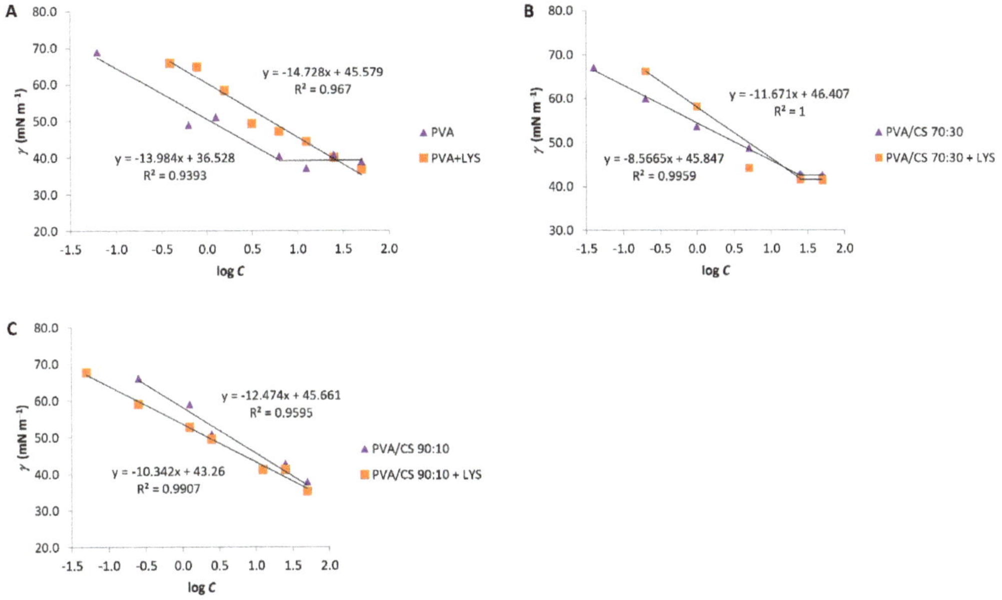

Figure 2. Surface tension profiles as a function of the logarithm of polymer concentration for PVA, PVA + Lys solutions (**A**); PVA/CS 70:30, PVA/CS 70:30 + Lys (**B**); and PVA/CS 90:10, PVA/CS 90:10 + Lys (**C**). SD (Standard Deviation) was ±0.01, and each point of the graphic was carried out in triplicates.

An increase in the conductivity of the solution promotes the formation of thinner fibers, contrary to the surface tension, which should not be too high, as it directly influences the shape of the formed structures, fibers, or drops.

It was not possible to produce nanofibers from the 2% CS gel solution despite having a higher conductivity (2324 S cm^{-1}) than the PVA/CS gel solutions. The results suggest an increase in this parameter by mixing the PVA/CS gel polymers. In the various attempts to produce fibers with this solution, the phenomenon of electrospray can be attributed to the low concentration of the gel polymer and a probable increase in surface tension. This gel solution was less viscous than the other solutions tested, which may hinder the production of nanofibers.

2.2. Morphology of the Nanofibers

The morphology of nanofibers is influenced by several parameters during the production process, namely temperature and humidity conditions, voltage, solution flow rate, the distance between the capillary end and the collector, and, in particular, the properties of the polymer(s) solution, including the concentration, viscosity, conductivity, surface tension, and nature of the solvent.

Not all attempts to produce PVA or PVA/CS nanofibers were effective, with the temperature of the electrospinning solution being the main parameter, with a consequent influence on viscosity. Thus, the optimal temperature range for electrospinning was 18–20 °C.

In the electrospinning, the gel solutions of 2% CS in 2% acetic acid, 2.5% CS in 50% acetic acid, and 4% and 3% CS in 90% acetic acid were tested, but in none of the cases, nanofibers were obtained. The production of pure CS nanofibers was very difficult due to the sensitivity of the process to humidity (<30%). However, the electrospinning process was improved by introducing the other gel polymer, PVA, and tested in different proportions (90:10 and 70:30).

The nanofibers obtained with PVA or PVA/CS, with and without Lys, presented, for the most part, a uniform appearance, white color, relatively thin and fragile, with a circular shape with a diameter of about 3.3 cm and a mass mean of 4.12 mg (Supplementary Material, Figure S1). Some reported studies where homogeneous fibers were obtained when chitosan was mixed with synthetic resins and nanoparticles to strengthen the gathered results [32].

PVA and PVA/CS nanofibers with different treatments were observed by optic microscopy and SEM with different magnifications (Supplementary Material, Figure S2). Figure 3 shows the images of PVA and PVA/CS nanofibers using SEM, respectively.

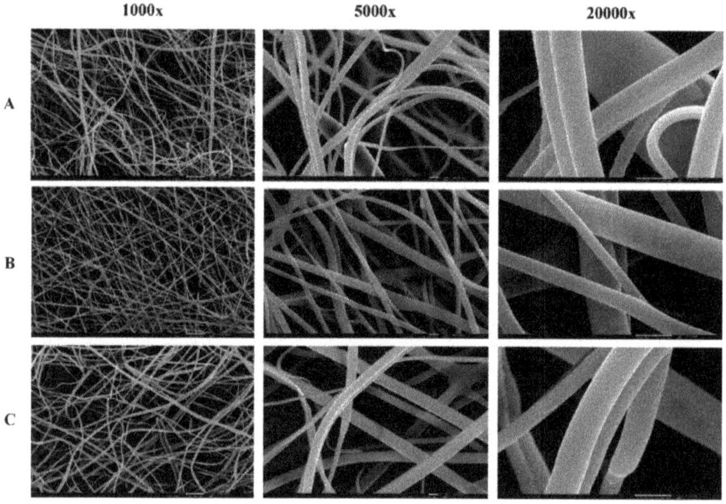

Figure 3. SEM images of the nanofibers of PVA/CS 70:30 (**A**); PVA/CS 90:10 (**B**), and PVA/CS 90:10 + Lys (**C**), magnification of 1000×, 5000× and 20,000×.

The mixture of PVA/CS polymers did not show significant differences in terms of morphology. SEM images show, for both PVA/CS 90:10 and 70:30 ratios, nanofibers with fibers of variable diameters, and some fibers appear folded on the surface, namely the fibers with Lys and in the PVA/CS 70:30 ratio (Figure 4). In terms of fiber diameter, compared with PVA fibers, they show similar diameters; however, there is greater porosity between fibers, especially the PVA/CS 70:30 nanofibers. PVA/CS 70:30 nanofibers also appear to be more flexible than PVA or PVA/CS 90:10 nanofibers.

The conjugation of PVA with CS in the formation of nanofibers seems to improve their chemical and resistance properties [33]. The interactions between these molecules are essentially based on the hydrophobic aggregation of the side chain and intermolecular and intramolecular hydrogen bonds [34].

After crosslinking with boronic acids, the nanofibers become less fragile. The immersion of these nanofibers in suspensions with iron oxide nanoparticles resulted in the sedimentation of black nanoparticles on the fiber, presenting a rough appearance with different colors (Supplementary Material, Figure S1).

All nanofibers produced presented randomly oriented fibers with variable diameters. Through the optical microscopy evaluation, it was observed that the nanofibers with Lys immobilized seem to have some drops on the fibers (Supplementary Material, Figure S2), which translates into roughness in the SEM images (Figure 3). Using optical microscopy, a deposition of brown drops on the fibers that have been crosslinked with boronic acids, namely the acids PBA, aPBA, and fPBA, was observed. SEM images confirm the deposition of inhomogeneous structures on the fibers, making them straighter and thicker, which may be due to the presence of boronic acids (Figure 3). Thus, the influence of these acids on the fibers can be qualitatively inferred based on the fiber diameter, which decreased as follows: fPBA > aPBA > PBA, while pore size decreased with aPBA > fPBA > PBA. The BTZ acid, through observation by optical microscopy, did not affect the nanofibers in a manner similar to the acids described, as it seems to affect the uniformity of the polymeric fibers, destroying their cohesion.

Figure 4. Electron microscopy images of nanofibers of PVA + PBA (**A**); PVA + PBA + IONPs (**B**); PVA + Lys (**C**); PVA + Lys + PBA + IONP (**D**); PVA + Lys + fPBA (**E**); PVA + Lys + aPBA (**F**) at 1000×, 5000× and 20,000×.

2.3. Release and Enzyme Activity Assays

The influence of pH and temperature on the release of Lys immobilized on PVA + Lys + PBA + IONPs nanofiber is shown in Figure 5.

Figure 5 shows the absorbance decay at 450 nm corresponding to the activity of *M. lysodeikticus* in contact with the nanofibers for 60 min for the different conditions tested, based on the RSM model. In the case of temperature and pH conditions, the nanofibers showed a controlled release profile compared with the profile corresponding to the free enzyme. After 60 min, a cell lysis rate was equivalent to the positive control with free enzyme (about 80%) (Figure 5B).

Based on the optimized temperature and pH conditions generated from the Response Surface Methodology (RSM) model, it was found that lysozyme showed a high release rate from the nanofiber at pH 6.74 and temperature 45.5 °C (Figure 5C). The model demonstrated excellent suitability as a tool to implement the lysozyme-tailored nanofibers to deliver the enzyme in a potential anticancer application since the pH in the cancer cells is lower than normal cells, and they are sensitive to temperature.

Figure 6A shows the cumulative release profiles as a function of time. The release pattern does not seem to be influenced by the polymers used, showing similar profiles between PVA fibers, PVA/CS 90:10 and PVA/CS 70:30. On the other hand, fibers with PBA acid crosslinks show a controlled release profile over time, with a moderate release up to 30 min [about 10% (*w*/*w*)]. Beyond 30 min, a highlighted release of the enzyme was observed until 48 h or 24 h for PVA or PVA/CS nanofibers, respectively. The presence of magnetic particles does not appear to influence the Lys release profile. In contrast, nanofibers without the crosslinks release Lys slowly, not showing a controlled release profile over time.

To study the enzymatic activity during the prolonged release assay, after reading by spectrophotometry (UV 260 nm), the samples collected during the Lys release assay from PVA and PVA/CS nanofibers 90:10 and 70:30, a solution of *M. lysodeikticus* was added to all samples, for about 8 min, to estimate the enzymatic activity as a function of the microbial lysis rate (Figure 6B).

The lysis profiles of *M. lysodeikticus* in contact with the Lys released from the different nanofibers were tested over time to confirm the results observed in the release assay. Solutions in contact with fibers with PBA acid crosslinks showed microbial lysis activity from 30 min, 1 h, and 48 h for PVA/CS 70:30, PVA, and PVA/CS 90:10 nanofibers, respectively. Contrary to the non-crosslinked fibers that show Lys release and antimicrobial activity from the first minutes, with a release profile similar to the free enzyme (Figure 6), the presence of magnetic particles appears not to influence both the release and the antimicrobial enzymatic activity.

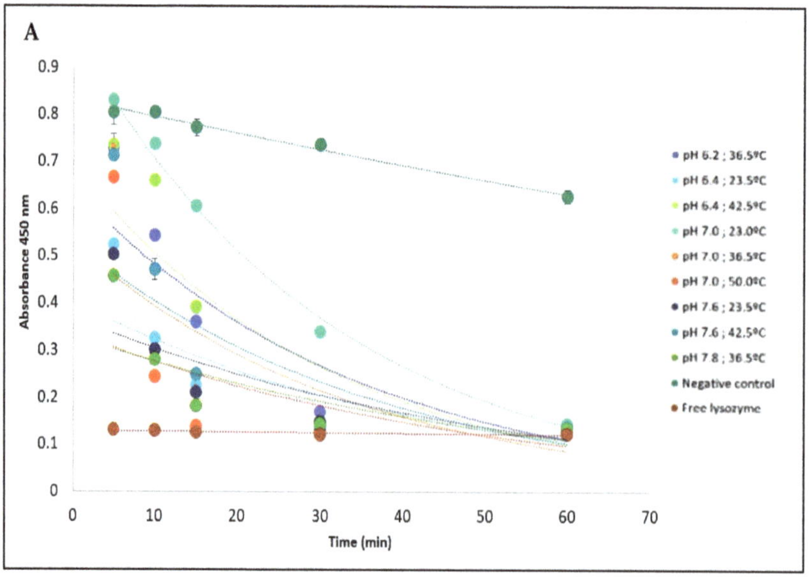

Figure 5. *Cont.*

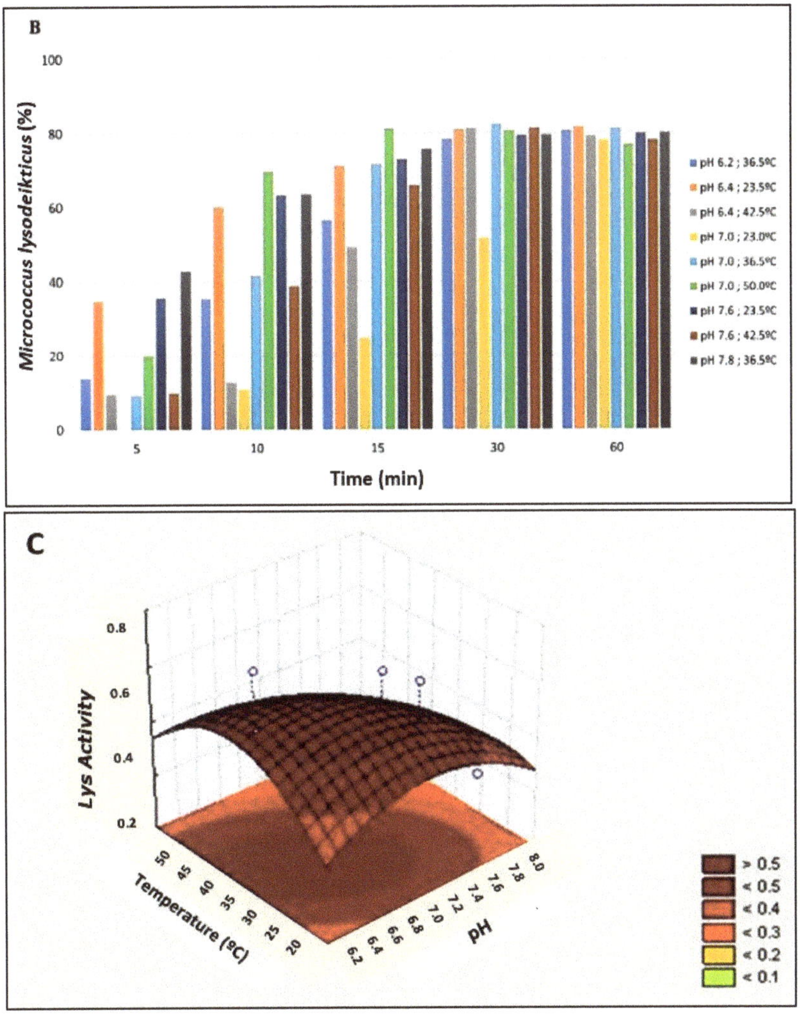

Figure 5. Lys release profile from PVA + Lys + PBA + IONPs fibers based on *Micrococcus lysodeikticus* (3 mg/mL) activity (absorbance at 450 nm) under different pH and temperature conditions (**A**); *Micrococcus lysodeikticus* lysis expressed as a percentage by the difference in absorbances at 450 nm after 60 min (**B**); Results of the CCD matrix to assess the pH and temperature at which lysozyme presented a higher release, based on the microbial reduction of *Micrococcus lysodeikticus* (**C**). SD (Standard Deviation) was ±0.005, and each point and column of the graphics (**A**,**B**) was carried out in triplicates.

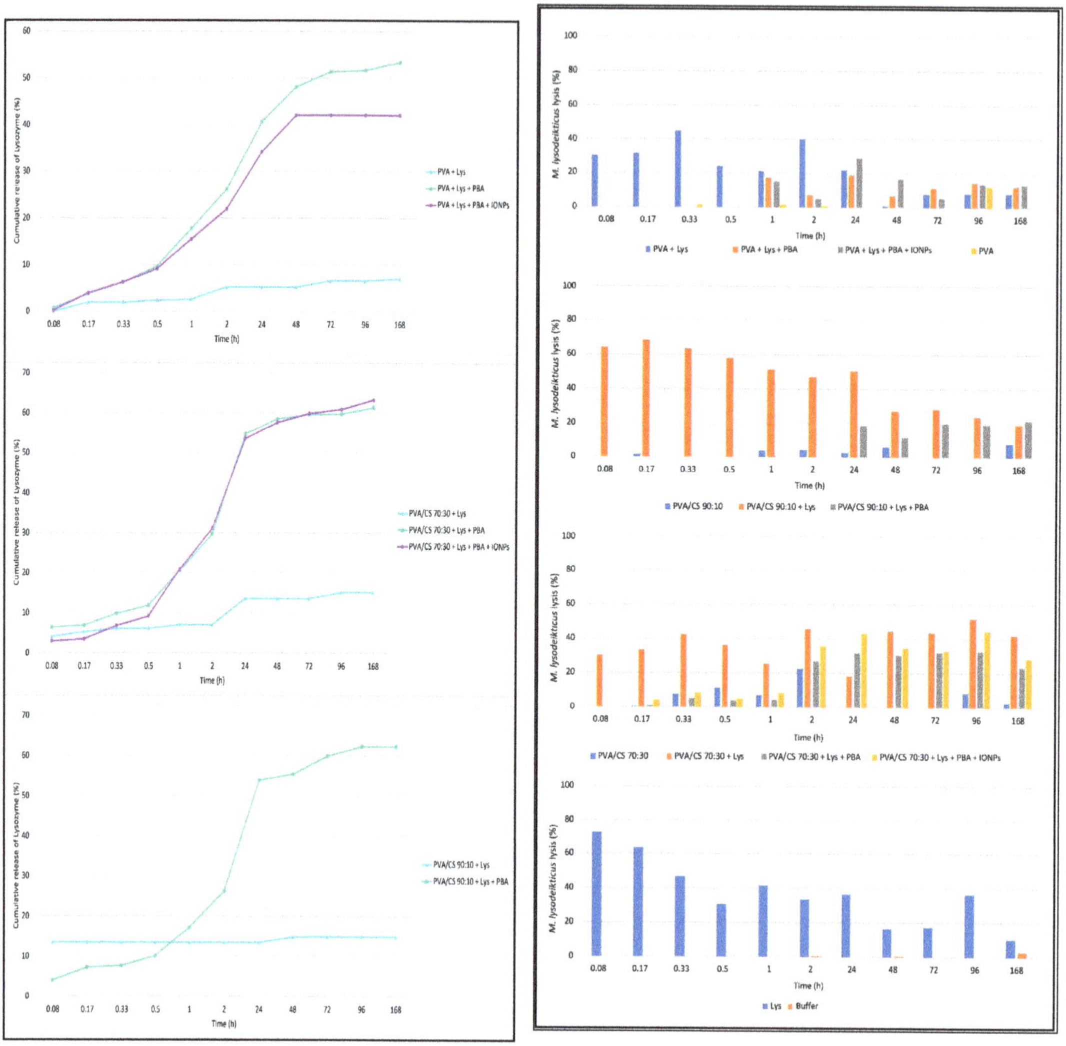

(A) **(B)**

Figure 6. (**A**) Cumulative release profiles of Lys from PVA (1), PVA/CS 70:30 (2), and PVA/CS 90:10 (3) nanofibers. These results are based on the Abs (Absorbance) at 260 nm and the average mass of the nanofibers used (5 mg). All samples were analyzed in triplicate; (**B**) Effect of the contact of nanofibers with the solvent (phosphate buffer) on the lysis of *Micrococcus lysodeikticus*. Results are expressed in percentages by the difference in absorbances at 450 nm. (1) PVA Nanofibers; (2) PVA/CS 90:10; (3) PVA/CS 70:30; (4) Positive (Lys) and Negative Controls in the buffer. SD (Standard Deviation) was ±0.01, and each point of the graphic was carried out in triplicates.

Interestingly, the fibers without Lys, namely the PVA/CS nanofibers in both proportions tested, showed some antimicrobial activity (on average 5% of the microbial lysis).

2.4. In Vitro Assays with Human Colon Adenocarcinoma (CaCo-2) Cell Line

The efficiency of the foreseen biosystems (PVA + LYS + PBA and PVA + LYS + PBA + ONPs) against Caco-2 cells, used as a model of colon cancer cells, was evaluated by placing them in contact for 10 days with cells. Caco-2 cell viability was evaluated using the MTT assay, and cells were visualized using SEM.

To assess the cytotoxicity of Lys and the studied boronic acids (PBA, fPBA, and aPBA), various concentrations of these components were tested in an in vitro assay with Caco-2 cells. The viability of Caco-2 cells after one week of incubation in the presence of these components and also when exposed to an aqueous mixture of Lys and boronic acids was evaluated according to the CCD matrix design.

Statistical significance compared with the control group (Caco-2 cells) was considered for * $p < 0.05$, ** $p < 0.001$ through the One ANOVA and post-Tukey tests, and the r^2 values represented the fit of the line to different experimental data. All boronic acids are cytotoxic to Caco-2 cells, mainly at concentrations above 0.313 mg/mL ($p < 0.001$). The fPBA acid showed the lowest rate of cell viability for all concentrations tested, showing about 70 ± 8.37% cell viability for the minimum concentration tested (0.005 mg/mL; $p < 0.005$) and an IC50 of 3.22 mg/mL ($r^2 = 0.9644$) for Caco-2 cells after 7 days of incubation. PBA acid showed a statistically significant difference in cell viability for concentrations less than 0.039 mg/mL and greater than 0.313 mg/mL, with an IC50 of 8.18 mg/mL ($r^2 = 0.8203$). In contrast, aPBA acid presented statistically significant values for concentration values greater than 0.078 mg/mL, with an IC50 of 7.10 mg/mL ($r^2 = 0.915$). They showed cytotoxicity values with statistical significance about the negative control for values greater than 1 mg/mL and less than 1.0 mg/mL, while for all concentrations of Lys tested, there was a reduction in cell viability with an IC_{50} of 13.27 mg/mL.

The cell viability of Caco-2 for the mixture of Lys with boronic acids at concentrations referring to the design of the RSM matrix showed similar results to those obtained with each boronic acid individually, while Lys combined with boronic acids showed a cumulative effect as a reduction in the rate of cell viability.

This study reports various types of nanofibers, including magnetized ones, and evaluates their cytotoxic effects on colon cancer cells. The impact of nanofibers produced with lysozyme and different gel polymers was evaluated on the cell viability of Caco-2 (Figure 7). MTT results indicated a reduction in cell viability for all nanofibers tested, regardless of the treatment or polymer used. The presence of the polymers studied showed cytotoxic activity in Caco-2 cells.

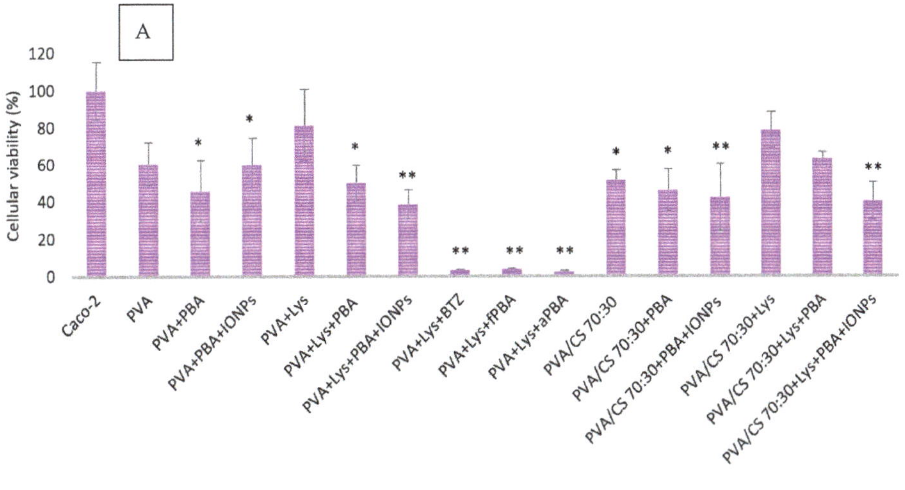

Figure 7. *Cont.*

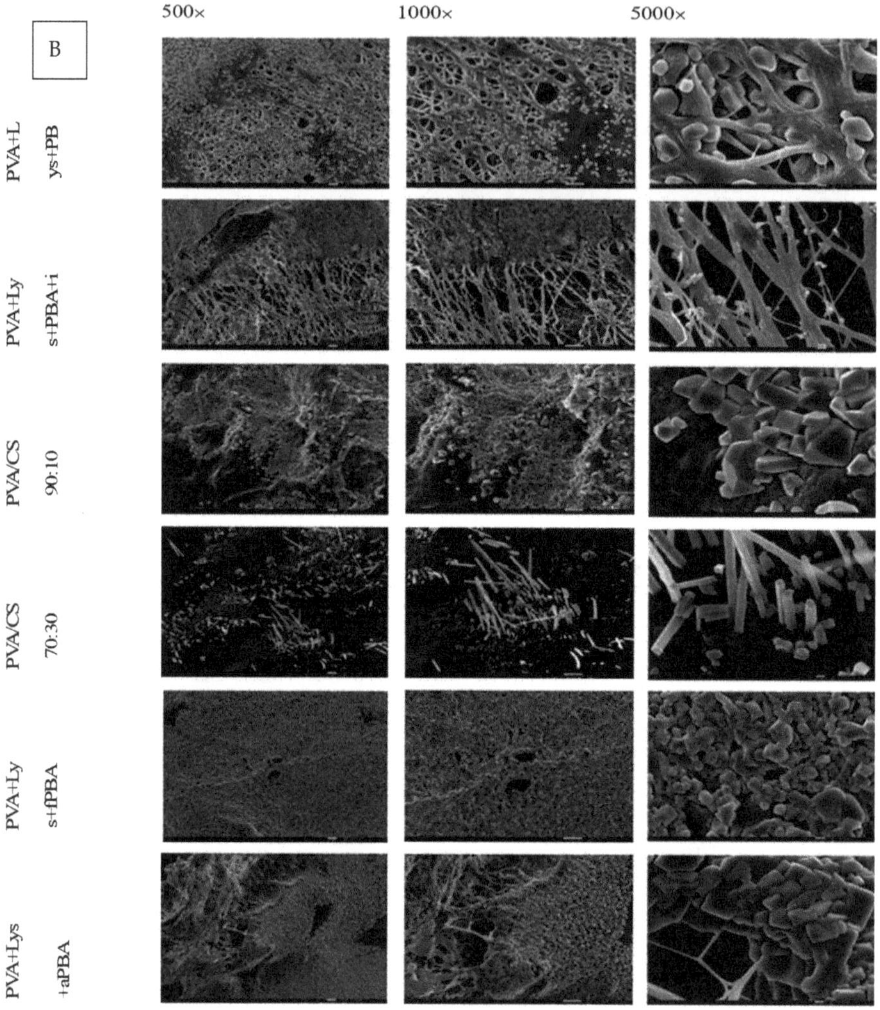

Figure 7. (**A**) Caco-2 cells viability after incubation with PVA and PVA/CS nanofibers by the MTT method (595 nm) at 37 °C for 7 days. Statistical significance compared with the control group (Caco-2). Statistical significance compared with the control group (Caco-2). * $p < 0.05$, ** $p < 0.001$ through One ANOVA and post Tukey tests. All data are expressed as mean ± standard deviation, $n = 3$; (**B**) Electron microscopy images of PVA + Lys + PBA nanofibers; from PVA + Lys + PBA + IONPs; PVA/CS 90:10; PVA/CS 70:30; PVA + Lys + fPBA and PVA + Lys + aPBA. After incubation with Caco-2 cells for 7 days. Magnifications of 500×, 1000× and 5000×.

Magnetized nanofibers, specifically PVA + Lys + PBA + IONPs and PVA/CS (70:30) + Lys + PBA + IONPs, exhibited high cytotoxicity compared with non-magnetized nanofibers. The cytotoxicity was observed to be significantly lower in non-magnetized nanofibers, such as PVA + Lys + PBA and PVA/CS (70:30) + Lys + PBA. Crosslinking treatment with different boronic acids resulted in a reduction in cell viability. A comparison of boronic acids (fPBA and aPBA) with positive control (crosslinking with BTZ acid) showed a similar effect on cell viability (<4%, $p < 0.001$).

Blending of polymers PVA and CS in nanofiber formation had a higher cytotoxic effect compared with PVA alone ($p < 0.05$). Figure 7B presents electron microscopy images of nanofibers used in in vitro assays with Caco-2 cells. Polyhedral structures were visible on all samples, with varying densities on different fibers. The structure of the PVA/CS 70:30 nanofibers showed "prismatic needles" that were distinct from other fibers, suggesting the possibility of Caco-2 cell adhesion. The structure of the polymeric nanofiber was noted to lose homogeneity and definition after incubation (Figure 7).

MTT results show that the nanofibers reduced the viability of Caco-2 cells. Fibers crosslinked with PBA and coated with IONPs have relatively high cytotoxicity comparable with bare PVA and PVA/CS fibers. This profile could favor the potential use of nanofibers as a co-adjuvant therapy for colon cancer.

The nanobiosystems of PVA and PVA/CS were successfully built using electrospinning. The nanofibers containing lysozyme encapsulated were produced, with increased stability by crosslinking with boronic acid derivates. Lysozyme was released gradually in different conditions of temperature and pH, namely in an acidic environment, simulated as a tumor microenvironment. The nanofibers with lysozyme encapsulated, crosslinked with fluorophenylboronic acid, and with IONPs adsorbed were able to reduce the viability of Caco-2 cells seeded on them.

Using a self-assembled nanostructured hen egg white lysozyme [35], it was able to induce 95% cell death in 24 h on MCF-7 breast cancer cells, mainly by inducing oxidative stress. The spherical nanosystem used by [35] consisted of partly folded monomeric lysozyme crosslinked with glutaraldehyde and functionalized with folic acid and was found to be stable at pH 7.4 and resistant to proteinase-K degradation. However, due to the preparation process, lysozyme was found to have lost part of its biological activity, meaning that cell death was not dependent on its enzymatic activity [35].

In vitro and in vivo studies confirmed the tumor-inhibitory activity of lysozyme. Examples of these human tumors include the uterus, colon, rectum, vulva, oral cavity, stomach, prostate, mammary carcinoma, lung carcinoma, small bowel reticulosarcoma, and multiple myeloma [36–39]. Different routes of lysozyme administration were tested, including mixing with tumor cells, peritumoral and intratumoral treatments, or indirectly through the systemic and oral routes [38,39]. The activity was dependent on the origin of the lysozyme, the type of tumor, and its degree of immunogenicity, with a potential effect on tumor lines that metastasize, suggesting that lysozyme influences the process of neoplastic dissemination [38].

One hypothesis of the antitumoral activity of lysozyme can be related to the bactericidal enzymatic action in the release of immunogenic substances, such as peptidoglycans, responsible for immunopotentiation and, consequently, antitumor activity [38–40].

Mahanta et al. [35] prepared, using the desolvation technique, a nanostructured self-organized lysozyme, which showed strong antiproliferative activity when tested in vitro against MCF-7 cells (breast cancer cell lines). When the antiproliferative activity of recombinant human lysozyme was tested on different stomach cancer cell lines (MGC803, MKN28, and MKN45), they showed positive results in inhibiting tumor evolution at concentrations of 100 and 1000 mg/L [39].

Wang et al. 2016 evaluated the effect of this enzyme on human lung carcinoma cells (A549 cell line) and observed that silencing lysozyme expression inhibits the invasiveness and migration of A549 lung carcinoma cells, suggesting that lysozyme is probably involved in the progression and metastasis of lung carcinoma, as a possible biomarker in the progression, prognosis and therapeutic effect of the disease.

3. Conclusions

Based on this study, some main conclusions can be addressed. Biodegradable nanofibers, with cytotoxic activity against Caco-2 cells, were produced adsorbed on magnetic nanoparticles, which make the system bio-responsive at higher temperatures, suggesting a potential in situ application in the treatment of tumors capable of metastasis, acting on the inhibition/reduction of the process.

4. Materials and Methods

4.1. Materials and Cells

The reagents used in this study were: poly(vinyl alcohol) (PVA, 99% degree of hydrolysis; average molecular weight 4441; Sigma Aldrich, Saint Louis, MO, USA); chitosan (CS, low molecular weight, Sigma Aldrich, Saint Louis, MO, USA); acetic acid (≥99%, VWR, Darmstadt, Germany), lysozyme obtained from hen's egg white (Lys, ~70,000 U/mg, Sigma Aldrich, Saint Louis, MO, USA), phosphate buffer [potassium dihydrogen phosphate (KH_2PO_4, Merk, Darmstadt, Germany) and di-potassium hydrogen phosphate (K_2HPO_4, VWR, Darmstadt, Germany)], phenylboronic acid (PBA, ≥97%, Sigma Aldrich, Saint Louis, MO, USA), 2-acetyl-phenylboronic acid (aPBA, 97%, Alfa Aesar, Kandel, Gemany), 2-formyl-phenylboronic acid (fPBA) and bortezomib (BTZ); RPMI-1640 medium (Sigma Aldrich, Saint Louis, MO, USA); fetal bovine serum (PBAS, VWR, Darmstadt, Germany); 100× antibiotic/antimycotic solution (Sigma Aldrich); 10× trypsin-EDTA (Sigma Aldrich, Saint Louis, MO, USA); Hank's Balanced Salt Solution (HBSS, Sigma Aldrich, Saint Louis, MO, USA); MTT [3-(4,5-dimethyliazol-2-yl)-2,5-diphenyltetrazolium bromide; Sigma Aldrich, Saint Louis, MO, USA) and dimethylsulfoxide (DMSO, Sigma Aldrich, Saint Louis, MO, USA).

Lyophilized *Micrococcus lysodeikticus* (*M. lysodeikticus*, ATCC No. 4698, Sigma-Aldrich, Saint Louis, MO, USA) was used.

A human colon adenocarcinoma cell line (Caco-2, ATCC No. HTB-37, Saint Louis, MO, USA) was used for cytotoxic evaluation.

All aqueous solutions were prepared with Milli-Q water.

4.2. Development of Electrospinning Polymer Solutions and Lys Loading

The gel polymers, PVA and CS, were used in the production of nanofibers by electrospinning.

According to the protocol described by Nunes et al. 2016 [27], aqueous solutions of PVA 10% (*m*/*v*) were prepared at about 90 °C for about 30 min under continuous magnetic stirring until complete dissolution of the polymer to a translucent gel and were subsequently stored at 4 °C.

For the CS gel solutions, a concentration of 2% (*m*/*v*) dissolved in 2% acetic acid was used, according to [28].

A solution-blending technique was employed in the production of nanofibers with both PVA-CS gel polymers. The aforementioned solutions were added in proportions 90/10 and 70/30 (*v*/*v*) with the aid of a 1 mL syringe and subsequently placed in magnetic stirring at 40–50 °C until complete dissolution of the polymers occurred.

Lysozyme immobilization was carried out at room temperature by adding the solid enzyme to the PVA or PVA-CS gel solutions at a concentration of 3% (*m*/*v*) under continuous magnetic stirring until complete dissolution (30–40 min).

4.3. Characterization of Gel Polymers Solutions

The electrical conductivity and surface tension of PVA and PVA-CS gel solutions (90/10 and 70/30) with and without lysozyme immobilization, used in the electrospinning assays, were evaluated.

4.3.1. Conductivity

For this procedure, it was necessary to prepare 10 mL of each gel solution, with subsequent dilutions with distilled water in different proportions (%): 100, 50, 25, 12.5, 6.25, 3.125, 1.56, 0.78; 0.39; 0.20 and 0.10.

The electrical conductivity and temperature of the solutions were measured using the multiparameter conductivity meter PC5 (ref. MPMT-005-001, XS Instruments). According to the device's instructions, the probe was introduced into the solution (10 mL), the solution was stabilized, and the conductivity value was recorded.

4.3.2. Surface Tension

Surface tensions were obtained using the Wilhelmy plate technique with a Krüss K12 tensiometer. The method consisted of lifting the container containing the solution to be analyzed until the liquid surface came into contact with the plate. The force applied to the plate is measured by a microbalance. The surface tension was calculated using Equation (1):

$$\gamma = F/(\rho \cos\theta) \quad (1)$$

where F is the force acting on the balance, ρ is the wetted length of the plate, and θ is the contact angle. The platinum plate was washed with water and acetone and flame-dried before each measurement.

4.4. Production of Nanofibers by Electrospinning

4.4.1. PVA Nanofibers Development

After the preparation of the polymeric solutions described in 2.2, nanofibers were produced in an in-house electrospinning system consisting of a 3.4 cm diameter plastic rotating cylindrical collector with a rotation controller motor; a continuous flow pump (New Era, NE-1000, Shropshire, UK) and a potential difference generator (73030DC, Genvolt, Shropshire, UK) with up to 30 kV.

All tests were performed under the following conditions: the plastic syringe (PIC, 1 mL) was filled with the solution to be tested, and a needle (Terumo, 23 G, 0.6 × 25 mm) was connected. They were fixed to the flow pump at a distance of 8 cm from the collector and, about 1 cm from the needle tip, the positive electrode of the potential difference generator was fixed. The applied electrical potential was 17.4 kV. The nanofibers were collected in the cylindrical collector previously lined with aluminum foil, with a rotation speed of 300 rpm. The flow rate varied between 50–70 μL/min, taking into account the most suitable conditions for each solution, namely in terms of viscosity (by extrapolation). The electrospinning process took place at a room temperature of 17–19 °C.

After the formation of the nanofibers, they were carefully removed with the aluminum foil base of the collector and placed in an oven at 50 °C for 15 min. Afterward, they were stored at room temperature. The encapsulation efficiency of lysozyme in the nanofibers was 95%.

4.4.2. Crosslinking and Magnetization

To improve the biomechanical properties of the fibers, they were crosslinked with boronic acids. This process includes the immersion of previously dried nanofibers in an aqueous boronic acid solution for ≈10 min. Subsequently, the nanofiber was removed and dried in an oven at 50 °C for 30–40 min. The boronic acids tested in this study are described in Supplementary Materials (Table S1).

4.4.3. Iron Oxide Nanoparticles (IONPs)

The effect of magnetization of the nanofibers after crosslinking with different boronic acids was evaluated. This process consisted of immersing the nanofibers in an aqueous suspension of 2 mg/mL of iron (II, III) oxide for 1 h with stirring (40 osc/min). Then, they were removed from the aqueous suspension, dried in an oven at 50 °C for 30–40 min, and stored at room temperature.

4.4.4. Characterization

The morphology of the nanofibers was evaluated using optical microscopy (OM) and electron microscopy (SEM). For SEM observation, 0.5 × 0.5 cm squares of nanofibers obtained in the different procedures were cut randomly and analyzed. The nanofibers were further characterized using a scanning electron microscopy-field emission gun (FEG-SEM) with a JEOL microscope, model JSM-7001F, operating at 5.0 kV. The surfaces were previously sputter-coated with a gold layer 20 nm thick to avoid charging effects during observation. Some fibers were cut to expose the interior.

4.5. Release and Enzyme Activity Assays

The Response Surface Methodology (RSM) is a statistical tool widely used in the modeling and optimization of various biotechnological processes based on factor analysis, which aims to obtain the maximum information about a process with a reduced number of trials [41]. One of the matrices used to determine quadratic models in RSM is the Central Composite Design (CCD). The design of a CCD matrix includes a 2 k factorial part, with two levels (minimum, −1 and maximum, 1) for each factor, these points being represented by (±1,. . ., ±1); an axial part (2 k points), at levels ± α (with α = $\sqrt{2}$) generally represented by (±α,. . ., 0) and (0,. . ., ±α); and central points, represented by (0,. . ., 0).

In order to study the influence of pH and temperature on the release of Lys immobilized on the nanofiber, a CCD matrix was designed with 2 factorial levels and triplicate of the central point, making up 11 tests: 22 factorial points; 3 repetitions of the central point and 4 axial points, positioned at a distance α from the central point, with α = 1.147. The pH and temperature values were defined based on previous studies. The range tested for pH was 6.15 (−α) to 7.85 (+α), and temperature of 23.5 °C (−α) and 50 °C (+α) (Supplementary Materials, Table S2).

The release of Lys from the nanofibers was evaluated by its activity on a culture of *M. lysodeikticus*. In each well of a 24-microplate (Orange Scientific, Braine-l'Alleud, Belgium), the nanofiber of PVA + Lys + PBA + IONPs was placed and submerged in 2 mL of a 2.4 mg/mL solution of *M. lysodeikticus* in phosphate buffer at different pH values and temperatures according to the experimental design described in Supplementary Materials (Table S2), for 60 min, at 200 rpm. At 5, 10, 15, 30, and 60 min, a 100 mL sample was removed. All samples were read in the spectrophotometer at 450 nm for 7 min in order to evaluate the lysis of *M. lysodeikticus* and consequently deduce the enzymatic activity. As a positive control, a free Lys solution at a concentration of 0.0125 mg/mL was used instead of the nanofiber, and the negative control was only phosphate buffer or PVA + PBA + IOPNs (without) nanofiber in contact with the *M. lysodeikticus* solution.

In order to obtain a release profile over time (168 h) in the different types of nanofibers produced, and under physiological conditions of pH and temperature, 7.4 and 37 °C, respectively. The following test was carried out: 0.5 × 0.5 cm of the nanofiber was cut out and placed in a 1.5 mL Eppendorf with 500 mL of phosphate buffer. Subsequently, it was placed in an oven at 37 °C, and 100 mL samples were collected at times 5, 10, 20, 30, 60, 120 min and 24, 48, 72, 96, and 168 h. The volume removed from each sample was replaced with phosphate buffer at 37 °C, with the final contact volume of the fiber being constant (500 mL). Each sample was collected on a 96-well microplate on ice to inactivate enzyme activity, and spectrophotometry at wavelengths of 260 and 280 nm was analyzed to estimate the amount of Lys over time. After UV reading, the physiological function of the enzyme was evaluated, and 50 mL of *M. lysodeikticus* (7.2 mg/mL) in phosphate buffer pH 7.4 was added to each sample, followed by a kinetic study of the enzyme activity in a microplate reading at 480 nm for 75 cycles (±24 min) or 100 cycles (±32 min), every ±19 s. For each type of fiber, triplicates were performed. As a negative control, a sample of phosphate buffer pH 7.4 was used. The positive control for kinetic evaluation was a solution of free Lys at a concentration of 3 mg/mL in phosphate buffer pH 7.4.

4.6. In Vitro Assay with Human Colon Adenocarcinoma (CaCo-2) Cell Line

4.6.1. Cell Culture

According to the protocol described by Frade et al. [29], the Caco-2 (human colorectal adenocarcinoma) cell line from ATCC (American Type Culture Collection) was cultured in RPMI-1640 medium supplemented with L-glutamine, 10% FBS (Fetal Bovine Serum) and 1% 100× antibiotic/antimycotic solution in 75 cm^2 cell culture flasks. Cultures were maintained in a humidified incubator at 37 °C and 5% CO_2 until they reached 80–90% confluence (8–10 days), approximately 10^6 cells/mL.

4.6.2. Influence of Lys, Boronic Acids, and IONPs on the Caco-2 Cell Viability

To evaluate the influence of Lys, the previously described boronic acids (PBA, fPBA, and aPBA) and the IONPs were tested on the cell viability of Caco-2. Two assays were performed: each one was tested individually in solutions of serial concentrations of Lys (0.00049–0.5 mg/mL), [IONPs] (0.00098–2 mg/mL), boronic acids (0.0049–10 mg/mL) (Supplementary Material, Table S2) and, a CCD array to assess the effect of both Lys and boronic acids both on the cell viability of Caco-2. Details about solutions of Lys, boronic acids, and IONPs at different concentrations, in distilled water are shown in Supplementary Materials (Table S3).

The CDD matrix was designed to study the combined effect of two variables: Lys and boronic acids (PBA, aPBA, fPBA, and BTZ), with a total of 11 trials, 22 trials for factorial points, triplicates of the central point, and 4 axial points, and α = 1.147 from the central point shown in Supplementary Materials (Table S3). The 11 trials were repeated 3 times for each combination of variables in order to assess the repeatability of results.

After reaching confluence, Caco-2 cells were trypsinized with 1% trypsin-EDTA (10 reagents in HBSS, diluted 1:5 in RPMI-1640 cell medium, previously supplemented with FBS and antibiotic/antimycotic solution, and plated 96 sterile wells microplates (Nunclon Surface, NuncTM), 100 mL per well. Cells were incubated with 100 mL of the compounds to be analyzed individually (Lys, boronic acids, and IONPs) in triplicate at the concentrations described previously. For the RSM assay, cells were incubated with 50 mL of each dependent variable (Lys and boronic acids) at the concentrations designated in the CCD matrix design (Supplementary Material, Table S4). The plates were incubated in a humid atmosphere at 37 °C and 5% CO_2 for 5 days, and the cell culture was monitored using light microscopy. In both assays, negative controls, in triplicates, of Caco-2 cells with RPMI-1640 medium alone (control cells) were used.

Cell viability was evaluated through the MTT (3-(4,5-dimethylthiazolyl-2)-2, 5-diphenyltetrazolium bromide) assay, a colorimetric test, which allowed measuring the number of viable cells after their incubation. First, the medium was carefully removed from each well, 200 mL of 1 mg/mL MTT solution in HBSS was added, and the plates were placed to incubate under the same conditions for 3–4 h. The medium was removed, and cells were washed with HBSS (100 mL per well). Finally, 100 mL of DMSO was added, which dissolved the purple formazan crystals. The plates were read on a spectrophotometer at 595 nm, and cell viability was calculated using the ratio (Equation (2)):

$$\%\ \text{cell viability} = (\text{Abs 595 nm treated cells})/(\text{Abs 595 nm control cells}) \times 100 \quad (2)$$

4.6.3. Effect of PVA and PVA/CS Nanofibers on Caco-2 Cell Viability

In addition to the in vitro assays with Lys, boronic acids and IONPs in their free form (in solution) were also tested when immobilized on PVA or PVA/CS nanofibers. In this assay, nanofibers, previously produced and treated as described previously, were cut into approximately 0.5 × 0.5 cm squares and placed in 24-well cell culture plates (Orange Scientific). Subsequently, the fibers already on the plates were sterilized by UV for 20 min. After the confluence of the Caco-2 cell culture, they were trypsinized and diluted 1:3 in RPMI-1640 cell medium, previously supplemented with FBS and antibiotic/antimycotic solution, and transferred (500 mL) directly on top nanofibers, and the cultures were incubated in a humid atmosphere, 37 °C, and 5% CO_2 for 7 days. During the incubation time, the culture was monitored. On day 7 of the incubation, the fibers were removed to a new plate, the culture medium was removed, and the cell viability was evaluated using the MTT assay in the plates with only cells. In this process, 500 mL of MTT and 200 mL of DMSO were used. The plates that only contained the fibers removed from the cell incubation were later observed using electron microscopy to assess whether there was adhesion and proliferation of cells in the fibers.

4.7. Statistical Analysis

All data referring to in vitro assays were repeated at least 3 times. The representativeness of these data was presented by the mean ± standard deviation. Statistical analyses were performed using IBM SPSS Statistics, version 24.0 55, considering a significance level, α, of 0.05.

Supplementary Materials: The following supporting information can be downloaded at https://www.mdpi.com/article/10.3390/gels9120968/s1, Figure S1: Nanofibers. (A) PVA; (B) PVA/CS 70:30 + Lys; (C) PVA + Lys + PBA + IONPs; (D) PVA/CS 90:10; Figure S2: Optical microscopy (400× magnification) and electron microscopy (500× magnification) images of the nanofibers of: PVA (A); PVA + PBA (B); PVA + PBA + Lys (C); PVA + Lys (D); PVA + Lys + PBA (E); PVA + Lys + PBA + IONPs (F); PVA + Lys + aPBA (G); PVA + Lys + fPBA (H); PVA + Lys + BTZ (I); PVA/CS 90:10 (J); PVA/CS 90:10 + Lys (K); PVA/CS 90:10 + Lys + PBA (L); PVA/CS 70:30 (M); PVA/CS 70:30 + Lys (N); Table S1: Boronic acids used in the crosslinking process; Table S2: CCD matrix delineation of the effect of pH and temperature on releasing Lys from the nanofiber; Table S3: CCD matrix design of the effect of blending Lys and boronic acids (PBA, aPBA, fPBA and BTZ) on the cell viability of Caco-2; Table S4: showed the different concentrations of lysozyme (Lys), boronic acids and IONPs tested on the Caco-2 cell viability.

Author Contributions: All authors have substantially contributed to the conceptualization, methodology, formal analysis, investigation, resources, data curation, writing—original draft preparation, writing—review, and editing. M.H.L.R., M.G. and F.F.F.G. conceived and designed research. M.G. and F.F.F.G. conducted experiments. M.E.R. contributed new analytic tools. M.G., C.F., M.E.R. and M.H.L.R. analyzed data. M.H.L.R. and M.G. wrote the manuscript. All authors have read and agreed to the published version of the manuscript.

Funding: This research was funded by FCT—FOUNDATION FOR SCIENCE AND TECHNOLOGY, I.P., through National Funds, under the projects UID/DTP/04138/2021 and UID/BIO/04565/2020. The authors also acknowledge funding from IT (UIDB/50008/2020) and project BioMaterARISES (10.54499/EXPL/CTM-CTM/0995/2021). The funders had no role in the design of the study, in the collection, analyses, or interpretation of data, in the writing of the manuscript, or in the decision to publish the results.

Institutional Review Board Statement: Not applicable.

Informed Consent Statement: This article does not contain any studies with human participants or animals performed by any of the authors.

Data Availability Statement: The data presented in this study are openly available in article.

Acknowledgments: Pedro Góis supplied some of the boronic acids, which is highly acknowledged.

Conflicts of Interest: The authors declare no conflict of interest.

References

1. Park, J.-M.; Kim, M.; Park, H.-S.; Jang, A.; Min, J.; Kim, Y.-H. Immobilization of lysozyme-CLEA onto electrospun chitosan nanofiber for effective antibacterial applications. *Int. J. Biol. Macromol.* **2013**, *54*, 37–43. [CrossRef] [PubMed]
2. Yang, J.M.; Yang, J.H.; Tsou, S.C.; Ding, C.H.; Hsu, C.C.; Yang, K.C.; Yang, C.C.; Chen, K.S.; Chen, S.W.; Wang, J.S. Cell proliferation on PVA/sodium alginate and PVA/poly(γ-glutamic acid) electrospun fiber. *Mater. Sci. Eng. C* **2016**, *66*, 170–177. [CrossRef] [PubMed]
3. Thambiliyagodage, C.; Jayanetti, M.; Mendis, A.; Ekanayake, G.; Liyanaarachchi, H.; Vigneswaran, S. Recent Advances in Chitosan-Based Applications—A Review. *Materials* **2023**, *16*, 2073. [CrossRef] [PubMed]
4. Gonçalves, R.P.; Ferreira, W.H.; Gouvêa, R.F.; Andrade, C.T. Effect of Chitosan on the Properties of Electrospun Fibers From Mixed Poly(Vinyl Alcohol)/Chitosan Solutions. *Mater. Res.* **2017**, *20*, 984–993. [CrossRef]
5. Pele, K.G.; Amaveda, H.; Mora, M.; Marcuello, C.; Lostao, A.; Alamán-Díez, P.; Pérez-Huertas, S.; Ángeles Pérez, M.; García-Aznar, J.M.; García-Gareta, E. Hydrocolloids of Egg White and Gelatin as a Platform for Hydrogel-Based Tissue Engineering. *Gels* **2023**, *9*, 505. [CrossRef] [PubMed]
6. Sava, G.; Benetti, A.; Ceschia, V.; Pacor, S. Lysozyme and cancer: Role of exogenous lysozyme as anticancer agent (review). *Anticancer. Res.* **1989**, *9*, 583–591. [PubMed]

7. Ulery, B.D.; Nair, L.S.; Laurencin, C.T. Biomedical Applications of Biodegradable Polymers. *J. Polym. Sci. B Polym. Phys.* **2011**, *49*, 832–864. [CrossRef] [PubMed]
8. Gao, Y.; Truong, B.Y.; Zhu, Y.; Kyratzis, L.I. Electrospun antibacterial nanofibers: Production, activity, and in vivo applications. *J. Appl. Polym. Sci.* **2014**, *131*, 9041–9053. [CrossRef]
9. Jin, L.; Bai, R. Mechanisms of Lead Adsorption on Chitosan/PVA Hydrogel Beads. *Langmuir* **2002**, *18*, 9765–9770. [CrossRef]
10. Mouro, C.; Gomes, A.P.; Gouveia, I.C. Emulsion Electrospinning of PLLA/PVA/Chitosan with *Hypericum perforatum* L. as an Antibacterial Nanofibrous Wound Dressing. *Gels* **2023**, *9*, 353. [CrossRef]
11. Xue, J.; Xie, J.; Liu, W.; Xia, Y. Electrospun Nanofibers: New Concepts, Materials, and Applications. *Acc. Chem. Res.* **2017**, *50*, 1976–1987. [CrossRef] [PubMed]
12. Thakkar, S.; Misra, M. Electrospun polymeric nanofibers: New horizons in drug delivery. *Eur. J. Pharm. Sci.* **2017**, *107*, 148–167. [CrossRef] [PubMed]
13. Ragland, S.A.; Criss, A.K. From bacterial killing to immune modulation: Recent insights into the functions of lysozyme. *PLoS Pathog.* **2017**, *13*, e1006512. [CrossRef] [PubMed]
14. Cegielska-Radziejewska, R.; Leśnierowski, G.; Kijowski, J.; Szablewski, T.; Zabielski, J. Effects of treatment with lysozyme and its polymers on the microflora and sensory properties of chilled chicken breast muscles. *Bull. Vet. Inst. Pulawy* **2009**, *53*, 455–461.
15. Lollike, K.; Kjeldsen, L.; Sengeløv, H.; Borregaard, N. Purification of lysozyme from human neutrophils, and development of an ELISA for quantification in cells and plasma. *Leukemia* **1995**, *9*, 206–209. [PubMed]
16. Ibrahim, H.; Higashiguchi, S.; Juneja, L.; Kim, M.; Yamamoto, T. A Structural Phase of Heat-Denatured Lysozyme with Novel Antimicrobial Action. *Agric. Food Chem.* **1996**, *44*, 1416–1423. [CrossRef]
17. Xue, F.; Chena, Q.; Lia, Y.; Liua, E.; Lia, D. Immobilized lysozyme onto 1,2,3,4-butanetetracarboxylic (BTCA)-modified magnetic cellulose microsphere for improving bio-catalytic stability and activities. *Enzym. Microb. Technol.* **2019**, *131*, 109425. [CrossRef]
18. Das, S.; Banerjee, S.; Gupta, J.D. Experimental Evaluation of Preventive and Therapeutic Potentials of Lysozyme. *Chemotherapy* **1992**, *38*, 350–357. [CrossRef]
19. Lee, H.-J.; Lee, S.J.; Uthaman, S.; Thomas, R.G.; Hyun, H.; Jeong, Y.Y.; Cho, C.-S.; Park, I.-K. Biomedical Applications of Magnetically Functionalized Organic/Inorganic Hybrid Nanofibers. *Int. J. Mol. Sci.* **2015**, *16*, 13661–13677. [CrossRef]
20. Alveroglu, E.; İlker, N.; Shah, M.T.; Rajar, K.; Gokceoren, A.T.; Koc, K. Effects of gel morphology on the lysozyme adsorption and desorption kinetics of temperature sensitive magnetic gel composites. *Colloids Surf. B Biointerfaces* **2019**, *181*, 981–988. [CrossRef]
21. Chen, Q.; Liu, D.; Wu, C.; Yao, K.; Li, Z.; Shi, N.; Wen, F.; Gates, I.D. Co-immobilization of cellulase and lysozyme on amino-functionalized magnetic nanoparticles: An activity-tunable biocatalyst for extraction of lipids from microalgae. *Bioresour. Technol.* **2018**, *263*, 317–324. [CrossRef] [PubMed]
22. Yu, Y.; Zhu, Z.-Y.; Ye, H.; Chen, G.-T.; Wu, M.-B.; Kong, X.; Militky, J.; Yao, J. Lysozyme@Fe_3O_4 composite adsorbents with enhanced adsorption rates and uranium/vanadium selectivity by photothermal property. *Compos. Comm.* **2022**, *36*, 101364. [CrossRef]
23. Erol, K.; Tatar, D.; Veyisoğlu, A.; Tokatlı, A. Antimicrobial magnetic poly(GMA) microparticles: Synthesis, characterization and lysozyme immobilization. *J. Polymer. Eng.* **2021**, *41*, 144–154. [CrossRef]
24. Li, Z.; Wang, C. Effects of Working Parameters on Electrospinning. In *One-Dimensional Nanostructures*; Springer Briefs in Materials; Springer: Berlin/Heidelberg, Germany, 2013; pp. 1323–1337.
25. Anastas, P.T.; Rodriguez, A.; de Winter, T.M.; Coish, P.; Zimmerman, J.B. A review of immobilization techniques to improve the stability and bioactivity of lysozyme. *Green Chem. Lett. Rev.* **2021**, *14*, 302–338. [CrossRef]
26. Winkler, R.; Ciria, M.; Ahmad, M.; Plank, H.; Marcuello, C. A Review of the Current State of Magnetic Force Microscopy to Unravel the Magnetic Properties of Nanomaterials Applied in Biological Systems and Future Direction for Quantum Technologies. *Nanomaterials* **2023**, *13*, 2585. [CrossRef] [PubMed]
27. Nunes, M.A.P.; Gois, P.M.P.; Rosa, M.E.; Martins, S.; Fernandes, P.C.B.; Ribeiro, M.H.L. Boronic acids as efficient cross linkers for PVA: Synthesis and application of tunable hollow microspheres in biocatalysis. *Tetrahedron* **2016**, *72*, 7293–7305. [CrossRef]
28. Alhosseini, S.N.; Moztarzadeh, F.; Mozafari, M.; Asgari, S.; Dodel, M.; Samadikuchaksaraei, F.; Kargozar, S.; Jalali, N. Synthesis and characterization of electrospun polyvinyl alcohol nanofibrous scaffolds modified by blending with chitosan for neural tissue engineering. *Int. J. Nanomed.* **2012**, *7*, 25–34.
29. Frade, R.F.M.; Simeonov, S.; Rosatella, A.A.; Siopa, F.; Afonso, C.A.M. Toxicological evaluation of magnetic ionic liquids in human cell lines. *Chemosphere* **2013**, *92*, 100–105. [CrossRef]
30. Osserman, E.F.; Klockars, M.; Halper, J.; Fischel, R.E. Effects of lysozyme on normal and transformed mammalian cells. *Nature* **1973**, *243*, 331–335. [CrossRef]
31. Li, W.; Wang, F.; Feng, S.; Wang, J.; Sun, Z.; Li, B.; Li, Y.; Yang, J.; Elzatahry, A.A.; Xia, Y.; et al. Sol–Gel Design Strategy for Ultradispersed TiO_2 Nanoparticles on Graphene for High-Performance Lithium Ion Batteries. *J. Am. Chem. Soc.* **2013**, *135*, 18300–18303. [CrossRef]
32. Saghafi, Y.; Baharifar, H.; Najmoddin, N.; Asefnejad, A.; Maleki, H.; Sajjadi-Jazi, S.M.; Bonkdar, A.; Shams, F.; Khoshnevisan, K. Bromelain- and Silver Nanoparticle- Loaded Polycaprolactone/Chitosan Nanofibrous Dressings for Skin Wound Healing. *Gels* **2023**, *9*, 672. [CrossRef] [PubMed]
33. Abrahama, A.; Solomanb, P.A.; Rejin, V.O. Poly (Vinyl Alcohol)-Chitosan Blends: Preparation, Mechanical and Physical Properties. *Procedia Technol.* **2016**, *24*, 741–748. [CrossRef]
34. Cho, Y.-W.; Han, S.-S.; Ko, S.-W. PVA containing chito-oligosaccharide side chain. *Polymer* **2000**, *41*, 2033–2039. [CrossRef]

35. Mahanta, S.; Paul, S.; Srivastava, A.; Pastor, A.; Kundu, B.; Chaudhuri, T.K. Stable self-assembled nanostructured hen egg white lysozyme exhibits strong anti-proliferative activity against breast cancer cells. *Colloids Surf. B Biointerfaces* **2015**, *130*, 237–245. [CrossRef] [PubMed]
36. Vizoso, F.; Plaza, E.; Vázquez, J.; Serra, C.; Lamelas, M.L.; González, L.O.; Merino, A.M.; Méndez, J. Lysozyme expression by breast carcinomas, correlation with clinicopathologic parameters, and prognostic significance. *Ann. Surg. Oncol.* **2001**, *8*, 667–674. [CrossRef] [PubMed]
37. Serra, C.; Vizoso, F.; Alonso, L.; Rodríguez, J.C.; González, L.O.; Fernández, M.; Lamelas, M.L.; Sánchez, L.M.; García-Muñiz, J.L.; Baltasar, A.; et al. Expression and prognostic significance of lysozyme in male breast cancer. *Breast Cancer Res.* **2002**, *4*, R16. [CrossRef] [PubMed]
38. Guo, T.; Zhao, X.; Xie, X.; Chen, Z.H.; Zhou, C.S.; Wei, L.; Zhang, H. The Anti-proliferative Effects of Recombinant Human Lysozyme on Human Gastric Cancer Cells. *J. Int. Med. Res.* **2007**, *35*, 353–360. [CrossRef]
39. Wang, H.; Guo, J.; Lu, W.; Feng, Z. Effect of Lysozyme on Invasion and Migration of Human Lung Carcinoma A549 Cells. *Int. J. Clin. Exp. Pathol.* **2016**, *9*, 8239–8246.
40. Marasovic, M.; Ivankovic, S.; Stojkovic, R.; Djermic, D.; Galic, B.; Milos, M. In vitro and in vivo antitumour effects of phenylboronic acid against mouse mammary adenocarcinoma 4T1 and squamous carcinoma SCCVII cells. *J. Enzym. Inhib. Med. Chem.* **2017**, *32*, 1299–1304. [CrossRef]
41. Montgomery, D.C. *Design and Analysis of Experiments*, 8th ed.; John Wiley & Sons, Inc.: Hoboken, NJ, USA, 2013; ISBN 978-1118-14692-7.

gels

MDPI

Review

Injectable Poloxamer Hydrogels for Local Cancer Therapy

Ana Camila Marques [1,2,*], Paulo Cardoso Costa [1,2], Sérgia Velho [3,4] and Maria Helena Amaral [1,2,*]

1 UCIBIO—Applied Molecular Biosciences Unit, MEDTECH, Laboratory of Pharmaceutical Technology, Department of Drug Sciences, Faculty of Pharmacy, University of Porto, R. Jorge Viterbo Ferreira 228, 4050-313 Porto, Portugal
2 Associate Laboratory i4HB—Institute for Health and Bioeconomy, Faculty of Pharmacy, University of Porto, R. Jorge Viterbo Ferreira 228, 4050-313 Porto, Portugal
3 i3S—Instituto de Investigação e Inovação em Saúde, University of Porto, R. Alfredo Allen 208, 4200-135 Porto, Portugal
4 IPATIMUP—Institute of Molecular Pathology and Immunology of the University of Porto, R. Júlio Amaral de Carvalho 45, 4200-135 Porto, Portugal
* Correspondence: amarques@ff.up.pt (A.C.M.); hamaral@ff.up.pt (M.H.A.)

Abstract: The widespread push to invest in local cancer therapies comes from the need to overcome the limitations of systemic treatment options. In contrast to intravenous administration, local treatments using intratumoral or peritumoral injections are independent of tumor vasculature and allow high concentrations of therapeutic agents to reach the tumor site with minimal systemic toxicity. Injectable biodegradable hydrogels offer a clear advantage over other delivery systems because the former requires no surgical procedures and promotes drug retention at the tumor site. More precisely, in situ gelling systems based on poloxamers have garnered considerable attention due to their thermoresponsive behavior, biocompatibility, ease of preparation, and possible incorporation of different anticancer agents. Therefore, this review focuses on the use of injectable thermoresponsive hydrogels based on poloxamers and their physicochemical and biological characterization. It also includes a summary of these hydrogel applications in local cancer therapies using chemotherapy, phototherapy, immunotherapy, and gene therapy.

Keywords: cancer therapy; injectable hydrogels; poloxamer; intratumoral administration

Citation: Marques, A.C.; Costa, P.C.; Velho, S.; Amaral, M.H. Injectable Poloxamer Hydrogels for Local Cancer Therapy. *Gels* **2023**, *9*, 593. https://doi.org/10.3390/gels9070593

Academic Editor: Christian Demitri

Received: 29 June 2023
Revised: 17 July 2023
Accepted: 21 July 2023
Published: 24 July 2023

1. Introduction

Local cancer therapy holds great potential to address the shortcomings of systemic treatment options, namely the lack of specificity for the target, low therapeutic efficiency, and drug resistance.

Different from intravenous (IV) administration, local treatments using intratumoral (IT) or peritumoral (PT) injections allow high concentrations of therapeutic agents to reach the tumor site, bypassing the bloodstream and non-specific interactions with healthy tissues [1]. Moreover, because local therapies are independent of tumor vasculature, delivery is not restricted to tumor regions with better perfusion. Besides increasing the stability of anticancer agents, local administration also allows for the use of novel combinations of co-solvents and polymers for solubilization, encapsulation, and incorporation of water-insoluble drugs.

In contrast to delivery systems based on implants (wafers, rods, and films) and particles, injectable biodegradable hydrogels require non-surgical procedures and promote retention of free or encapsulated drugs at the tumor site [2]. Indeed, treatment with injectable cisplatin (CDDP)/epinephrine gel has been proven practicable by direct injection into superficially accessible tumors or endoscopically for esophageal cancer [3]. Regarding the distribution dynamics, nanoparticles embedded in hydrogels were observed to cover larger areas of the tumor than free nanoparticles upon IT administration [4].

Injectable gels include pre-formed gels with shear-thinning and self-healing properties [5] and in situ-forming gels [6]. After injection as free-flowing polymer solutions, in situ gelling systems transform into gels at the injection site, acting as drug depots for sustained drug release. Generally, the mechanism of depot formation is an in situ phase transition triggered by external stimuli such as changes in temperature.

In this review, we emphasize the use of poloxamers in the development of injectable thermoresponsive hydrogels and their characterization. The extensive application of injectable poloxamer hydrogels in local cancer therapy, including chemotherapy, phototherapy, immunotherapy, and gene therapy, is also summarized.

2. Poloxamer-Based Thermoresponsive Hydrogels

Hydrogels are thermoresponsive if they respond to changes in temperature. A case in point is thermoresponsive polymer solutions exhibiting a sol–gel phase transition upon the shift from room to body temperature (37 °C). The temperature at which the transition from one phase to two phases (polymer-enriched and aqueous phases) takes place is termed the "critical solution temperature" of the polymer [7].

Not many thermoresponsive polymers show an upper critical solution temperature (UCST) in aqueous media under relevant conditions. Conversely, most thermoresponsive polymers having a phase transition temperature near 37 °C present a lower critical solution temperature (LCST) in water. When raising the temperature above the LCST, these polymers become insoluble and separate from the solution, yielding a hydrogel. During this process, polymer–solvent interactions based on hydrogen bonds are weakened, which leads to partial dehydration and aggregation of polymer chains. The progressive exclusion of water molecules with heating will expose the hydrophobic polymer domains and thus facilitate the establishment of polymer–polymer hydrophobic interactions (hydrophobic effect). Gelation is reversible at temperatures below the LCST because polymers become miscible with water again [8,9].

The basic requirements of in situ gelling systems for local administration are injectability and gelation under physiological conditions. Therefore, in addition to allowing the incorporation of therapeutic agents, thermoresponsive polymer solutions should have low viscosity to flow easily during administration and rapidly form a gel once injected [1]. Although other in situ-forming thermoresponsive hydrogels based on chitosan [10,11], polyacrylamides [12], and polyesters [13,14] have been developed for this purpose, poloxamers have gained significant attention from researchers over the past two decades.

This family of poly(ethylene oxide)-poly(propylene oxide)-poly(ethylene oxide) (PEO PPO-PEO) triblock copolymers (Figure 1A) is synthesized by sequential ring-opening polymerization of propylene oxide and ethylene oxide monomers, using an alkaline catalyst (i.e., potassium or sodium hydroxide) [15]. A series of poloxamers diverging in molecular weight and PPO/PEO ratio is available with different trade names like Pluronic, Lutrol, or Synperonic [16].

The individual PEO-PPO-PEO block copolymers (unimers) self-assemble into micelles above a "critical micelle concentration" because of their amphiphilic character in aqueous solutions. Structurally, the micellar core comprising hydrophobic PPO blocks is surrounded by a hydrophilic PEO shell that forms hydrogen bonds with the water molecules. Due to the nanosized and core-shell structure of poloxamer micelles, it is feasible to entrap hydrophobic drugs in the core and carry them into an aqueous environment [15]. Increasing temperature to the "critical micellization temperature" also induces micelle formation by favoring the dehydration of PPO groups and their interaction via van der Waals forces [17]. At a fixed PPO/PEO ratio, the longer the PPO block, the lower the temperature and the poloxamer concentration needed to form micelles in water [18]. The effect of temperature on micellization is reversible because cooling causes an equilibrium shift from micelles back to unimers. Upon further heating above the LCST, micelles aggregate and entangle to become a gel (Figure 1B).

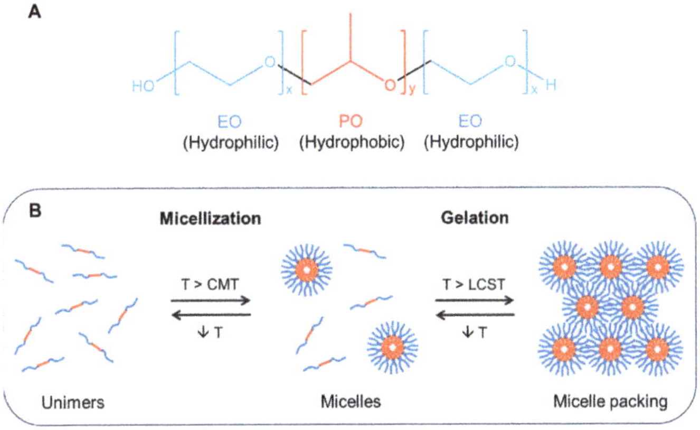

Figure 1. Schematic of the structure and reversible thermoresponsive behavior of poloxamers. (**A**) Poloxamers comprise a central block of "y" numbers of propylene oxide (PO) units and two side blocks of "x" numbers of ethylene oxide (EO) units. (**B**) Heating above the critical micellization temperature (CMT) and above the lower critical solution temperature (LCST) of poloxamer induces micellization and gel formation by micelle packing, respectively. Reprinted from [17], Copyright 2019, with permission from Elsevier.

The rich phase behavior of poloxamer dispersions makes them a versatile platform for drug delivery in the form of micelles or hydrogels, which have enduring popularity in cancer therapy [19–21]. However, before considering the development of pure poloxamer hydrogels, researchers should be aware of their limitations under high compression or in aqueous media due to low mechanical strength and fast erosion, respectively [22]. Notwithstanding that thermosensitivity is appealing in the biomedical field, this property depends on concentrated aqueous dispersions of poloxamers [15]. Moreover, to impart biodegradability to the hydrogel, modifications in the polymer structure can be considered [23].

Among poloxamers, poloxamer 407 (P407) or Pluronic® F127 (PF127) solutions are the most formulated for IT or PT injection on account of their reversible thermoresponsive nature that allows gelation close to the physiological temperature (at concentrations of 20% (*w*/*w*) or less) and residence at the site of implantation as a sustainable carrier [18,24]. Accordingly, P407 exists in a sol state under colder conditions, during which it can be loaded with therapeutic agents for posterior release from its gel form. Moreover, the matrix structure of P407—and thus, the release kinetics of therapeutics—can be modified with other polymers by altering the hydrogel degradation rate [25]. P407, with a molecular weight of ~12.6 kDa, has been approved as an excipient by the U.S. Food and Drug Administration (FDA) for pharmaceutical applications and is listed in the US and European Pharmacopoeia [15,26]. The molecular weight and end group identity of P407 can also be tailored to adjust gelling and adhesiveness properties [27].

3. Characterization of Injectable Poloxamer Hydrogels

Typically, poloxamer hydrogels are prepared according to the "cold method", consisting of adding an appropriate amount of polymer to water or phosphate-buffered saline (PBS) solution for blank hydrogels, or to nanoparticle dispersions for nanocomposite hydrogels, under stirring at 4 °C until a clear dispersion is obtained [28,29]. It is also possible to mix free drugs or drug-loaded nanoparticles with preformed polymer dispersions cooled to 4 °C. Then, the obtained poloxamer dispersions are usually characterized regarding sol–gel transition, rheological properties, in vitro degradation, and release profile.

3.1. Gelation Temperature

Different techniques have been used for measuring the sol–gel transition temperature ($T_{sol\text{-}gel}$). The simplest is the test tube inversion method, where each formulation is sealed in a test tube and heated slowly in a water bath [30]. The temperature at which no flow is observed as inverting the tube is considered the gelation temperature. Alternatively, poloxamer formulations are gradually heated under stirring in a beaker containing a small magnetic bar. Once the magnetic bar stops moving, the recorded temperature is referred to as $T_{sol\text{-}gel}$ [31]. Inasmuch as observation of gelation is straightforward and requires minimal equipment, visual methods lack precision and accuracy [17]. More reliable methods for investigating phase transition include rheological and differential scanning calorimetry (DSC) measurements. Although temperature-controlled rotary viscometers can be used [32], $T_{sol\text{-}gel}$ is determined more precisely by oscillatory shear rheology, measuring the elastic (or storage) modulus (G') and viscous (or loss) modulus (G'') of the hydrogel. In oscillation mode, gelation can be detected as a function of temperature or time, with the G'/G'' crossover point indicating the gelation point. Contrastingly, the gelation process is evidenced by a secondary endothermic peak in the DSC thermogram with no insight into changes in the rheological behavior [17]. To be suitable for in vivo applications, gelation temperature should be near body temperature but never above 37 °C. Otherwise, the sol–gel transition might not occur at the injection site, resulting in leakage of the poloxamer formulation to the surrounding tissues. The $T_{sol\text{-}gel}$ values are also expected to be higher than room temperature to avoid premature gelation that impedes injection [31]. It is established that poloxamer concentration and gelation temperature vary inversely [33]. Therefore, the combination of PF407 with poloxamer 188 (Pluronic® F68, PF68) is often recommended to obtain an acceptable gelation temperature, which also increases gel strength versus the polymers used alone [34].

3.2. Rheological Behavior, Mechanical Strength, and Injectability

As expected, the rheological properties of poloxamer dispersions are temperature-dependent, exhibiting Newtonian behavior at low temperature and non-Newtonian, shear-thinning behavior at higher temperatures [31,35]. The mechanical strength of the hydrogels is essential for maintaining their integrity in the body, but is usually overlooked during characterization. Nevertheless, in vitro degradation and release studies provide evidence of low gel strength. As a result of rapid erosion, there is a burst release behavior, meaning the fast release of a considerable fraction of payload into a hydrolytic medium that simulates human physiological conditions (PBS at 37 °C) [36,37]. Of note, as to delivery to tumor cells, the tumor microenvironment is more closely imitated if acidity (pH 6.8) and the presence of enzyme hyaluronidase are considered [38]. Some researchers sought to improve the mechanical properties of poloxamer hydrogels by introducing a chain extender (hexamethylene diisocyanate, abbreviated to HDI) into the polymer [39] or adding bioadhesive polymers such as N,N,N-trimethyl chitosan [40], alginate [41], or xanthan gum [38]. Ju et al. [42] upgraded this strategy and prepared a P407 hydrogel interpenetrated by a network of carboxymethyl chitosan crosslinked with glutaraldehyde but losing thermosensitivity. Instead, chitosan can be crosslinked with genipin to form an interpenetrating scaffold within P407 hydrogels [43].

Notwithstanding that injectability is a critical parameter of the injection performance, not many authors assess the force required to perform the administration via a syringe [44]. Those who do either evaluate the easiness of passing the hydrogel through a needle in qualitative terms [45] or conduct uniaxial tensile testing using a mechanical testing machine with a syringe fitted with a needle or catheter [46].

3.3. Biocompatibility and Sterilization

The safety and non-toxicity of poloxamer hydrogels for local tumor administration have been observed in vitro [47,48] and in vivo through histological examination for signs of inflammation after subcutaneous implantation [39,49]. In addition to biocompatibility, sterility is a requirement for considering the potential clinical use of any material intended to be in close contact with the human body [50]. However, the impact of sterilization on poloxamer hydrogels has been understudied. To date, steam sterilization (121 °C, 15–20 min) was the most investigated, which was found to cause a slight decrease in gelation temperature [25,51]. This observation can be explained by an increase in polymer weight fraction due to water evaporation during autoclaving. It was suggested that autoclaving at a lower temperature for a longer time (e.g., 105 °C for 30 min) would allow the poloxamer hydrogels to be more like the non-sterilized ones [52].

4. Local Tumor Administration of Poloxamer Hydrogels

The proof-of-concept of poloxamer hydrogels for local tumor administration has been demonstrated by the growth inhibition of several tumors in different mouse models. Nevertheless, it is noteworthy that most tumor models were established in mice through the subcutaneous inoculation of cancer cells. Subcutaneous (or ectopic) tumors might be advantageous to monitoring tumor growth and performing local injections, but fail to mimic the tumor microenvironment. Despite being more clinically relevant because tumor xenografts are placed in the tissue/organ of origin, orthotopic mouse models still do not reflect the size of tumors that develop naturally in patients. Moreover, potential adverse effects in cancer patients with intact immunity may go unnoticed if studies in immunodeficient animals are the case [53].

The application of injectable poloxamer hydrogels in local cancer therapy is depicted in Figure 2 and discussed below, with several examples organized by therapeutic modality.

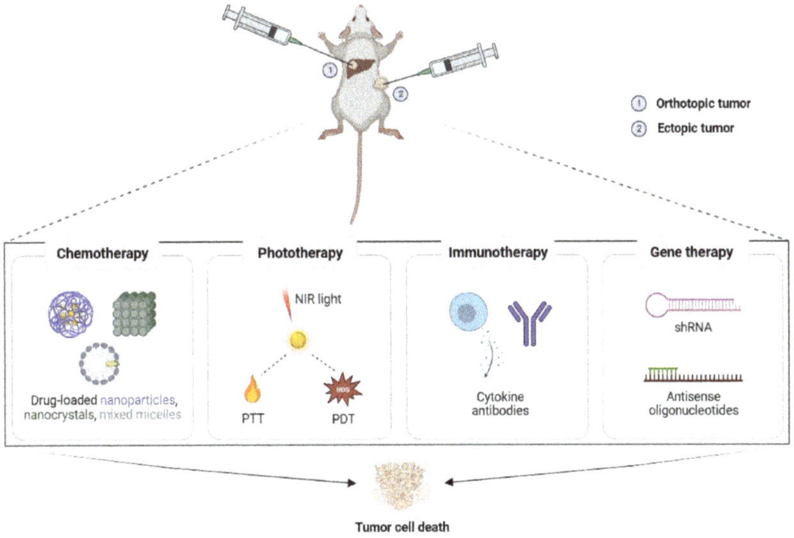

Figure 2. The in vivo administration of poloxamer hydrogels for local cancer therapies including chemotherapy, photothermal therapy (PTT), photodynamic therapy (PDT), immunotherapy, and gene therapy.

4.1. Potential Applications in Cancer Chemotherapy

For local cancer chemotherapy, P407 or PF127 solutions were mixed with free drugs, such as paclitaxel (PTX) [32], topotecan [54], doxorubicin (DOX) [55], and salinomycin [47]. Still, most PF127 hydrogels reported for IT or PT injection accommodate anticancer drugs encapsulated in nanoparticles [56,57], nanocrystals [28,58,59], cyclodextrin inclusion complexes [60,61], hyaluronic acid-based nanocomplexes [62,63], and mixed micelles [63,64]. The literature also contains several examples of in situ-forming gels using mixtures of PF127 and PF68 for the local delivery of free [45,65] and encapsulated [30,31,49,66] drugs.

The direct incorporation of free PTX into a P407 solution at the final concentration of 0.5 mg/mL, albeit simple, resulted in a very slow in vitro drug release from P407 hydrogel because of the poor water solubility of PTX. Moreover, although it was completely dissolved at lower concentrations, PTX formed a suspension when the final concentration was doubled (1.0 mg/mL) [32]. This reflects the low solubilization of hydrophobic drugs in P407, which generally imparts limited drug loading and physical instability to poloxamer micelles. Therefore, hybrid systems integrating drug-loaded nanoparticles and thermoresponsive hydrogels have been intensively studied to improve drug release and increase drug loading capacity [28,67].

The combination of liposomes and poloxamer hydrogels was proposed to stabilize the lactone form of 7-ethyl-10-hydroxycamptothecin [66] and prolong the release of PTX [49,56]. Whereas PF127/PF68 hydrogels enhanced the retention of drug-loaded liposomes at the tumor site [49,66], the use of liposomes made of 0.21–1.25% soybean phospholipids was suggested to allow a 3–9 wt% decrease in the poloxamer concentration required for an in situ-forming PF127 gel [56]. Notwithstanding the evidence from studies in MCF-7 breast cancer cells supporting the higher cytotoxic activity of tamoxifen citrate-loaded niosomes compared to the free drug, the low viscosity of niosomal suspensions prompted their dispersion into poloxamer hydrogels [31]. In a very interesting approach to the treatment of melanoma, Yu et al. [57] prepared a PF127 hydrogel to intratumorally deliver CDDP-loaded poly(α-L-glutamate)-g-mPEG nanoparticles and microspheres entrapping losartan potassium that exerts antifibrotic effects, namely by inhibiting the production of collagen I in tumors. The incorporation of both microspheres and nanoparticles into the gel enabled most losartan to be released first and reduce the collagen content prior to the release of CDDP, which occurred in the following days after the nanoparticles have penetrated more deeply into the tumor. Differently, Shen et al. [68] combined nanotechnology and active targeting with thermoresponsive polymers for IT administration of PTX in pancreatic tumors. For that, they prepared a PF127/PF68/hydroxypropyl methylcellulose gel bearing PTX-loaded mPEG-poly(D,L-lactide-co-glycolide)-poly(L-lysine) nanoparticles functionalized with a cyclic peptide, which specifically binds to $\alpha v\beta 3$ integrin overexpressed on the endothelial tumor cells. Later, Xie et al. [69] also developed a PF127/PF68/hydroxypropyl methylcellulose hydrogel to improve the efficacy and safety of norcantharidin (NCTD) for treating hepatic cancer. In another work, Gao et al. [29] took into consideration that NCTD has poor solubility in water, thereby preparing NCTD-loaded polymeric nanoparticles before incorporating them into a DOX-containing PF127 hydrogel to treat hepatocellular carcinoma via IT administration.

The formulation and dispersion of drug nanocrystals into PF127 hydrogels deserved some attention, considering that nanocrystals provide higher drug loading than other nanocarriers [28,58]. Further, Gao et al. [58] dissolved D-α-tocopherol PEG 1000 succinate in PF127 solutions to impair drug efflux and reverse drug resistance of P-glycoprotein-overexpressing liver cancer cells. Together with lapatinib-loaded microparticles, PTX nanocrystals were incorporated into PF127 hydrogel for PT injection to imitate the slow and fast release of these two drugs in clinical use [59].

Attempts to increase the water solubility of the anticancer agent β-lapachone involve the formation of inclusion complexes with cyclodextrins. Intending to design injectable thermoresponsive hydrogels containing β-lapachone, Landin's group used Artificial Neural Network modeling to understand the interactions between the polymer (PF127) and the solubilizing agent (cyclodextrin) and obtain the optimal formulation [60]. A significant decrease in cell viability and tumor volume was observed following the treatment of MCF-7 cells and in the breast xenograft mouse model with this ternary system [61]. When studying the effect of methylated β-cyclodextrin and ethanol on the β-lapachone solubility and gel properties, these authors confirmed that both additives promote drug solubilization [70]. However, the addition of ethanol as a co-solvent may render Pluronic® (F127 and P123) dispersions inappropriate for IT administration. Data from rheological characterization suggested that autoclaving may not affect the gelation temperature and gel strength of Pluronic® systems with β-lapachone [70].

An injectable PF127 hydrogel containing DOX complexed with HA and $MgCl_2$ was developed by Jhan et al. [62] and was demonstrated to cause the growth inhibition of C26 colon cancer cells in a mouse model. This drug delivery system was patented (US9364545B2) [71] and then ameliorated by adding a mixed micellar formulation composed of PF127 and Pluronic® L121 for carrying a second chemotherapeutic drug (DTX) [63]. Mixed micelles consisting of PF127 and another surfactant, such as Solutol® HS15 [30], Tween® 80 [64], or D-α-tocopherol PEG 1000 succinate [72], have been incorporated into PF127 hydrogels to deliver hydrophobic drugs, namely DTX [30,64] and PTX [72].

By synthesizing the dalteparin-P407 copolymer, Li et al. [73] repositioned low-molecular-weight heparin as an anticancer agent and fabricated a novel thermosensitive and injectable hydrogel carrying DOX-loaded laponite nanoparticles.

Only one of the articles reviewed [74] reported the use of poloxamer hydrogels for local chemoradiotherapy. The concurrent IT administration of chemotherapeutics and radiation was achieved by using PF127 hydrogels co-loaded with DOX and gold nanoparticles.

4.2. Potential Applications in Cancer Phototherapy

Considering the mechanisms of light conversion, phototherapy includes photothermal therapy (PTT) and photodynamic therapy (PDT). Phototherapy based on PTT or PDT can eliminate cancer cells by generating hyperthermia or reactive species of oxygen (ROS) [75,76].

PTT involves the laser activation of photothermal agents, followed by near-infrared (NIR) light conversion into heat. Despite great progress in cancer PTT, most photothermal agents are made of heavy metals and given intravenously, causing safety concerns to arise. To reduce putative systemic toxicity and enhance local retention, Fu and colleagues indicated PF127 hydrogels embedding copper sulfide nanodots [77] or Prussian blue nanospheres [78] for PT administration. Given that seaweed polysaccharides have good biocompatibility, biodegradability, and non-toxicity, Chen et al. [79] prepared an injectable photothermal hydrogel using iota carrageenan-capped gold-silver nanoparticles and PF127. The in vivo results pointed to a multifunctional hydrogel that could prevent tumor growth and recurrence and promote post-surgical wound healing without chemotherapeutic drugs and antibiotics [79]. An alternative to metal nanoparticles as photothermal agents is organic agents (i.e., indocyanine green), but their intrinsic instability limits their therapeutic effects. As a result, organic–inorganic hybrid nanomaterials such as titanium carbide (Ti_3C_2) nanoparticles have received attention and have been combined with PF127 through a simple mixture to form an injectable hydrogel for local PTT [80].

The combination of chemotherapy with other therapeutic modalities, namely phototherapy, has gained momentum in recent years. One such example is the work by Zhang et al. [81], which was aimed at achieving complete tumor ablation via IT injection of HDI-PF127 nanocomposite hydrogel incorporating PTX-loaded chitosan micelles and PEGylated gold nanorods. Qin et al. [82] chose PF127 as the hydrogel matrix and black phosphorus nanosheets as photothermal agents because of their broad absorption in the NIR region and extinction coefficient larger than other 2D materials. While investigating the in vitro release profile of gemcitabine, it was observed that black phosphorus nanosheets accelerated drug release from PF127 hydrogel under NIR irradiation (808 nm, 2.0 W/cm^2, 10 min). Compared to chemotherapy alone, this hydrogel exhibited a superior antitumor effect and good photothermal effect in BALB/c mice bearing 4T1 xenograft tumors [82]. In another paper [83], the application of NIR light induced on-demand release for up to 14 days after a single administration of PF127 hydrogel with liposomes incorporating DOX and gold-manganese oxide nanoparticles.

In addition to the analyses described in Section 3, the photothermal properties of these poloxamer hydrogels are usually assessed in terms of photo–heat conversion ability and photothermal stability under repeated 808 nm laser irradiation.

Tumor destruction by conventional PDT relies on the photochemical reaction between a light-activated photosensitizer and molecular oxygen to produce ROS, resulting in cell death [84]. However, PDT often fails to completely eradicate tumors due to the limited penetration of currently available photosensitizers into the tissue. Building upon the use of two-photon excitation to improve light penetration, Luo et al. [85] proposed the co-encapsulation of a two-photon absorption compound (T1) and a photosensitizer (pyropheophorbide a) into polymeric micelles combined with PF127. In 4T1 xenograft mice, the obtained hydrogel was shown to inhibit tumor growth in more than 50% (under 1 cm-thick muscle tissue) after IT administration and NIR irradiation. By capitalizing on the synergistic effects of chemotherapy and PDT, Li et al. [86] employed DTX-loaded micelles and black phosphorus nanosheets as a hydrophobic model drug and photosensitizer, respectively, incorporating them into a PF127/PF68 hydrogel. The photodynamic performance of these hydrogels is evaluated by singlet oxygen detection.

4.3. Potential Applications in Cancer Immunotherapy

The most recent revolutionary wave in cancer therapy came with immunotherapy and its improvements for patients in terms of survival and quality of life [87]. However, two of the most used classes of immunotherapeutics, cytokines and checkpoint inhibitors, face similar and appreciable delivery challenges. For example, the use of Toll-like receptor (TLR) 7/8 agonists is often limited to IT administration because IV administration can lead to systemic toxicity by stimulating the entire immune system [88]. Therefore, local delivery of TLR 7/8 agonists, such as MEDI9197 [89] and imiquimod [90], is preferred, which can be attained by mixing them with P407 aqueous solutions. Fakhari et al. [89] demonstrated significant antitumor activity of P407 thermogel with MEDI9197 after two IT injections in a B16-OVA melanoma tumor model. In another work [90], imiquimod was first encapsulated in 1,2-dipalmitoyl-sn-glycero-3-phosphatidylcholine liposomes before being incorporated into PF127 hydrogel, with the final delivery system producing promising results in a breast cancer model.

Cytotoxic T-lymphocyte-associated protein 4 (CTLA-4), belonging to the class of checkpoint inhibitors, can also be explored to generate antitumor immune responses. To control the release of anti-CTLA-4 antibodies, Chung et al. [48] pioneered the optimization of CTLA-4 therapy using P407-based injectable hydrogels. The authors observed a significant reduction in serum anti-CTLA-4 levels and effective tumor growth inhibition in CT26 tumor-bearing mice receiving the hydrogel peritumorally. A major feature of the tumor microenvironment is extracellular acidosis, which seems to antagonize the efficacy of immune checkpoint inhibitors. One prominent strategy to alleviate extracellular tumor acidity capitalizes on sodium bicarbonate therapy, but can cause metabolic alkalosis. As such, Jin et al. [91] employed $NaHCO_3$-loaded PF127 hydrogel for precise delivery into the tumor, rendering its microenvironment immunologically favorable. Indeed, tumor clearance was improved when treating MC-38-bearing mice with a low dose of immune checkpoint inhibitors after local tumor neutralization with the gel. Very recently, the overall goal of maximizing the therapeutic index of vemurafenib and antagonistic programmed cell death protein 1 antibody used in combination to treat BRAF-mutated melanoma was achieved with PF127-g-gelatin hydrogel and IT administration [92].

Although dendritic cells were initially recognized for their role in antiviral immunity, recent attention has been directed toward their potential to boost the patients' immune system in the fight against cancer [93]. Still, considering their short viability and low in vivo migration capacity, treatment with adjuvants for the recruitment and maturation of dendritic cells is of great interest. To illustrate, Lemdani et al. [94] designed a mucoadhesive hydrogel consisting of P407 and xanthan gum for IT co-delivery of granulocyte-macrophage colony-stimulating factor and heat-killed *Mycobacterium tuberculosis* to refine the local antitumor immune response. Though administering a solution of these immunomodulatory agents elicited minimal therapeutic effects, their IT injection in the gel led to the infiltration of T cells in the tumor, as well as growth inhibition.

4.4. Potential Applications in Cancer Gene Therapy

To date, the utilization of poloxamer hydrogels in local cancer gene therapy is scarce, with only two works being reported.

After coupling conjugated linoleic acid (CLA) with P407, Guo et al. [95] demonstrated that CLA-coupled poloxamer hydrogel could be a local delivery system for PTX with enhanced antitumor efficacy. The evidence of apoptotic cell death inspired these authors to use the obtained hydrogel for combination therapy with PTX and Akt1 shRNA [96]. Knowing that the phosphoinositide 3-kinase/Akt1 signaling pathway has emerged as a target for breast cancer therapy, it is no surprise that the inhibition of Akt1 warrants special attention. In addition to synergistic inhibitory effects in vitro (MDA-MB-231 and MCF-7 cells) and in vivo (MDA-MB-231 xenograft), local treatment with PTX and Akt1 via CLA-coupled PF127 hydrogel was confirmed to decrease Akt1 phosphorylation levels and inhibit angiogenesis [96].

Another promising target for breast cancer is survivin, whose inhibition merits in situ injection to ensure tissue and cell specificity. Taking advantage of electrostatic interactions between a cationic polymer (poly[(R)-3-hydroxybutyrate]-b-poly(2-dimethylamino) ethyl methacrylate) and negatively charged survivin antisense oligonucleotide, Zhao et al. [97] developed a gene delivery nanocomplex subsequently incorporated into injectable PF127 hydrogels for local retention. A single injection was enough to achieve a sustained gene release for up to 16 days and counteract PTX-induced multidrug resistance by silencing up-regulated survivin.

At the end of this subsection, Table 1 summarizes the described injectable poloxamer hydrogels published for the period 2018–2023.

Table 1. Injectable poloxamer hydrogels for intratumoral or peritumoral administration.

Cancer Therapy	Injection Type	Hydrogel Composition	Cancer Cell (In Vitro)	Cancer (In Vivo)	Ref.
Chemotherapy	IT	P407, Topotecan	-	Retinoblastoma (Y79)	[54]
Chemotherapy	-	P407, DOX	MC-38 (colon)	-	[55]
Chemotherapy	IT	P407, Salinomycin	U251	GBM	[47]
Chemotherapy	IT	P407, CDDP NP, LP microspheres	-	Melanoma (B16)	[57]
Chemotherapy	IT	P407, P188, SN-38 liposomes	-	HCC (H22)	[66]
Chemotherapy	IT	P407, P188, DTX micelles	-	Colon (HT-29)	[30]
Chemotherapy	IT	P407, P188, HPMC, NCTD	-	HCC (H22)	[69]
Chemotherapy	IT	P407, DOX, NCTD NP	HepG2	HCC (H22)	[29]
Chemotherapy	IT	P407, HA, HPMC K_4M, DOX, PTX micelles	-	-	[72]
Chemotherapy	IT	P407, P188, Alginate, 5-FU	-	Colon (CT26-luc)	[41]
Chemotherapy	IT	P407, P188, Xanthan gum, PTX NP	MCF-7	Breast	[38]
Chemotherapy	PT	Heparin-P407, DOX laponite NP	S180	Sarcoma	[73]
PTT	PT	P407, CuS nanodots	4T1	Breast	[77]

Table 1. *Cont.*

Cancer Therapy	Injection Type	Hydrogel Composition	Cancer Cell (In Vitro)	Cancer (In Vivo)	Ref.
PTT	PT	P407, Prussian blue nanospheres	4T1	Breast	[78]
PTT	IT	P407, CA-AuAg NP	4T1 B16F10	Breast Melanoma	[79]
PTT	IT	P407, Ti_3C_2 NP	4T1	Breast	[80]
PTT + Chemotherapy	IT	P407, BP nanosheets, Gemcitabine	-	Breast (4T1)	[82]
PDT	IT	P407, T1 and PPa co-encapsulated micelles	4T1	Breast	[85]
PDT + Chemotherapy	IT	P407, P188, BP nanosheets, DTX micelles	-	Breast (4T1)	[86]
Immunotherapy	IT	P407, Imiquimod liposomes	4T1	Breast	[90]
Immunotherapy	PT	P407, CTLA-4 Ab	MC-38 (colon)	Colon (CT26)	[48]
Immunotherapy	IT	P407, $NaHCO_3$	-	Colon (MC-38)	[91]
Immunotherapy	IT	P407-g-gelatin, Vemurafenib, PD-1 mAb	D4M B16F10	Melanoma	[92]
Immunotherapy	IT	P407, Xanthan gum, GM-CSF, HKMT	B16 CT26 3LL (Lewis lung carcinoma)	Colon (CT26)	[94]
Gene therapy	IT	P407, Sur-ASON, PHB-b-PDMAEMA	MCF-7/PDR	Breast	[97]

5-FU: 5-fluorouracil; Ab: antibody; BP: black phosphorus; CA-AuAg: iota carrageenan-capped gold-silver; CDDP: cisplatin; CTLA-4: cytotoxic T-lymphocyte-associated protein 4; DOX: doxorubicin; DTX: docetaxel; GBM: glioblastoma multiforme; GM-CSF: granulocyte-macrophage colony stimulating factor; HA: hyaluronic acid; HCC: hepatocellular carcinoma; HKMT: heat-killed Mycobacterium tuberculosis; HPMC: hydroxypropyl methylcellulose; IT: intratumoral; LP: losartan potassium; NCTD: norcantharidin; NP: nanoparticles; P188: poloxamer 188; P407: poloxamer 407; PD-1 mAb: programmed cell death protein 1 monoclonal antibody; PDT: photodynamic therapy; PHB-b-PDMAEMA: poly[(R)-3-hydroxybutyrate]-b-poly(2-dimethylamino) ethyl methacrylate; PPa: pyropheophorbide a; PT: peritumoral; PTT: photothermal therapy; PTX: paclitaxel; SN-38: 7-ethyl-10-hydroxycamptothecin; Sur-ASON: survivin antisense oligonucleotide.

5. Conclusions

A myriad of in situ gelling systems triggered by temperature changes has been developed for the IT or PT administration of different therapeutic agents. Among thermoresponsive polymers, poloxamer-based hydrogels are in the spotlight due to their low cost, simplicity of preparation, and compatibility with biological systems [15,98].

Over the past two decades, poloxamer dispersions have been proposed as injectable formulations to assist local cancer treatment using chemotherapy, phototherapy, immunotherapy, and gene therapy. However, the application of injectable poloxamer hydrogels for local tumor administration remains in the proof-of-concept stage, despite promising preclinical (in vitro and in vivo) outcomes. First, it is highly recommended that poloxamers are modified or used in novel combinations of polymers to reduce the erosion rate of conventional poloxamer hydrogels and ensure precise delivery. Moreover, the developed hydrogels are more likely to reach the clinical testing phase if researchers evaluate therapeutic efficacy in larger animals (e.g., monkeys, pigs, and dogs) instead of using rodent models [75]. In addition to clinical translation, the scale-up process from laboratory to industry is also very effortful, the first step being a thorough characterization of the hydrogels including insights into the morphology and thermal properties, and not only rheological and biological analyses [99].

In the future, injectable poloxamer hydrogels are expected to remain an exciting research topic not only for drug delivery but also for tissue engineering [100] and cartilage repair [101,102]. Regarding local cancer therapy, one can envision phototherapy and immunotherapy succeeding chemotherapy as the most applied therapeutic modalities, with increasingly frequent reports. Concurrently, more researchers will follow the trend of combination therapy as a new direction for cancer treatment and establish injectable poloxamer-based hydrogel as a key element in upcoming therapeutic strategies [103].

Author Contributions: Conceptualization, A.C.M.; writing—original draft preparation, A.C.M.; writing—review and editing, M.H.A., P.C.C. and S.V.; supervision, M.H.A., P.C.C. and S.V. All authors have read and agreed to the published version of the manuscript.

Funding: This work was financed by national funds from FCT—Fundação para a Ciência e a Tecnologia, I.P., in the scope of the project UIDP/04378/2020 and UIDB/04378/2020 of the Research Unit on Applied Molecular Biosciences—UCIBIO and the project LA/P/0140/2020 of the Associate Laboratory Institute for Health and Bioeconomy—i4HB. Ana Camila Marques gratefully acknowledges FCT for financial support (grant reference: 2020.06766.BD).

Institutional Review Board Statement: Not applicable.

Informed Consent Statement: Not applicable.

Data Availability Statement: Not applicable.

Conflicts of Interest: The authors declare no conflict of interest.

References

1. Fakhari, A.; Subramony, J.A. Engineered in-situ depot-forming hydrogels for intratumoral drug delivery. *J. Control. Release* **2015**, *220*, 465–475. [CrossRef] [PubMed]
2. Marques, A.C.; Costa, P.J.; Velho, S.; Amaral, M.H. Stimuli-responsive hydrogels for intratumoral drug delivery. *Drug Discov. Today* **2021**, *26*, 2397–2405. [CrossRef] [PubMed]
3. Burris, H.A., 3rd; Vogel, C.L.; Castro, D.; Mishra, L.; Schwarz, M.; Spencer, S.; Oakes, D.D.; Korey, A.; Orenberg, E.K. Intratumoral cisplatin/epinephrine-injectable gel as a palliative treatment for accessible solid tumors: A multicenter pilot study. *Otolaryngol. Head Neck Surg.* **1998**, *118*, 496–503. [CrossRef] [PubMed]
4. Brachi, G.; Ruiz-Ramírez, J.; Dogra, P.; Wang, Z.; Cristini, V.; Ciardelli, G.; Rostomily, R.C.; Ferrari, M.; Mikheev, A.M.; Blanco, E.; et al. Intratumoral injection of hydrogel-embedded nanoparticles enhances retention in glioblastoma. *Nanoscale* **2020**, *12*, 23838–23850. [CrossRef]
5. Chen, M.H.; Wang, L.L.; Chung, J.J.; Kim, Y.-H.; Atluri, P.; Burdick, J.A. Methods To Assess Shear-Thinning Hydrogels for Application As Injectable Biomaterials. *ACS Biomater. Sci. Eng.* **2017**, *3*, 3146–3160. [CrossRef]
6. Thambi, T.; Li, Y.; Lee, D.S. Injectable hydrogels for sustained release of therapeutic agents. *J. Control. Release* **2017**, *267*, 57–66. [CrossRef]
7. Chakraborty, D.D.; Nath, L.K.; Chakraborty, P. Recent Progress in Smart Polymers: Behavior, Mechanistic Understanding and Application. *Polym. Plast. Technol. Eng.* **2017**, *57*, 945–957. [CrossRef]
8. Zhang, Q.; Weber, C.; Schubert, U.S.; Hoogenboom, R. Thermoresponsive polymers with lower critical solution temperature: From fundamental aspects and measuring techniques to recommended turbidimetry conditions. *Mater. Horiz.* **2017**, *4*, 109–116. [CrossRef]
9. Fan, R.; Cheng, Y.; Wang, R.; Zhang, T.; Zhang, H.; Li, J.; Song, S.; Zheng, A. Thermosensitive Hydrogels and Advances in Their Application in Disease Therapy. *Polymers* **2022**, *14*, 2379. [CrossRef]
10. Bragta, P.; Sidhu, R.K.; Jyoti, K.; Baldi, A.; Jain, U.K.; Chandra, R.; Madan, J. Intratumoral administration of carboplatin bearing poly (ε-caprolactone) nanoparticles amalgamated with in situ gel tendered augmented drug delivery, cytotoxicity, and apoptosis in melanoma tumor. *Colloids Surf. B Biointerfaces* **2018**, *166*, 339–348. [CrossRef]
11. Pesoa, J.I.; Rico, M.J.; Rozados, V.R.; Scharovsky, O.G.; Luna, J.A.; Mengatto, L.N. Paclitaxel delivery system based on poly(lactide-co-glycolide) microparticles and chitosan thermo-sensitive gel for mammary adenocarcinoma treatment. *J. Pharm. Pharmacol.* **2018**, *70*, 1494–1502. [CrossRef]
12. Fong, Y.T.; Chen, C.-H.; Chen, J.-P. Intratumoral Delivery of Doxorubicin on Folate-Conjugated Graphene Oxide by In-Situ Forming Thermo-Sensitive Hydrogel for Breast Cancer Therapy. *Nanomaterials* **2017**, *7*, 388. [CrossRef]
13. Zhou, X.; He, X.; Shi, K.; Yuan, L.; Yang, Y.; Liu, Q.; Ming, Y.; Yi, C.; Qian, Z. Injectable Thermosensitive Hydrogel Containing Erlotinib-Loaded Hollow Mesoporous Silica Nanoparticles as a Localized Drug Delivery System for NSCLC Therapy. *Adv. Sci.* **2020**, *7*, 2001442. [CrossRef]

14. Babaei, M.; Davoodi, J.; Dehghan, R.; Zahiri, M.; Abnous, K.; Taghdisi, S.M.; Ramezani, M.; Alibolandi, M. Thermosensitive composite hydrogel incorporated with curcumin-loaded nanopolymersomes for prolonged and localized treatment of glioma. *J. Drug Deliv. Sci. Technol.* **2020**, *59*, 101885. [CrossRef]
15. Russo, E.; Villa, C. Poloxamer Hydrogels for Biomedical Applications. *Pharmaceutics* **2019**, *11*, 671. [CrossRef]
16. Giuliano, E.; Paolino, D.; Fresta, M.; Cosco, D. Mucosal Applications of Poloxamer 407-Based Hydrogels: An Overview. *Pharmaceutics* **2018**, *10*, 159. [CrossRef]
17. Soliman, K.A.; Ullah, K.; Shah, A.; Jones, D.S.; Singh, T.R. Poloxamer-based in situ gelling thermoresponsive systems for ocular drug delivery applications. *Drug Discov. Today* **2019**, *24*, 1575–1586. [CrossRef]
18. Bodratti, A.M.; Alexandridis, P. Formulation of Poloxamers for Drug Delivery. *J. Funct. Biomater.* **2018**, *9*, 11. [CrossRef]
19. Valenzuela-Oses, J.K.; García, M.C.; Feitosa, V.A.; Pachioni-Vasconcelos, J.A.; Gomes-Filho, S.M.; Lourenço, F.R.; Cerize, N.N.; Bassères, D.S.; Rangel-Yagui, C.O. Development and characterization of miltefosine-loaded polymeric micelles for cancer treatment. *Mater. Sci. Eng. C Mater. Biol. Appl.* **2017**, *81*, 327–333. [CrossRef]
20. Vaidya, F.U.; Sharma, R.; Shaikh, S.; Ray, D.; Aswal, V.K.; Pathak, C. Pluronic micelles encapsulated curcumin manifests apoptotic cell death and inhibits pro-inflammatory cytokines in human breast adenocarcinoma cells. *Cancer Rep.* **2018**, *2*, e1133. [CrossRef]
21. Kotta, S.; Aldawsari, H.M.; Badr-Eldin, S.M.; Nair, A.B.; Kaleem, M.; Dalhat, M.H. Thermosensitive Hydrogels Loaded with Resveratrol Nanoemulsion: Formulation Optimization by Central Composite Design and Evaluation in MCF-7 Human Breast Cancer Cell Lines. *Gels* **2022**, *8*, 450. [CrossRef]
22. Xu, N.; Xu, J.; Zheng, X.; Hui, J. Preparation of Injectable Composite Hydrogels by Blending Poloxamers with Calcium Carbonate-Crosslinked Sodium Alginate. *Chemistryopen* **2020**, *9*, 451–458. [CrossRef]
23. Salehi, S.; Naghib, S.M.; Garshasbi, H.R.; Ghorbanzadeh, S.; Zhang, W. Smart stimuli-responsive injectable gels and hydrogels for drug delivery and tissue engineering applications: A review. *Front. Bioeng. Biotechnol.* **2023**, *11*, 1104126. [CrossRef] [PubMed]
24. Zhang, K.; Shi, X.; Lin, X.; Yao, C.; Shen, L.; Feng, Y. Poloxamer-based in situ hydrogels for controlled delivery of hydrophilic macromolecules after intramuscular injection in rats. *Drug Deliv.* **2014**, *22*, 375–382. [CrossRef] [PubMed]
25. Beard, M.C.; Cobb, L.H.; Grant, C.S.; Varadarajan, A.; Henry, T.; Swanson, E.A.; Kundu, S.; Priddy, L.B. Autoclaving of Poloxamer 407 hydrogel and its use as a drug delivery vehicle. *J. Biomed. Mater. Res. Part B: Appl. Biomater.* **2020**, *109*, 338–347. [CrossRef]
26. Fakhari, A.; Corcoran, M.; Schwarz, A. Thermogelling properties of purified poloxamer 407. *Heliyon* **2017**, *3*, e00390. [CrossRef]
27. Ci, L.; Huang, Z.; Liu, Y.; Liu, Z.; Wei, G.; Lu, W. Amino-functionalized poloxamer 407 with both mucoadhesive and thermosensitive properties: Preparation, characterization and application in a vaginal drug delivery system. *Acta Pharm. Sin. B* **2017**, *7*, 593–602. [CrossRef] [PubMed]
28. Lin, Z.; Gao, W.; Hu, H.; Ma, K.; He, B.; Dai, W.; Wang, X.; Wang, J.; Zhang, X.; Zhang, Q. Novel thermo-sensitive hydrogel system with paclitaxel nanocrystals: High drug-loading, sustained drug release and extended local retention guaranteeing better efficacy and lower toxicity. *J. Control. Release* **2014**, *174*, 161–170. [CrossRef]
29. Gao, B.; Luo, J.; Liu, Y.; Su, S.; Fu, S.; Yang, X.; Li, B. Intratumoral Administration of Thermosensitive Hydrogel Co-Loaded with Norcantharidin Nanoparticles and Doxorubicin for the Treatment of Hepatocellular Carcinoma. *Int. J. Nanomed.* **2021**, *16*, 4073–4085. [CrossRef]
30. Xu, M.; Mou, Y.; Hu, M.; Dong, W.; Su, X.; Wu, R.; Zhang, P. Evaluation of micelles incorporated into thermosensitive hydrogels for intratumoral delivery and controlled release of docetaxel: A dual approach for in situ treatment of tumors. *Asian J. Pharm. Sci.* **2018**, *13*, 373–382. [CrossRef]
31. Shaker, D.S.; Shaker, M.A.; Klingner, A.; Hanafy, M.S. In situ thermosensitive Tamoxifen citrate loaded hydrogels: An effective tool in breast cancer loco-regional therapy. *J. Drug Deliv. Sci. Technol.* **2016**, *35*, 155–164. [CrossRef]
32. Amiji, M.M.; Lai, P.-K.; Shenoy, D.B.; Rao, M. Intratumoral Administration of Paclitaxel in an In Situ Gelling Poloxamer 407 Formulation. *Pharm. Dev. Technol.* **2002**, *7*, 195–202. [CrossRef] [PubMed]
33. Sguizzato, M.; Valacchi, G.; Pecorelli, A.; Boldrini, P.; Simelière, F.; Huang, N.; Cortesi, R.; Esposito, E. Gallic acid loaded poloxamer gel as new adjuvant strategy for melanoma: A preliminary study. *Colloids Surf. B Biointerfaces* **2020**, *185*, 110613. [CrossRef] [PubMed]
34. Shastri, D.H.; Prajapati, S.T.; Patel, L.D. Design and Development of Thermoreversible Ophthalmic In Situ Hydrogel of Moxifloxacin HCl. *Curr. Drug Deliv.* **2010**, *7*, 238–243. [CrossRef]
35. Chen, Y.; Lee, J.-H.; Meng, M.; Cui, N.; Dai, C.-Y.; Jia, Q.; Lee, E.-S.; Jiang, H.-B. An Overview on Thermosensitive Oral Gel Based on Poloxamer 407. *Materials* **2021**, *14*, 4522. [CrossRef]
36. Jung, Y.-S.; Park, W.; Park, H.; Lee, D.-K.; Na, K. Thermo-sensitive injectable hydrogel based on the physical mixing of hyaluronic acid and Pluronic F-127 for sustained NSAID delivery. *Carbohydr. Polym.* **2017**, *156*, 403–408. [CrossRef]
37. Bhattacharjee, S. Understanding the burst release phenomenon: Toward designing effective nanoparticulate drug-delivery systems. *Ther. Deliv.* **2021**, *12*, 21–36. [CrossRef]
38. Jeswani, G.; Chablani, L.; Gupta, U.; Sahoo, R.K.; Nakhate, K.T.; Taksande, A.G. Ajazuddin Exploration of hemocompatibility and intratumoral accumulation of paclitaxel after loco-regional administration of thermoresponsive hydrogel composed of poloxamer and xanthan gum: An application to dose-dense chemotherapy. *Int. J. Biol. Macromol.* **2023**, *226*, 746–759. [CrossRef]
39. Chen, Y.-Y.; Wu, H.-C.; Sun, J.-S.; Dong, G.-C.; Wang, T.-W. Injectable and Thermoresponsive Self-Assembled Nanocomposite Hydrogel for Long-Term Anticancer Drug Delivery. *Langmuir* **2013**, *29*, 3721–3729. [CrossRef]

40. Turabee, H.; Jeong, T.H.; Ramalingam, P.; Kang, J.H.; Ko, Y.T. N,N,N-trimethyl chitosan embedded in situ Pluronic F127 hydrogel for the treatment of brain tumor. *Carbohydr. Polym.* **2018**, *203*, 302–309. [CrossRef]
41. Al Sabbagh, C.; Seguin, J.; Agapova, E.; Kramerich, D.; Boudy, V.; Mignet, N. Thermosensitive hydrogels for local delivery of 5-fluorouracil as neoadjuvant or adjuvant therapy in colorectal cancer. *Eur. J. Pharm. Biopharm.* **2020**, *157*, 154–164. [CrossRef] [PubMed]
42. Ju, C.; Sun, J.; Zi, P.; Jin, X.; Zhang, C. Thermosensitive Micelles–Hydrogel Hybrid System Based on Poloxamer 407 for Localized Delivery of Paclitaxel. *J. Pharm. Sci.* **2013**, *102*, 2707–2717. [CrossRef] [PubMed]
43. Kelly, H.; Duffy, G.; Rossi, S.; Hastings, C. A Thermo-Responsive Hydrogel for Intratumoral Administration as a Treatment in Solid Tumor Cancers. WO2019092049A1, 7 November 2018.
44. Phogat, K.; Ghosh, S.B.; Bandyopadhyay-Ghosh, S. Recent advances on injectable nanocomposite hydrogels towards bone tissue rehabilitation. *J. Appl. Polym. Sci.* **2022**, *140*, e53362. [CrossRef]
45. Soni, G.; Yadav, K.S. High encapsulation efficiency of poloxamer-based injectable thermoresponsive hydrogels of etoposide. *Pharm. Dev. Technol.* **2014**, *19*, 651–661. [CrossRef]
46. Rossi, S.M.; Murray, T.E.; Cassidy, J.; Lee, M.J.; Kelly, H.M. A Custom Radiopaque Thermoresponsive Chemotherapy-Loaded Hydrogel for Intratumoural Injection: An In Vitro and Ex Vivo Assessment of Imaging Characteristics and Material Properties. *Cardiovasc. Interv. Radiol.* **2018**, *42*, 289–297. [CrossRef]
47. Norouzi, M.; Firouzi, J.; Sodeifi, N.; Ebrahimi, M.; Miller, D.W. Salinomycin-loaded injectable thermosensitive hydrogels for glioblastoma therapy. *Int. J. Pharm.* **2021**, *598*, 120316. [CrossRef]
48. Chung, C.K.; Fransen, M.F.; van der Maaden, K.; Campos, Y.; García-Couce, J.; Kralisch, D.; Chan, A.; Ossendorp, F.; Cruz, L.J. Thermosensitive hydrogels as sustained drug delivery system for CTLA-4 checkpoint blocking antibodies. *J. Control. Release* **2020**, *323*, 1–11. [CrossRef]
49. Mao, Y.; Li, X.; Chen, G.; Wang, S. Thermosensitive Hydrogel System With Paclitaxel Liposomes Used in Localized Drug Delivery System for In Situ Treatment of Tumor: Better Antitumor Efficacy and Lower Toxicity. *J. Pharm. Sci.* **2016**, *105*, 194–204. [CrossRef]
50. Galante, R.; Pinto, T.D.J.A.; Colaco, R.; Serro, A.P. Sterilization of hydrogels for biomedical applications: A review. *J. Biomed. Mater. Res. Part B Appl. Biomater.* **2017**, *106*, 2472–2492. [CrossRef]
51. Ferreira, I.; Marques, A.C.; Costa, P.C.; Amaral, M.H. Effects of Steam Sterilization on the Properties of Stimuli-Responsive Polymer-Based Hydrogels. *Gels* **2023**, *9*, 385. [CrossRef]
52. Burak, J.; Grela, K.P.; Pluta, J.; Karolewicz, B.; Marciniak, D.M. Impact of sterilisation conditions on the rheological properties of thermoresponsive pluronic F-127-based gels for the ophthalmic use. *Acta Pol. Pharm.—Drug Res.* **2018**, *75*, 471–481. [CrossRef] [PubMed]
53. He, H.; Liu, L.; Morin, E.E.; Liu, M.; Schwendeman, A. Survey of Clinical Translation of Cancer Nanomedicines—Lessons Learned from Successes and Failures. *Accounts Chem. Res.* **2019**, *52*, 2445–2461. [CrossRef] [PubMed]
54. Huo, Y.; Wang, Q.; Liu, Y.; Wang, J.; Li, Q.; Li, Z.; Dong, Y.; Huang, Y.; Wang, L. A temperature-sensitive phase-change hydrogel of topotecan achieves a long-term sustained antitumor effect on retinoblastoma cells. *OncoTargets Ther.* **2019**, *12*, 6069–6082. [CrossRef]
55. Chung, C.K.; García-Couce, J.; Campos, Y.; Kralisch, D.; Bierau, K.; Chan, A.; Ossendorp, F.; Cruz, L.J. Doxorubicin Loaded Poloxamer Thermosensitive Hydrogels: Chemical, Pharmacological and Biological Evaluation. *Molecules* **2020**, *25*, 2219. [CrossRef] [PubMed]
56. Yang, Z.; Nie, S.; Hsiao, W.W.; Pam, W. Thermoreversible Pluronic® F127-based hydrogel containing liposomes for the controlled delivery of paclitaxel: In vitro drug release, cell cytotoxicity, and uptake studies. *Int. J. Nanomed.* **2011**, *6*, 151–166. [CrossRef]
57. Yu, M.; Zhang, C.; Tang, Z.; Tang, X.; Xu, H. Intratumoral injection of gels containing losartan microspheres and (PLG-g-mPEG)-cisplatin nanoparticles improves drug penetration, retention and anti-tumor activity. *Cancer Lett.* **2018**, *442*, 396–408. [CrossRef]
58. Gao, L.; Wang, X.; Ma, J.; Hao, D.; Wei, P.; Zhou, L.; Liu, G. Evaluation of TPGS-modified thermo-sensitive Pluronic PF127 hydrogel as a potential carrier to reverse the resistance of P-gp-overexpressing SMMC-7721 cell lines. *Colloids Surf. B Biointerfaces* **2016**, *140*, 307–316. [CrossRef]
59. Hu, H.; Lin, Z.; He, B.; Dai, W.; Wang, X.; Wang, J.; Zhang, X.; Zhang, H.; Zhang, Q. A novel localized co-delivery system with lapatinib microparticles and paclitaxel nanoparticles in a peritumorally injectable in situ hydrogel. *J. Control. Release* **2015**, *220*, 189–200. [CrossRef]
60. Díaz-Rodríguez, P.; Landin, M. Smart design of intratumoral thermosensitive β-lapachone hydrogels by Artificial Neural Networks. *Int. J. Pharm.* **2012**, *433*, 112–118. [CrossRef]
61. Seoane, S.; Díaz-Rodríguez, P.; Sendon-Lago, J.; Gallego, R.; Perez-Fernandez, R.; Landin, M. Administration of the optimized β-Lapachone–poloxamer–cyclodextrin ternary system induces apoptosis, DNA damage and reduces tumor growth in a human breast adenocarcinoma xenograft mouse model. *Eur. J. Pharm. Biopharm.* **2013**, *84*, 497–504. [CrossRef] [PubMed]
62. Jhan, H.-J.; Liu, J.-J.; Chen, Y.-C.; Liu, D.-Z.; Sheu, M.-T.; Ho, H.-O. Novel injectable thermosensitive hydrogels for delivering hyaluronic acid–doxorubicin nanocomplexes to locally treat tumors. *Nanomedicine* **2015**, *10*, 1263–1274. [CrossRef]
63. Sheu, M.-T.; Jhan, H.-J.; Su, C.-Y.; Chen, L.-C.; Chang, C.-E.; Liu, D.-Z.; Ho, H.-O. Codelivery of doxorubicin-containing thermosensitive hydrogels incorporated with docetaxel-loaded mixed micelles enhances local cancer therapy. *Colloids Surf. B Biointerfaces* **2016**, *143*, 260–270. [CrossRef]

64. Yang, Y.; Wang, J.; Zhang, X.; Lu, W.; Zhang, Q. A novel mixed micelle gel with thermo-sensitive property for the local delivery of docetaxel. *J. Control. Release* **2009**, *135*, 175–182. [CrossRef] [PubMed]
65. Gao, M.; Xu, H.; Zhang, C.; Liu, K.; Bao, X.; Chu, Q.; He, Y.; Tian, Y. Preparation and characterization of curcumin thermosensitive hydrogels for intratumoral injection treatment. *Drug Dev. Ind. Pharm.* **2013**, *40*, 1557–1564. [CrossRef]
66. Bai, R.; Deng, X.; Wu, Q.; Cao, X.; Ye, T.; Wang, S. Liposome-loaded thermo-sensitive hydrogel for stabilization of SN-38 via intratumoral injection: Optimization, characterization, and antitumor activity. *Pharm. Dev. Technol.* **2017**, *23*, 106–115. [CrossRef] [PubMed]
67. Basso, J.; Miranda, A.; Nunes, S.; Cova, T.; Sousa, J.; Vitorino, C.; Pais, A. Hydrogel-Based Drug Delivery Nanosystems for the Treatment of Brain Tumors. *Gels* **2018**, *4*, 62. [CrossRef]
68. Shen, M.; Xu, Y.-Y.; Sun, Y.; Han, B.-S.; Duan, Y.-R. Preparation of a Thermosensitive Gel Composed of a mPEG-PLGA-PLL-cRGD Nanodrug Delivery System for Pancreatic Tumor Therapy. *ACS Appl. Mater. Interfaces* **2015**, *7*, 20530–20537. [CrossRef]
69. Xie, M.-H.; Ge, M.; Peng, J.-B.; Jiang, X.-R.; Wang, D.-S.; Ji, L.-Q.; Ying, Y.; Wang, Z. In-vivo anti-tumor activity of a novel poloxamer-based thermosensitive in situ gel for sustained delivery of norcantharidin. *Pharm. Dev. Technol.* **2018**, *24*, 623–629. [CrossRef]
70. Cunha-Filho, M.S.S.; Alvarez-Lorenzo, C.; Martínez-Pacheco, R.; Landin, M. Temperature-Sensitive Gels for Intratumoral Delivery of β-Lapachone: Effect of Cyclodextrins and Ethanol. *Sci. World J.* **2012**, *2012*, 126723. [CrossRef]
71. Jhan, H.J.; Ho, H.O.; Sheu, M.T.; Shen, S.C.; Ho, Y.S.; Liu, J.J. Thermosensitive Injectable Hydrogel for Drug Delivery. U.S. Patent 9,364,545, 14 June 2016.
72. Emami, J.; Rezazadeh, M.; Akbari, V.; Amuaghae, E. Preparation and characterization of an injectable thermosensitive hydrogel for simultaneous delivery of paclitaxel and doxorubicin. *Res. Pharm. Sci.* **2018**, *13*, 181–191. [CrossRef] [PubMed]
73. Li, J.; Pan, H.; Qiao, S.; Li, Y.; Wang, J.; Liu, W.; Pan, W. The utilization of low molecular weight heparin-poloxamer associated Laponite nanoplatform for safe and efficient tumor therapy. *Int. J. Biol. Macromol.* **2019**, *134*, 63–72. [CrossRef] [PubMed]
74. Li, T.; Zhang, M.; Wang, J.; Wang, T.; Yao, Y.; Zhang, X.; Zhang, C.; Zhang, N. Thermosensitive Hydrogel Co-loaded with Gold Nanoparticles and Doxorubicin for Effective Chemoradiotherapy. *AAPS J.* **2015**, *18*, 146–155. [CrossRef] [PubMed]
75. Xie, Z.; Shen, J.; Sun, H.; Li, J.; Wang, X. Polymer-based hydrogels with local drug release for cancer immunotherapy. *Biomed. Pharmacother.* **2021**, *137*, 111333. [CrossRef] [PubMed]
76. Overchuk, M.; Weersink, R.A.; Wilson, B.C.; Zheng, G. Photodynamic and Photothermal Therapies: Synergy Opportunities for Nanomedicine. *ACS Nano* **2023**, *17*, 7979–8003. [CrossRef]
77. Fu, J.-J.; Zhang, J.-Y.; Li, S.-P.; Zhang, L.-M.; Lin, Z.-X.; Liang, L.; Qin, A.-P.; Yu, X.-Y. CuS Nanodot-Loaded Thermosensitive Hydrogel for Anticancer Photothermal Therapy. *Mol. Pharm.* **2018**, *15*, 4621–4631. [CrossRef]
78. Fu, J.; Wu, B.; Wei, M.; Huang, Y.; Zhou, Y.; Zhang, Q.; Du, L. Prussian blue nanosphere-embedded in situ hydrogel for photothermal therapy by peritumoral administration. *Acta Pharm. Sin. B* **2018**, *9*, 604–614. [CrossRef]
79. Chen, X.; Tao, J.; Zhang, M.; Lu, Z.; Yu, Y.; Song, P.; Wang, T.; Jiang, T.; Zhao, X. Iota carrageenan gold-silver NPs photothermal hydrogel for tumor postsurgical anti-recurrence and wound healing. *Carbohydr. Polym.* **2022**, *298*, 120123. [CrossRef]
80. Yao, J.; Zhu, C.; Peng, T.; Ma, Q.; Gao, S. Injectable and Temperature-Sensitive Titanium Carbide-Loaded Hydrogel System for Photothermal Therapy of Breast Cancer. *Front. Bioeng. Biotechnol.* **2021**, *9*, 791891. [CrossRef]
81. Zhang, N.; Xu, X.; Zhang, X.; Qu, D.; Xue, L.; Mo, R.; Zhang, C. Nanocomposite hydrogel incorporating gold nanorods and paclitaxel-loaded chitosan micelles for combination photothermal–chemotherapy. *Int. J. Pharm.* **2016**, *497*, 210–221. [CrossRef]
82. Qin, L.; Ling, G.; Peng, F.; Zhang, F.; Jiang, S.; He, H.; Yang, D.; Zhang, P. Black phosphorus nanosheets and gemcitabine encapsulated thermo-sensitive hydrogel for synergistic photothermal-chemotherapy. *J. Colloid Interface Sci.* **2019**, *556*, 232–238. [CrossRef]
83. Huang, S.; Ma, Z.; Sun, C.; Zhou, Q.; Li, Z.; Wang, S.; Yan, Q.; Liu, C.; Hou, B.; Zhang, C. An injectable thermosensitive hydrogel loaded with a theranostic nanoprobe for synergistic chemo–photothermal therapy for multidrug-resistant hepatocellular carcinoma. *J. Mater. Chem. B* **2022**, *10*, 2828–2843. [CrossRef] [PubMed]
84. Correia, J.H.; Rodrigues, J.A.; Pimenta, S.; Dong, T.; Yang, Z. Photodynamic Therapy Review: Principles, Photosensitizers, Applications, and Future Directions. *Pharmaceutics* **2021**, *13*, 1332. [CrossRef] [PubMed]
85. Luo, L.; Zhang, Q.; Luo, Y.; He, Z.; Tian, X.; Battaglia, G. Thermosensitive nanocomposite gel for intra-tumoral two-photon photodynamic therapy. *J. Control. Release* **2019**, *298*, 99–109. [CrossRef] [PubMed]
86. Li, R.; Shan, L.; Yao, Y.; Peng, F.; Jiang, S.; Yang, D.; Ling, G.; Zhang, P. Black phosphorus nanosheets and docetaxel micelles co-incorporated thermoreversible hydrogel for combination chemo-photodynamic therapy. *Drug Deliv. Transl. Res.* **2021**, *11*, 1133–1143. [CrossRef]
87. Esfahani, K.; Roudaia, L.; Buhlaiga, N.; Del Rincon, S.V.; Papneja, N.; Miller, W.H., Jr. A Review of Cancer Immunotherapy: From the Past, to the Present, to the Future. *Curr. Oncol.* **2020**, *27* (Suppl. S2), S87–S97. [CrossRef]
88. Riley, R.S.; June, C.H.; Langer, R.; Mitchell, M.J. Delivery technologies for cancer immunotherapy. *Nat. Rev. Drug Discov.* **2019**, *18*, 175–196. [CrossRef]
89. Fakhari, A.; Nugent, S.; Elvecrog, J.; Vasilakos, J.; Corcoran, M.; Tilahun, A.; Siebenaler, K.; Sun, J.; Subramony, J.A.; Schwarz, A. Thermosensitive Gel–Based Formulation for Intratumoral Delivery of Toll-Like Receptor 7/8 Dual Agonist, MEDI9197. *J. Pharm. Sci.* **2017**, *106*, 2037–2045. [CrossRef]

90. Tsai, H.-C.; Chou, H.-Y.; Chuang, S.-H.; Lai, J.-Y.; Chen, Y.-S.; Wen, Y.-H.; Yu, L.-Y.; Lo, C.-L. Preparation of Immunotherapy Liposomal-Loaded Thermal-Responsive Hydrogel Carrier in the Local Treatment of Breast Cancer. *Polymers* **2019**, *11*, 1592. [CrossRef]
91. Jin, H.-S.; Choi, D.-S.; Ko, M.; Kim, D.; Lee, D.-H.; Lee, S.; Lee, A.Y.; Kang, S.G.; Kim, S.H.; Jung, Y.; et al. Extracellular pH modulating injectable gel for enhancing immune checkpoint inhibitor therapy. *J. Control. Release* **2019**, *315*, 65–75. [CrossRef]
92. Kim, J.; Archer, P.A.; Manspeaker, M.P.; Avecilla, A.R.; Pollack, B.P.; Thomas, S.N. Sustained release hydrogel for durable locoregional chemoimmunotherapy for BRAF-mutated melanoma. *J. Control. Release* **2023**, *357*, 655–668. [CrossRef]
93. Salah, A.; Wang, H.; Li, Y.; Ji, M.; Ou, W.-B.; Qi, N.; Wu, Y. Insights Into Dendritic Cells in Cancer Immunotherapy: From Bench to Clinical Applications. *Front. Cell Dev. Biol.* **2021**, *9*, 686544. [CrossRef] [PubMed]
94. Lemdani, K.; Seguin, J.; Lesieur, C.; Al Sabbagh, C.; Doan, B.-T.; Richard, C.; Capron, C.; Malafosse, R.; Boudy, V.; Mignet, N. Mucoadhesive thermosensitive hydrogel for the intra-tumoral delivery of immunomodulatory agents, in vivo evidence of adhesion by means of non-invasive imaging techniques. *Int. J. Pharm.* **2019**, *567*, 118421. [CrossRef] [PubMed]
95. Guo, D.-D.; Xu, C.-X.; Quan, J.-S.; Song, C.-K.; Jin, H.; Kim, D.-D.; Choi, Y.-J.; Cho, M.-H.; Cho, C.-S. Synergistic anti-tumor activity of paclitaxel-incorporated conjugated linoleic acid-coupled poloxamer thermosensitive hydrogel in vitro and in vivo. *Biomaterials* **2009**, *30*, 4777–4785. [CrossRef] [PubMed]
96. Guo, D.-D.; Hong, S.-H.; Jiang, H.-L.; Minai-Tehrani, A.; Kim, J.-E.; Shin, J.-Y.; Jiang, T.; Kim, Y.-K.; Choi, Y.-J.; Cho, C.-S.; et al. Synergistic effects of Akt1 shRNA and paclitaxel-incorporated conjugated linoleic acid-coupled poloxamer thermosensitive hydrogel on breast cancer. *Biomaterials* **2012**, *33*, 2272–2281. [CrossRef]
97. Zhao, D.; Song, H.; Zhou, X.; Chen, Y.; Liu, Q.; Gao, X.; Zhu, X.; Chen, D. Novel facile thermosensitive hydrogel as sustained and controllable gene release vehicle for breast cancer treatment. *Eur. J. Pharm. Sci.* **2019**, *134*, 145–152. [CrossRef]
98. Van Hemelryck, S.; Dewulf, J.; Niekus, H.; van Heerden, M.; Ingelse, B.; Holm, R.; Mannaert, E.; Langguth, P. In vitro evaluation of poloxamer in situ forming gels for bedaquiline fumarate salt and pharmacokinetics following intramuscular injection in rats. *Int. J. Pharm. X* **2019**, *1*, 100016. [CrossRef]
99. Alonso, J.M.; del Olmo, J.A.; Gonzalez, R.P.; Saez-Martinez, V. Injectable Hydrogels: From Laboratory to Industrialization. *Polymers* **2021**, *13*, 650. [CrossRef]
100. Cui, N.; Dai, C.-Y.; Mao, X.; Lv, X.; Gu, Y.; Lee, E.-S.; Jiang, H.-B.; Sun, Y. Poloxamer-Based Scaffolds for Tissue Engineering Applications: A Review. *Gels* **2022**, *8*, 360. [CrossRef]
101. Li, Y.; Cao, J.; Han, S.; Liang, Y.; Zhang, T.; Zhao, H.; Wang, L.; Sun, Y. ECM based injectable thermo-sensitive hydrogel on the recovery of injured cartilage induced by osteoarthritis. *Artif. Cells Nanomed. Biotechnol.* **2018**, *46*, 152–160. [CrossRef]
102. Chen, I.-C.; Su, C.-Y.; Chen, P.-Y.; Hoang, T.C.; Tsou, Y.-S.; Fang, H.-W. Investigation and Characterization of Factors Affecting Rheological Properties of Poloxamer-Based Thermo-Sensitive Hydrogel. *Polymers* **2022**, *14*, 5353. [CrossRef] [PubMed]
103. Liu, Y.; Zhang, J.; Guo, Y.; Wang, P.; Su, Y.; Jin, X.; Zhu, X.; Zhang, C. Drug-grafted DNA as a novel chemogene for targeted combinatorial cancer therapy. *Exploration* **2022**, *2*, 20210172. [CrossRef] [PubMed]

gels

Review

Hydrogels in Cutaneous Wound Healing: Insights into Characterization, Properties, Formulation and Therapeutic Potential

Mariana Ribeiro [1,2,3], Marco Simões [2,4], Carla Vitorino [1,3,4,*] and Filipa Mascarenhas-Melo [5,6,*]

1 Faculty of Pharmacy, University of Coimbra, Pólo das Ciências da Saúde, Azinhaga de Santa Comba, 3000-548 Coimbra, Portugal; marianasmribeiro99@gmail.com
2 CISUC—Center for Informatics and Systems, University of Coimbra, Pinhal de Marrocos, 3030-290 Coimbra, Portugal; msimoes@dei.uc.pt
3 Coimbra Chemistry Centre, Institute of Molecular Sciences—IMS, Department of Chemistry, University of Coimbra, 3000-535 Coimbra, Portugal
4 CIBIT—Coimbra Institute for Biomedical Imaging and Translational Research, University of Coimbra, Pólo das Ciências da Saúde, Azinhaga de Santa Comba, 3000-548 Coimbra, Portugal
5 Higher School of Health, Polytechnic Institute of Guarda, Rua da Cadeia, 6300-307 Guarda, Portugal
6 REQUIMTE/LAQV, Department of Pharmaceutical Technology, Faculty of Pharmacy, University of Coimbra, Azinhaga de Santa Comba, 3000-548 Coimbra, Portugal
* Correspondence: csvitorino@ff.uc.pt (C.V.); filipamelo@ff.uc.pt (F.M.-M.); Tel.: +351-239-488-400 (C.V.); +351-271-205-220 (F.M.-M.)

Citation: Ribeiro, M.; Simões, M.; Vitorino, C.; Mascarenhas-Melo, F. Hydrogels in Cutaneous Wound Healing: Insights into Characterization, Properties, Formulation and Therapeutic Potential. *Gels* **2024**, *10*, 188. https://doi.org/10.3390/gels10030188

Academic Editors: Ana Paula Serro, Ana Isabel Fernandes and Diana Silva

Received: 1 February 2024
Revised: 26 February 2024
Accepted: 5 March 2024
Published: 8 March 2024

Abstract: Hydrogels are polymeric materials that possess a set of characteristics meeting various requirements of an ideal wound dressing, making them promising for wound care. These features include, among others, the ability to absorb and retain large amounts of water and the capacity to closely mimic native structures, such as the extracellular matrix, facilitating various cellular processes like proliferation and differentiation. The polymers used in hydrogel formulations exhibit a broad spectrum of properties, allowing them to be classified into two main categories: natural polymers like collagen and chitosan, and synthetic polymers such as polyurethane and polyethylene glycol. This review offers a comprehensive overview and critical analysis of the key polymers that can constitute hydrogels, beginning with a brief contextualization of the polymers. It delves into their function, origin, and chemical structure, highlighting key sources of extraction and obtaining. Additionally, this review encompasses the main intrinsic properties of these polymers and their roles in the wound healing process, accompanied, whenever available, by explanations of the underlying mechanisms of action. It also addresses limitations and describes some studies on the effectiveness of isolated polymers in promoting skin regeneration and wound healing. Subsequently, we briefly discuss some application strategies of hydrogels derived from their intrinsic potential to promote the wound healing process. This can be achieved due to their role in the stimulation of angiogenesis, for example, or through the incorporation of substances like growth factors or drugs, such as antimicrobials, imparting new properties to the hydrogels. In addition to substance incorporation, the potential of hydrogels is also related to their ability to serve as a three-dimensional matrix for cell culture, whether it involves loading cells into the hydrogel or recruiting cells to the wound site, where they proliferate on the scaffold to form new tissue. The latter strategy presupposes the incorporation of biosensors into the hydrogel for real-time monitoring of wound conditions, such as temperature and pH. Future prospects are then ultimately addressed. As far as we are aware, this manuscript represents the first comprehensive approach that brings together and critically analyzes fundamental aspects of both natural and synthetic polymers constituting hydrogels in the context of cutaneous wound healing. It will serve as a foundational point for future studies, aiming to contribute to the development of an effective and environmentally friendly dressing for wounds.

Keywords: hydrogels; wound healing; natural polymers; synthetic polymers; critical attributes; critical analysis

1. Introduction

The skin is the largest organ of the human body, accounting for almost 10% of the total body mass [1,2]. It serves as a fundamental anatomical barrier against pathogens and protects the external environment. The skin performs several important functions for maintaining the balance between the biological system and the surrounding environment, such as controlling the thermoregulation process. Furthermore, it is the human organ most frequently injured [1–3].

Thousands of years ago, ancient civilizations like the Greeks and Egyptians used tree bark, turmeric, aloe vera, and honey to treat wounds. The increased perception that injured skin is susceptible to contamination and dehydration boosted the development of both synthetic and natural dressings [4].

Since the 1960s, wound dressings have been considered favorable for wound healing because they create an environment conducive to skin regeneration [2]. The application of wound dressings aims to cover the wound, promote re-epithelialization, prevent mechanical trauma, and protect it from infections [5].

The ideal dressing should ensure a moist environment and have the capacity to absorb tissue exudate while allowing gaseous exchange, which is related to its porosity. It must protect the wound against microorganisms and stimulate tissue regeneration. Additionally, it should be rigid enough to allow for fixation on the wound, while remaining flexible and elastic to adapt to body movements. Moreover, it must be biocompatible and biodegradable, ensuring that its by-products are safe. The dressing should provide mechanical stability, and be widely available and cost-effective [1,2,4–9].

Due to their intrinsic properties, hydrogels fulfill various requirements for an ideal wound dressing [10]. They offer protection against microorganisms and new lesions [3]. Additionally, they can absorb large amounts of water, up to thousands of times their dry weight [5,11]. Therefore, the highly hydrated three-dimensional (3D) polymeric network allows for the maintenance of a high level of moisture in the wound bed [11]. Moreover, they adhere to the wound but are also easily removable. Their transparency facilitates visual inspection of the wound, and they are customizable and easily adapt to the contours of the wound, promoting autolytic debridement (removal of debris and necrotic tissues), and intrinsically stimulating healing through various mechanisms [5]. These mechanisms include promoting angiogenesis (formation and growth of blood vessels) in wounds with poor perfusion, modulating the immune cells within the wound, or enhancing the migration of keratinocytes and fibroblasts in wound healing [12–15]. Hydrogels overcome some limitations of traditional treatments, such as prolonged healing, limited body movement, traumatic removal, and poor regeneration of skin attachments [6].

Depending on the type of polymer that constitutes the hydrogel, it can be classified as natural or synthetic. Natural polymers offer better biocompatibility, while synthetic polymers exhibit improved mechanical strength and adjusted properties [3,9]. Regenerative medicine takes advantage of natural polymers, especially as dressings for wound treatment, due to their intrinsic characteristics of biocompatibility and biodegradability. They easily induce tissue repair and skin regeneration because of their interconnected 3D networks embedded in water or biological fluids, as well as their similarity to the extracellular matrix (ECM) [16].

This review provides a comprehensive overview of dressings developed exclusively from hydrogels, which also identifies the critical attributes of an ideal dressing for wound healing. A detailed and in-depth description of various polymers, both natural and synthetic, is also presented, outlining their origins, and alluding to their structure and intrinsic properties. In addition, a critical analysis is provided, which brings together all the critical attributes identified for polymers when used for wound healing. In addition, experimental studies related to wound healing that have used hydrogels in different approaches will be presented, taking advantage of the potential that hydrogels can offer in this context and for different types of applications. This manuscript is a distinctive review that, in addition to presenting detailed information on polymers and hydrogels, also provides a distinct and high-quality discussion about their critical attributes that are key tools for the successful development of wound-healing dressings.

2. Wound Healing Process

Wound healing comprises four overlapping phases: hemostasis (blood clotting, stopping bleeding), inflammation (inhibition of microbial growth and wound bed preparation), cellular proliferation (stimulation of fibroblast proliferation and migration to cover the wound surface), and matrix remodeling (strengthening of the tissue and collagen synthesis). These phases cooperate to restore and recover the injured tissue [2,9,17].

A high level of angiogenesis is a critical factor for tissue repair and normal wound healing, as an adequate blood supply ensures the transport of oxygen and nutrients to the wound site [18,19]. Simultaneously, the survival of the new tissue and its integration into the surrounding tissue depends on proper vascularization [20]. It has been observed that fibroblasts cultivated in poly (vinyl alcohol) hydrogels with incorporated arginine, a precursor of nitric oxide, which, in turn, is a key signaling molecule in the regulation of angiogenesis and vasodilation, exhibit increased ECM production compared to fibroblasts cultivated in hydrogels without arginine [4].

The focus of most studies is on the development of absorbent dressings for the treatment of wet or exudative wounds. However, not all wounds require the same care. For instance, burns require continuous hydration and not total moisture absorption [6]. Conventional beliefs advocated for keeping wounds in a dry environment. However, George Winter introduced the concept of moist healing [21]. The wet process accelerates wound healing 3 to 5 times because dehydration (dry process) disrupts the ideal conditions necessary to trigger the wound healing process [1]. The "wet wound healing theory" suggests that a moist healing environment increases the activities of cells and enzymes, facilitates skin cell division, and is advantageous for the formation of granulation tissue, thereby promoting the healing process [21,22]. Therefore, it is crucial to keep the wound environment moist [1,22]. This theory has revolutionized the area of wound management, redirecting research focus from traditional passive drying materials to active moisturizing products [21].

Wound healing can be facilitated by cells, namely fibroblasts, which aid in the formation of the granulation layer, while keratinocytes are involved in re-epithelialization [23,24]. It is well established that fibroblasts and keratinocytes interact synergistically. Previous studies have proven that treatment is more advantageous when applying both types of cells and not just one type. Paracrine signaling provides mechanical, biochemical, and structural properties superior to those of isolated cells. Fibroblasts secrete growth factors that stimulate the growth and differentiation of keratinocytes. Keratinocytes, in turn, secrete molecules that promote the proliferation of fibroblasts [8].

Several growth factors regulate the wound healing process, namely the fibroblast growth factor (FGF), which promotes cell migration and proliferation, as well as angiogenesis; epidermal growth factor (EGF), which improves migration and proliferation of fibroblasts, stimulates epithelization, angiogenesis, and induces the secretion of growth factors by fibroblasts; vascular endothelial growth factor (VEGF), the main growth factor responsible for stimulating angiogenesis; and transforming growth factor (TGF-β), which, in addition to promoting angiogenesis, proliferation, and cell migration, induces the secretion of ECM proteins [25].

3. Hydrogel Wound Dressings

The ideal dressing must meet certain requirements to create a conducive environment for wound healing. It should prevent infections, promote the body's self-healing process, adhere to the wound site without causing damage during dressing changes, and adapt to the body's movements, among other characteristics. The key properties, which we will refer to as critical quality attributes that an ideal wound dressing should possess, have been outlined in Table 1 [1,2,4–9,11].

Table 1. Critical attributes of an ideal wound dressing.

Critical Attributes	Justification
Moisture	Ensuring a balanced and moist environment to promote cell migration and proliferation. It is considered as key factor for wound healing. (supplying moisture to the dry wounds and absorbing moisture and exudates from wet wounds)
Absorption	Controlling the level of wound exudate and preventing tissue maceration by effectively absorbing excess fluid
Permeability	Facilitating gaseous exchange (water vapor, O_2, CO_2) to the wound bed to accelerate cellular activity
Protection	Preventing microbial infections that could impede the wound healing process and prolong its duration
Transparency	Enabling clinicians to visualize the wound and monitor the healing process
Mechanical robustness	Mimicking the structure of native skin while being rigid enough to allow fixation on the wound
Flexibility and elasticity	Adapting to body movements and minimizing patient discomfort and pain during application and dressing replacement
Adhesiveness	Providing good adhesion to healthy skin without sticking to the wound itself, facilitating easy removal after re-epithelialization, and preventing secondary injuries in the newly formed tissue
Biocompatibility	Minimizing the risk of immune reactions or side effects
Safety	Ensuring the absence of toxic substances that could cause damage or result in dire consequences; ensuring that products resulting from degradation are safe and follow the normal metabolic pathway
Cost-effectiveness	Supplying a cost-effective wound dressing solution
Availability	Widely available to all patients and healthcare centers

Hydrogels can play various roles in the field of tissue engineering, such as filling spaces, functioning as wound dressings, or serving as drug delivery systems [8]. Interestingly, the first hydrogel was developed in 1960 by Wichterle and Lim to be used as contact lenses [3].

Hydrogels are promising materials for clinical applications, particularly in the treatment of wounds, due to their intrinsic characteristics that align with the essential aspects an ideal dressing should possess. The primary goal is to stimulate and accelerate healing while enhancing the quality of life for patients [2].

As the prefix "hydro" (water) suggests, hydrogels contain water in their composition, up to 96%. However, the hydrogel sheet itself is not wet. Its hydrophilicity creates a microenvironment in the area of the lesion with an adequate moisture content, which is a critical factor for the rapid healing of the wound. Moreover, moisture is essential to ensure cell viability and proper physiological functioning [2–4,8].

The presence of hydrophilic groups (–OH, –SO_3H, –NH_2, –COOH, –$CONH_2$) is what enables hydrogels to bind to water molecules [3]. The high water content facilitates the transmission of water vapor and oxygen [10,26]. The water vapor transmission rate allows for evaluating the ability of dressings to ensure an adequate level of moisture in the wound. In the case of commercial dressings, it should be between 426 and 2047 $g/m^2/day$ [27].

In addition to moisturizing tissues, the 3D structure of hydrogels enables them to absorb and retain excess exudate from the wound surface, as well as remove toxins and water-soluble waste from the wound [2,3,10].

The swelling capacity of hydrogels is intriguing because this behavior increases the pore size, facilitating the adhesion and proliferation of cells within the 3D structure. However, this property is only favorable to a certain extent, as it can negatively affect the mechanical properties and integrity of the surrounding tissues [4].

Their dense network per se and their ability to adhere firmly and evenly to the wound surface prevent bacteria from reaching and invading the wound. Hydrogels are very pliable and soft, and their degradation occurs through hydrolysis. They are available in various shapes and sizes, and their flexibility and elasticity allow them to adapt to different parts of the body, making them atraumatic [2,3,10]. In the clinical context, the mechanical stiffness of wound dressings should fall within the range of normal healthy skin, where Young's modulus varies between 0.42 and 0.85 MPa, allowing for painless movement. If the stiffness falls outside this range, it may compromise the fixation of the dressing to the wound or cause discomfort [6].

Also, the role of the hydrogel depends on the healing phase. For example, during the inflammatory phase, they can intrinsically stimulate the cleaning of the wound bed through the induction of autolytic debridement of the necrotic eschar tissue [2,3,8,10].

Despite the valuable properties of hydrogels and the fact that the polymer network hinders the entry of pathogens, they do not have inherent antimicrobial properties. One of the simplest alternatives is the direct incorporation of an antimicrobial agent, such as minocycline, gentamicin, colistin, or 1% sulfadiazine, either onto the surface or within the hydrogel structure itself (acting as a vector). One problem with these dressings is the rapid release of the drug [3]. In addition to antibacterial drugs, they can effectively transport other bioactive molecules to the wound site. However, the water content decreases as active substances are added [2,3].

Additionally, their structure allows for the deposition and organization of cells, which will differentiate according to specific stimuli and form specific tissue [8]. Moreover, they can release growth factors to promote cell proliferation and stimulate vascular regeneration, aiding in the proliferative phase [5]. Hydrogels also promote the process of keratinization, and their hypoxic and slightly acidic environment promotes angiogenesis at the wound site, inhibits bacterial growth, and attracts cells involved in the wound repair process, such as fibroblasts [2,3]. In addition to participating in the formation of granulation tissue during the healing process, human fibroblasts secrete growth factors, soluble cytokines, and components of the ECM, such as fibronectin, collagen, and hyaluronic acid, that stimulate the proliferation of epithelial cells [8,22].

A study on hydrogels incorporating arginine, a precursor of nitric oxide, which is an important signaling molecule in the regulation of angiogenesis and vasodilation, also demonstrated an increase in ECM production [4,28]. Hydrogels can also be loaded with growth factors. Xiong et al. [29] studied the influence of FGF2 on fibroblast proliferation and found a 40 to 75% increase in the proliferation rate when 100 ng/mL of FGF2 was added. FGF, one of the main promoters of cell proliferation with chemotactic activity, plays an important role in skin healing [4]. Several studies on the incorporation of VEGF into hydrogels have shown improvements in cell proliferation and wound tissue remodeling [30].

Some hydrogel polymers contain RGD (arginine–glycine–aspartic acid) peptide sequences, responsible for interacting with fibronectin and integrin, acting as anchors [6,31]. These sequences are involved in the mechanism of cell adhesion to the ECM and improve cell survival [32], promote migration and proliferation of cells such as keratinocytes and fibroblasts [33], and induce the secretion of growth factors and angiogenic cytokines crucial for ECM remodeling [19]. The type of cell adhesion ligand, particularly RGD peptides, their spatial arrangement within the hydrogel, the combination of several ligands, or the association of ligands and soluble factors can regulate the phenotype and cellular function of the formed tissues [34]. In addition to the described effects, in vitro and in vivo models have demonstrated that RGD peptide enhances the formation of the keratinocyte layer, production of granulation tissue, and strengthening of the wound matrix, thereby improving wound healing [35].

Moreover, their transparency allows for the assessment of even small changes, visual inspection of the wound, and monitoring of the healing process without removing the dressing [2,3,36]. Additionally, removing the dressing negatively interferes with the healing process and should be discouraged [36].

The properties of hydrogel dressings can be enhanced. For instance, the addition of NaCl to the solution during the preparation of type I collagen hydrogels can improve their mechanical properties and optical performance. Similarly, increasing the pH (near the isoelectric point) can enhance their transparency and linear viscoelastic properties [37].

Because hydrogels are designed to come into direct contact with the wound, they should be immunologically neutral [2,3,38]. Hydrogels have shown low rates of adverse effects and low irritation rates [3].

4. Polymers

Natural polymers (Figure 1) are biocompatible and often equivalent to macromolecules recognized by the human body [11,17,20]. However, they have relatively low mechanical strength compared to synthetic polymers [11,16,20]. Additionally, they are susceptible to batch-to-batch variations, which may result in slight differences in physicochemical characteristics [11].

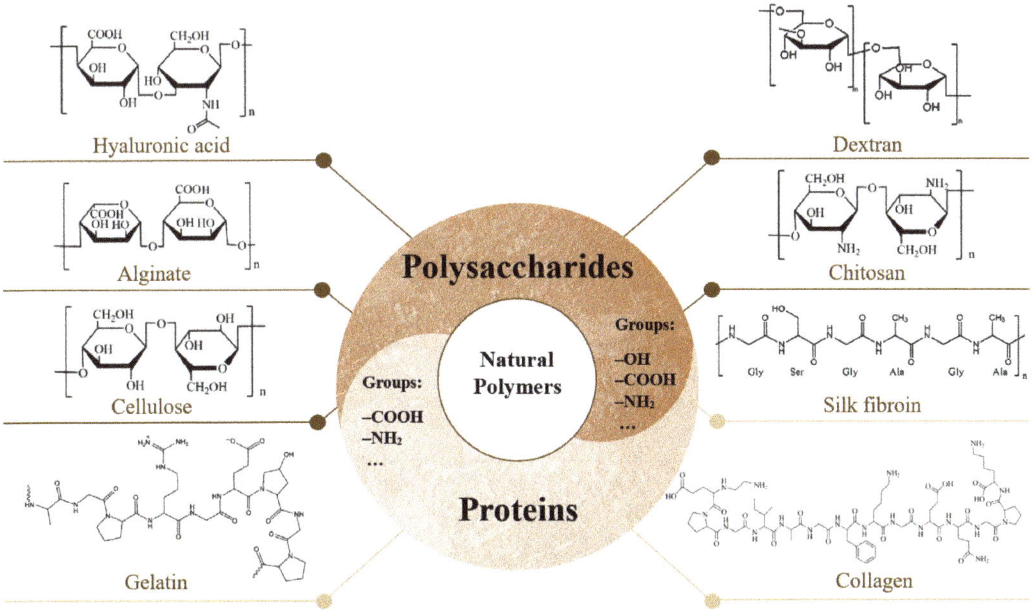

Figure 1. Representation of the chemical structures of some natural polymers classified into proteins (collagen, gelatin, and silk fibroin) and polysaccharides (alginate, hyaluronic acid, cellulose, dextran, and chitosan).

Synthetic polymers (Figure 2) are chemically synthesized and exhibit controllable and predictable properties. They maintain constant and homogeneous physicochemical properties, often displaying excellent mechanical properties and controlled degradation [5,11,17,39,40]. Some synthetic polymers, such as polyesters, are biodegradable and generally more cost-effective compared to natural polymers [40] and have more abundant sources of raw materials [5]. However, unlike natural polymers, which are biologically inert, synthetic polymers carry the associated toxicity risk and may present biocompatibility issues [17,39,40].

Moreover, these materials often require surface treatment or combination with natural polymers to improve cytocompatibility, as they exhibit weak cellular interactions. Surface treatments aim to reduce hydrophobicity through chemical alterations or enhance cell adhesion by adding adhesion peptides, for example [40]. The most commonly used strategy involves combining synthetic and natural polymers [5,9,40,41], as the latter contribute to their natural biological activity due to their similarity to tissues and the native ECM [20,40]. They mimic the natural microenvironment of cells in the human body, facilitating processes such as cell adhesion, proliferation, migration, and differentiation [20]. Blending polymers improves the mechanical properties of natural polymers. However, their biocompatibility may be somewhat affected [16].

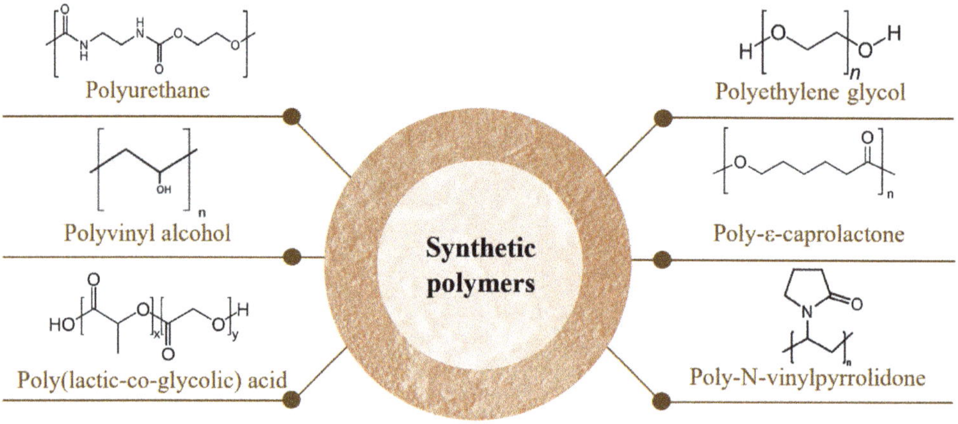

Figure 2. Synthetic polymers and their respective chemical structures.

The selection of the polymeric composition of hydrogels should consider the most suitable characteristics according to the type of wound to be treated and the patient's clinical conditions [2].

Table 2 gathers the main sources for obtaining natural polymers and the primary pathways for the chemical synthesis of synthetic polymers.

Table 2. Main sources for obtaining natural polymers and chemical synthesis pathways of synthetic polymers.

	Polymers	Main Natural Sources/Chemical Synthesis Pathways	References
Natural polymers	Collagen	Bovine, porcine, red algae, fish, from species such as *Prionace glauca*, *Oreochromis niloticus*, and *Lophius litulon*, octopuses, starfish, jellyfish such as *Rhopilema esculentum*, polar bears, whales, seals, and marine sponges	[4,42]
	Gelatin	Bovine, porcine, caprine, mammalian tissues, squid, sponges, jellyfish, and snails	[17]
	Silk Fibroin	Cocoons of Mulberry silkworm *Bombyx mori* and the non-mulberry silkworm *Antheraea assama*	[43]
	Alginate	Brown marine algae such as *Laminaria*, *Ascophyllum*, and *Macrocystis*, red and green marine macroalgae, or bacteria like *Pseudomonas* and *Azotobacter*	[1,34,38,44]
	Hyaluronic acid	Human vitreous humor, joints in the umbilical cord, connective tissue	[45]
	Bacterial cellulose	Bacteria strains such as *Gluconacetobacter*, *Rhizobium*, *Agrobacterium*, *Sarcina*, and *Rhodobacter*	[1,16,46,47]
	Dextran	Bacteria, such as *Leuconostoc*, *Weissella*, *Lactobacillus*, and *Streptococcus*	[48,49]
	Chitosan	Insects, marine invertebrates, and crustaceans such as crabs, shrimp, and lobster, as well as in the cell walls of fungi	[9,16,50–53]

Table 2. *Cont.*

	Polymers	Main Natural Sources/Chemical Synthesis Pathways	References
Synthetic polymers	**Polyurethane**	Result from the polyaddition polymerization between polyols (molecules containing hydroxyl groups) and isocyanates, in the presence of a chain extender and a catalyst	[54–56]
	Polyethylene glycol	Result from the condensation of ethylene oxide and water	[57,58]
	Polyvinyl alcohol	Results from the hydrolysis, aminolysis, or alcoholysis of vinyl acetate	[59]
	Poly-ε-caprolactone	Results from the ring-opening polymerization of ε-caprolactone	[60]
	Poly-N-vinylpyrrolidone	Results from high-pressure free-radical polymerizations	[61]
	Poly(lactic-co-glycolic) acid	Results from the blend of polylactic acid (derives from lactic acid and results from renewable sources such as corn, starch, or sugarcane) and polyglycolic acid (results from the hydrating carbonylation of formaldehyde, with H_2SO_4 as a catalyst, or from the pyrolysis of renewable sources such as sugar beets, sugarcane, cantaloupe, and grapes)	[9,54,60,62]

4.1. *Natural Polymers*

Within natural polymers, we can classify them as proteins (collagen, gelatin, and silk fibroin) and polysaccharides, which can be further subdivided into acidic polysaccharides (alginate and hyaluronic acid), neutral polysaccharides (cellulose and dextran), or basic polysaccharides (chitosan) [16]. Figure 1 schematically represents some of the natural polymers.

4.1.1. Collagen

Collagen is considered the most abundant protein in the human body [16], constituting approximately 30% of the overall protein content [22]. The main types of collagens present in the body are types I, II, III, IV, and V. This amino acid is ubiquitously present in the ECM [4]. Native collagen is insoluble and needs to undergo a pre-treatment where non-covalent bonds are broken to solubilize and extract it [63].

The main source of commercially used collagen isolation is from marine sources [42], representing an abundant and environmentally friendly reservoir of collagen [64]. The process of obtaining collagen from the transformation of by-products of fish skin and scale is inexpensive and has minimal environmental impact [65].

Although different types of collagen have various structural organizations, they all share the fibrillar triple helix structure [66], composed of the tripeptide sequence (Gly-X-Y)n, where Gly (glycine) accounts for 30% of the total amino acids, and X is typically proline while Y is hydroxyproline [37]. Within the three polypeptide chains, there are repetitive sequences of specific amino acids, such as RGD (arginine–glycine–aspartic acid), which interact with integrins present in cells and promote the adhesion and migration of cells such as keratinocytes and fibroblasts [33,35].

This polymer is highly biocompatible [33,38], eco-friendly [67], and bio-stable. It is a promising candidate for hydrogel applications due to its water absorption capacity, as well as its abundance, and plain structure [4]. They are commonly employed as biomimetic materials for the skin matrix, effectively simulating the natural microenvironment, particularly concerning skin elasticity [68]. Collagen dressings are semi-flexible [69], easily applied and removed, and serve as scaffolds for cytokines and growth factors [4]. Collagen-based hydrogels have demonstrated desirable biodegradability, excellent shape consistency at physiological temperature (37 °C), and good formation of micro and macropores, which are important for cell adhesion and proliferation [20,70]. Collagen-based scaffolds exhibit a high level of hydration and transparency, and provide an appropriate surface area for cellular adhesion, as well as the necessary conditions to ensure their viability, migration, and proliferation [70]. Collagen can polymerize in vitro into a fibrillar hydrogel at physiological temperature, ionic strength, and pH [71].

Collagen stimulates the molecular and cellular cascade of wound healing [16], promotes debridement of necrotic tissue [4,16], leads to the synthesis of growth factors that stimulate angiogenesis, and provides a hydrophilic environment that favors re-epithelialization [4]. Collagen acts through a hemostatic effect mechanism [4,46], inducing platelet activation and aggregation, resulting in the release of coagulation factors that lead to the deposition of a fibrin clot at the wound site [33] and increasing the availability of fibronectin, which plays an important role in cell endurance and is essential for the succession of the cell cycle. Collagen plays a crucial role in maintaining cell viability and preserves immune system cells such as macrophages and leukocytes [4].

Collagen dressings inhibit the activity of metalloproteinases (stagnate wound healing in the inflammatory phase), allowing the healing process to proceed [4,46]. In the inflammatory phase, the activation of immune cells stimulates the secretion of pro-inflammatory cytokines, recruiting fibroblasts, endothelial, and epithelial cells. Fibroblasts contribute to collagen synthesis [4,33].

At the same time, collagen and its degradation products are responsible for various cellular processes crucial for wound healing, including differentiation, migration, proliferation, and protein synthesis, including collagen itself [35]. Collagen fragments, in turn, release molecules that attract keratinocytes to the wound area [4] and recruit fibroblasts and endothelial cells to regulate granulation and re-epithelialization [4,33,35], aiding in the formation of new tissue [16], which is more resistant to remodeling [35]. Collagen also acts in the remodeling phase (balance between the degradative activity of metalloproteinases and the formation of the new matrix) determining tensile strength [33] and scar formation [33].

Although marine collagen exhibits a high degree of bio-adhesion, biocompatibility, biodegradability, and low immunogenicity [4], it has a lower amount of hydroxyproline amino acids, resulting in lower thermal resistance [72]. When subjected to chemical or enzymatic hydrolysis, it forms marine collagen peptides with low molecular weight, which have higher hydrophilicity, strong calcium affinity, and are more easily absorbed [25]. These peptides possess physiological characteristics that provide activities such as antioxidant and wound-healing properties [25]. Collagen sponges are commonly used in wound treatment. They are highly porous materials and absorb a significant amount of water [4,9]. Their elastic nature and absorbent behavior make them resistant to bacterial infections [4]. However, their rapid degradation limits in vivo use [9].

Collagen-based dressings of avian, bovine, or porcine origin are recommended for the treatment of full- or partial-thickness wounds with minimal to moderate exudate, but they are contraindicated in allergic or sensitive patients and third-degree burns [16].

Collagen-based hydrogel has some limitations, particularly weak mechanical properties [73,74] and tissue adhesion [75]. Furthermore, when in contact with living tissues, it may induce inflammation [74]. On the other hand, the lack of intrinsic antibacterial activity is also a disadvantage, as it alone cannot protect the wound area from infections [76]. In addition to the high cost of pure collagen, which makes it economically unfeasible for large-scale approaches, collagen degradation results in amino acids with the ability to activate the coagulation cascade and with thrombogenic potential, limiting its application as a biomaterial [37]. It can quickly lose its shape and stability due to enzymatic degradation [77]. Moreover, the long gelation time and low mechanical strength of printed collagen-based hydrogels pose additional challenges [78,79].

In a neutral solvent, at a temperature close to physiological, collagen molecules can autonomously assemble into collagen fibers. The hydrogel is then created through the interaction among these collagen fibers [22]. The in vitro degradation of collagen hydrogels occurs by incubating them with a collagenase solution (5 U/mL) at 37 °C, resulting in complete hydrogel disintegration within 3 h [80].

Almeida Cruz et al. [42] compiled the results of several in vivo (animal models) studies on marine-derived collagen. Overall, there was an increase in the amount of collagen in the wound bed, increased granulation tissue, angiogenesis, and promotion of re-epithelialization in the animals with skin wounds, burns, or injuries treated with collagen or collagen peptides extracted from different marine species.

In vitro studies demonstrated that marine collagen peptides isolated from the skin of *Nibea japonica* increased the proliferation and migration of NIH-3T3 fibroblast cells and were classified as non-cytotoxic and hypoallergenic [25,81]. These peptides also showed antioxidant activity against superoxide anion, hydroxyl, DPPH, and ABTS radicals. In vitro studies (scratch wound assay) showed that collagen peptides significantly improved scratch closure rate, and according to another in vitro test (Western blotting), they increased the expression of proteins such as nuclear factor kB (NF-kB) p65, inhibitory-κB kinase (IKK) α, and β, from the NF-kB signaling pathway, which regulates the transcription of genes associated with cellular functions such as adhesion, migration, proliferation, and cell survival. Moreover, they also increased in vitro (Western blotting) the levels of growth factors such as EGF, FGF, VEGF, and TGF-β, all related to wound healing [25]. These results were similar to those obtained with marine collagen hydrolysate regarding skin repair and tissue regeneration capacity. Collagen isolated from *Chum salmon* and *Nile tilapia* skin also demonstrated excellent wound-healing properties [22].

Feng et al. [18] studied aminated collagen hydrogel, isolated from fish scales, in the healing of full-thickness wounds. Aminated collagen is highly biocompatible, minimally immunogenic, eco-friendly, sustainable, and low-cost. The results demonstrated that the hydrogel improved angiogenesis (in vitro) and was effective in wound healing in vivo (rat model). Mature and organized collagen deposition was observed at the wound site, along with re-epithelialization after 14 days, with the formation of tight junctions between the dermis and epidermis, which are crucial for tissue functional recovery.

Previous studies also indicate that hydrogels with higher collagen concentrations are stable, enhance cell viability, and allow for the expression of genes related to matrix macromolecules and cytokines involved in neovascularization and re-epithelialization. This suggests that concentrated collagen hydrogels could be a novel option for cellular therapy in the treatment of chronic skin wounds [82].

4.1.2. Gelatin

Gelatin is a natural polymer resulting from the hydrolysis and controlled denaturation of collagen at high temperatures [6,11,61,83]. It is ubiquitous [84], eco-friendly [67,85,86], sustainable [85], and recyclable [84].

Similar to collagen, gelatin possesses repetitive sequences of amino acids (Gly-X-Y)n, where X and Y are typically proline and hydroxyproline, respectively [17]. Additionally, it contains RGD motifs (arginine–glycine–aspartic acid), which promote cell adhesion, migration, and proliferation. By providing binding sites for integrins, it facilitates keratinocyte migration and improves tissue remodeling [27,83,87,88].

Moreover, the structure of gelatin is flexible and contains numerous free functional groups, such as hydroxyl, amino, and carboxyl groups [31,89], enabling the modification of its structure through chemical conjugation [27].

Gelatin is highly biocompatible and biodegradable [17,27,83,87]. It is hydrophilic, capable of absorbing large quantities of water and exudate [27,90], and mimics the native ECM, making it interesting for wound treatment applications [11,20,38,83,90]. Moreover, its extraction and synthesis are relatively easy [11,91], and dressings based on gelatin exhibit good transparency [91–94].

Derived from collagen, the main protein of the dermal ECM, gelatin provides the necessary ingredients for dermal regeneration and exerts a positive effect on the biological response [6], without triggering immune responses [6,88]. Compared to collagen, gelatin is cheaper and has lower antigenicity and immunogenicity, as it is partially denatured [27,83]. Gelatin can be used as a drug delivery system or as a scaffold for cell proliferation [17].

This biopolymer positively affects cell viability, mainly due to the RGD peptide sequences, which are responsible for interacting with fibronectin and integrin, acting as anchors and enhancing cell adhesion to the ECM and cell proliferation [6,31]. Despite its cellular adhesion properties, and even though Ma et al.'s study [89] demonstrates that adding gelatin to the polyacrylic acid and polyacrylamide hydrogel enhances the hydrogel's viscosity and adhesion to surfaces like glass and plastic, the tissue adhesive properties of gelatin are insufficient [90]. Therefore, it is often functionalized with dopamine, imparting adhesive properties to the hydrogel due to its structural resemblance to mussel adhesive proteins [88,95,96]. The key to the good adhesive properties of mussels lies in the abundance of catechol groups [88]. In Wang et al.'s study [96], the addition of 0.06% polydopamine to the gelatin–polyacrylamide hydrogel improved the adhesion of the hydrogels on porcine skin without compromising the ease of removal.

Gelatin undergoes easy and low-temperature processing and exhibits a thermoreversible gelation process [84]. At ambient temperature, gelatin forms hydrogels, but at temperatures above 35 °C, it dissolves in water and forms a transparent gel, due to the loss of hydrogen bonds that connect the chains in a triple helix [17,90]. Furthermore, it has a high degradation rate and is not very viscous above 27 ± 1 °C, which limits its printability. To overcome this low formability, gelatin can be combined with other biopolymers such as alginate and silk fibroin or undergo specific modifications such as methacrylation [11,20]. Gelatin methacrylate, in addition to being a biocompatible and biodegradable polymer, has high thermal sensitivity and facilitates cell migration, making it attractive for wound healing applications. Moreover, its ability for UV radiation-induced photopolymerization provides it with mechanical stability, resulting in high shape fidelity. It also exhibits in situ and rapid gelling [20,87].

Gelatin hydrogels lack antibacterial activity [97], and their stability decreases at high temperatures [98], while they exhibit poor mechanical properties [90,95,97,99]. One solution to improve the mechanical properties of gelatin and reduce its water solubility is through crosslinking. The strength and stability of the crosslinking depend on the crosslinking agent, as well as the water absorption capacity. Recently, crosslinking agents such as lactose, citric acid, and genipin have gained attention due to their biocompatibility. Lactose reacts through the Maillard reaction and results in non-enzymatic glycation of the gelatin chains, while citric acid and genipin react with the amino group, but genipin forms a heterocyclic compound [27]. Lactose has a carbonyl group that reacts with the amino group of gelatin. The resulting structure is more compact, reducing the water absorption capacity. Citric acid has three carboxyl groups. Those that do not react can form hydrogen bonds with polar groups of gelatin. The resulting structure is looser, increasing the water absorption capacity [27].

Ren et al. [100] prepared a biodegradable, recyclable, sustainable, and environmentally friendly gelatin hydrogel, avoiding secondary pollution.

Ionescu et al. [90] prepared a film based on a gelatin derivative containing norbornene functionalities and evaluated its wound healing potential in vivo (rat model). The results showed that there was no significant degradation during the 3-week study period. There was a considerable improvement in the wound healing process. Additionally, a high percentage of wound contraction (80% ± 0.3) was observed at the end of 12 days.

4.1.3. Silk Fibroin

Silk fibroin (SF) is a natural protein-based biopolymer [54,60,101–112]. Silk fibers are primarily composed of proteins derived from silkworms, namely fibroin, sericin, and 18 different amino acids [56,113]. Among these, alanine, glycine, and serine residues make up the majority of the amino acid sequence [54,56,101].

Chemically, this fibrous bioproduct possesses abundant amino groups (primary amines), hydroxyl, and carboxyl groups along its molecular chain that are accessible for chemical modifications [101,107,112,114].

Furthermore, this structured amphiphilic copolymer is composed of alternatively repeating units of hydrophilic light chains and disulfide-connected hydrophobic heavy chains [60,101,115]. In solution, SF adopts metastable amorphous conformations such as random coil and α-helix structure (Silk I). When exposed to physical and chemical stimuli, such as shear forces, high silk concentration, low pH, high temperature, vortexing, sonication, cross-linking agents, blending with other polymers, or organic solvent treatment, the amorphous portion of SF can be transformed into stable and insoluble anti-parallel crystallized β-sheets (Silk II) [106,114–118].

In physiological conditions, the SF protein adopts a low-energy β-sheet conformation and tends to aggregate, resulting in hydrogel formation through self-aggregation. This self-assembly approach is relatively simple. However, it can be time-consuming [104,110]. Another technique used to induce hydrogel formation is the ultrasonication technique. The formation of β-sheets occurs through alterations in hydrophobic hydration. The application of ultrasound waves promotes intermolecular interactions of fibers, leading to a structural rearrangement of the protein. This process involves conformational changes from a random coil to a β-sheet, culminating in hydrogel formation [104,110,115]. Previous studies have shown that this technique reduced the gelation time of SF from days to minutes, with the formation of hydrogels reported within 15 to 45 s [110,115].

In general, the SF hydrogel state is preferred for its adaptability and ease of use [111]. The conformation of the SF chain determines the morphological structure and solubility of the hydrogel. The higher the concentration of SF, the greater the content of the β-sheets (Silk II) [117]. In turn, the crystalline Silk II structure imparts greater mechanical resistance and long-term stability to the hydrogel [104,113,116]. The hydrophilic blocks enhance toughness and elasticity [113].

SF possesses excellent biological properties, including outstanding biocompatibility both in vitro and in vivo, and tunable biodegradability, forming non-inflammatory by-products. It also exhibits high tensile strength and robust mechanical properties, along with excellent flexibility, elasticity, and malleability [16,17,54,57,60,104–108,110,112–114,116,117,119,120]. Moreover, it is naturally abundant, bio-sustainable [121], and eco-friendly [121–123].

In addition to addressing biodegradation issues associated with non-biological materials [115], SF can act as a natural strengthening agent, providing the desired mechanical strength and hardness, instead of relying on synthetic polymers and chemicals [106,110,113]. Studies have shown that the addition of small amounts of SF improved the physical and mechanical properties of a collagen scaffold without affecting its biological nature [106].

The SF hydrogel has become a promising biomimetic dressing, given its similarity to Young's modulus of the skin and its ability to adjust mechanical properties to match the desired resilience and elasticity of native tissues. In addition to supporting cell proliferation, it allows for the controlled release of antibacterial molecules and bioactive compounds for skin wound regeneration [110,112,114,117,120].

Furthermore, its low immunogenicity, non-toxicity, good accessibility, low cost (especially when compared with collagen due to the rigorous processing associated with collagen extraction), easy processability, outstanding stability, and chemical modifiability significantly contribute to its application in the biomedical field [54,57,102,108,110,112,115–117,120,124,125].

Moreover, due to the presence of hydrophilic groups on SF backbones that are easily hydrated, such as hydroxyl and carboxyl groups capable of forming hydrogen bonds with water molecules [101,112,114], SF hydrogels can absorb exudate and maintain a moist environment. This helps keep the wound area hydrated, promoting cell adhesion and migration [16,17,107,109,112,114,116].

SF hydrogel dressings exhibit excellent inherent potential in wound healing [17,60,102,107,110,112,116–118,124]. They demonstrate a good water vapor transmission rate, water retention capacity, gelation behavior, and proper oxygen permeation. Additionally, SF hydrogel dressings possess hemostatic properties and support the recruitment of various cell types, including endothelial cells, macrophages, neutrophils, fibroblasts, and keratinocytes. Furthermore, they promote cell proliferation and migration, facilitating re-epithelialization and the formation of granulation tissue [17,104–106,108,109,112,119,126].

In addition to numerous studies reporting the effectiveness of SF hydrogels in drug delivery and the regulation of growth factors [60,104], the literature documents indicate that SF can induce the release of EGF [107].

SF has excellent cell adhesion [125]. In vitro studies have demonstrated that SF-based hydrogels provide mechanical support and act as a matrix for tissue formation, significantly enhancing cell adhesion and proliferation. This effect may be attributed to their capacity to provide nutrients, promote cell–cell interactions, and facilitate the spreading of normal human keratinocytes and fibroblasts under in vitro conditions [17,54,101,107,109,115].

Moreover, SF-based scaffolds downregulate the expression of pro-inflammatory cytokines, suppressing inflammation [115]. Low-molecular-weight SF can activate the intrinsic coagulation cascade, acting during the hemostasis phase [56].

Furthermore, previous studies have shown that SF is used due to its capacity to stimulate collagen synthesis and contribute to the production and deposition of ECM components. This makes a direct contribution to the wound repair and tissue regeneration process [101,107,108,110,112,119].

NF-kB is an important mediator of inflammatory responses. The NF-kB pathway is crucial in the activation and differentiation of innate immune cells and inflammatory T cells [102]. SF can activate the NF-kB signaling pathway through the expression of cyclin D1, vimentin, fibronectin, and VEGF. In addition to modulating cellular activities such as cell adhesion, differentiation, and proliferation, this activation promotes ECM deposition and regulates inflammation and ROS elimination. This signaling pathway is closely related to wound healing and constitutes the underlying mechanism in the healing process of SF-based dressings [102,104,110,116].

Despite the highlighted promising inherent properties, SF hydrogels present some limitations, such as the lack of angiogenic activity [115], and antibacterial and antioxidant properties [110,115,116,124]. Additionally, they have a high gelation time [98].

As a single component, SF may not be sufficient for skin regeneration due to the lack of cell-specific binding sites [109]. Furthermore, the high cutaneous affinity of SF hydrogels allows them to adhere easily to the skin surface without the need for biological adhesives [105,117]. However, in the case of wounds with irregular contours, additional auxiliary fixation is required since they cannot adapt to deformation [111].

The transparency of SF hydrogels sparks controversy. Some authors acknowledge their good transparency [121,125], while others assert that the production of hydrogels leads to microstructures lacking in transparency [127].

The degradation of silk protein is carried out through proteases present in the biological system. The degradability and degradation rate depend on the type of silk, the concentration of SF, and the content of β-sheets [106].

Guan et al. [118] developed an SF-based hydrogel and investigated its mechanism for skin repair and wound healing in a second-degree burn mouse model. The results showed that the SF hydrogel provided a moist environment conducive to wound healing. It improved adhesion and migration of fibroblast and keratinocyte cells in vitro through the Talin 1 pathway, which is closely related to the healing process, and associated proteins (vinculin, paxillin, p-FAK, and FAK). The interconnected porous morphology of the hydrogel, suitable for cell growth, facilitated nutrient delivery to cells and promoted the absorption of cellular debris or metabolic waste. In the in vivo study, the treated wound appeared smooth and moist, with no apparent inflammatory reactions. After 12 days, the burn skin treated with SF hydrogel exhibited morphology and histology similar to normal skin, showing complete wound regeneration without edema or granulation tissue. The hydrogel-treated group demonstrated faster re-epithelialization and wound closure compared to the control groups.

Li et al. [117] investigated the therapeutic effect of an SF-based hydrogel on hypertrophic scars in vivo using rabbit ears. Hypertrophic scars, occurring in 33% to 91% of burn victims, lack an established optimal treatment method. The findings of the study suggest that the SF hydrogel exhibits outstanding inhibitory effects on hypertrophic scars, reducing their thickness and lightening their color. These effects were attributed by the authors to the downregulation of α-smooth muscle actin (α-SMA) expression levels. Additionally, the hydrogel-treated group demonstrated lower collagen fiber density and more organized arrangements than the blank control and scar removal cream groups.

4.1.4. Alginate

Alginate is a natural polysaccharide primarily found in the cell walls of marine algae (macroalgae) in the form of alginic acid. Additionally, it can be obtained from bacteria [34,44]. Bacterial biosynthesis enables the production of alginates with more defined physical properties and chemical structures compared to those derived from marine algae. Moreover, the ability to manipulate bacteria has allowed for the customization of alginate characteristics [34].

Alginate is an unbranched linear anionic polymer composed of repeating units of (1,4)-linked α-L-guluronic acid and β-D-mannuronic acid, referred to as G blocks and M blocks, respectively. Depending on the source from which alginate is extracted, these two types of monomeric acids may be present in varying proportions and assume diverse structural arrangements, thereby influencing the properties of alginate [1,21,34,44].

It is believed that only the G blocks are involved in the intermolecular binding with di and trivalent cations [34,44]. A higher content of G blocks enables the formation of stiffer hydrogels with improved mechanical properties [21] due to the cationic interaction they establish with ions such as calcium, forming a structure resembling an "egg-box" [1,38]. On the other hand, a hydrogel with a higher content of M blocks tends to be less adhesive and is capable of stimulating human monocytes to produce cytokines, making it applicable in wound healing processes [1].

Alginate is an insoluble salt, but it can be extracted through treatment with aqueous alkaline solutions, typically NaOH, resulting in the formation of water-soluble sodium alginate [1,34].

It is a highly porous, hydrophilic, moisture-permeable, biostable, highly biocompatible, biodegradable, and non-toxic polymer. Furthermore, it is widely recognized as safe, readily available in nature, inherently non-adhesive, easy to process, and possesses good resistance in acidic media [1,21,38,39,44,53,128,129] and transparency [78,92]. Moreover, alginate is renewable [130,131], has an eco-friendly nature [131–134], and is cost-effective [1,39,44,129,130,132,134]. However, it still has some shortcomings, including weak mechanical properties and cell adhesion [1,21,44,128], the inability to promote angiogenesis, inhibit microbial infections, slow hemostasis, and difficulty in reducing scar formation [129].

While some researchers have found little to no immune response, others have reported that alginates with a high content of M blocks were immunogenic. This observed immunogenic response was attributed to residual impurities in the alginate itself, such as endotoxins, polyphenolic compounds, and heavy metals, due to its natural origin [34]. Orive et al. [135] demonstrated that purified alginates had fewer impurities and did not induce any significant reaction when implanted in animals (mice).

Alginate is widely used either in its hydrated gel form to provide moisture to dry wounds or in its dry form for wound fluid absorption [34]. It is highly absorbent, particularly in a lyophilized form, capable of absorbing water or body fluids up to twenty times its weight [16,21]. It is especially useful in wounds with moderate to high exudate levels [1,44,77]. Alginate can channel wound exudate towards the surrounding healthy skin, a phenomenon known as "lateral capillarity". However, research indicates that a high content of M blocks hinders this effect [21].

Calcium alginate-based biomaterials (insoluble in water) tend to partially dissolve when in contact with body fluids. As they absorb wound exudate, an ion exchange occurs between calcium (from the alginate) and sodium (from body fluids), releasing calcium ions. This results in the formation of a solubilized gel (sodium alginate), which is conducive to wound healing, providing a moist microenvironment in the wound area that stimulates re-epithelialization, assists in granulation tissue formation and minimizes bacterial infections. They promote autolytic debridement and do not adhere to tissues, making their removal almost painless, accelerating wound closure, and preventing secondary injuries during dressing changes [1,6,16,21,34,38,53,128]. However, the ion exchanges that occur under physiological conditions limit the long-term stability of ionically crosslinked hydrogels [34].

Alginate biopolymers can be applied as drug delivery systems [1,21,34] and are commonly used in biomedical products for the treatment of bleeding/exudating skin wounds and burns due to their hemostatic properties [16,53,77]. On one hand, calcium ions released under physiological conditions promote hemostasis, and on the other hand, the hydrogel itself allows for the aggregation of platelets and red blood cells within its matrix [34]. Alginate stimulates macrophage activation and induces the production of interleukin-6 and tumor necrosis factor α by monocytes, initiating the second phase of the wound healing process, the inflammatory phase [21,128].

The removal of alginate-based dressings causes less pain than traditional dressings and will not induce additional harm to the wound site [1,21]. However, its weak adhesive properties require a secondary dressing to protect and attach it to the wound area [16,53]. Since the inherent cell adhesion properties of alginate are deficient [20,98], recently, alginate derivatives have been prepared by chemically introducing adhesive cell peptides, including the RGD sequence and others like YIGSR (Tyr-Ile-Gly-Ser-Arg) and DGEA (Asp-Gly-Glu-Ala) [34].

Alginate exhibits relevant rheological properties such as viscosity, rapid gelation, and the ability to stabilize dispersions [1,98]. However, it has weak mechanical properties and low tensile strength, which limits its application in wound healing [98,128]. Bahadoran et al. [136] prepared a sodium alginate/polyvinyl alcohol hydrogel and observed that an increase in alginate concentration resulted in more porous structures with enhanced swelling capacity, improved elasticity, and degradation rate.

Increasing the molecular weight is one way to enhance the physical properties of the resulting hydrogel. However, an increase in molecular weight raises the viscosity of the alginate solution, which may compromise the viability of cells or proteins incorporated into the mixture due to generated shear forces. The viscosity of the pre-gelation solution and post-gelation stiffness can be controlled independently, so a combination of low- and high-molecular-weight polymers can be used to increase hydrogel elasticity with a minimal increase in solution viscosity [34]. The viscosity of alginate solutions remains unchanged within pH values between 5 and 11. When the pH falls below 5, protonation of the $–COO^-$ groups in alginate occurs, reducing electrostatic repulsion between alginate chains, facilitating their approximation, and forming hydrogen bonds, leading to increased viscosity. When the pH exceeds 11, depolymerization occurs, decreasing viscosity [34].

The functional groups of alginate enable it to easily cross-link with other biopolymers, forming a network that enhances the physical stability of the dressing. The ionic nature of alginate promotes the formation of bioactive interpolymeric complexes with compounds such as chitosan, a cationic polysaccharide [1].

4.1.5. Hyaluronic Acid

Hyaluronic acid (HA) is an animal-derived glycosaminoglycan and is a structural component abundantly present in the ECM found in embryonic mesenchymal tissues [7]. This linear polyanionic heteropolysaccharide consists of alternating units of β-1,3-N-acetyl-D-glucosamine and α-1,4-D-glucuronic acid [17,137].

HA is a highly biocompatible, biodegradable, non-toxic polymer that can be easily chemically modified [7,9,19,38,137,138]. It is a non-adhesive, non-immunogenic [139], and eco-friendly glycosaminoglycan [140,141] with high transparency [142–144]. In the architecture of the ECM, HA plays regulatory roles in water homeostasis, ensuring proper tissue hydration during wound repair [137,138]. It is a semi-flexible [145] and water-soluble macromolecule. HA is one of the most hygroscopic molecules, capable of absorbing and retaining large amounts of water, increasing its volume by up to 1000 times [7,142]. It has several carboxyl and hydroxyl groups in its structure, which impart a highly hydrophilic character to it, enabling it to absorb exudate [45]. It provides a moist environment and allows for oxygen permeation [142,146]. At physiological pH, HA carries a negative charge. The acetamido and carboxyl groups in its structure form hydrogen bonds with water molecules, stabilizing the polymer's secondary structure [45].

Most approaches to obtaining HA hydrogel dressings involve high costs and tedious multi-step reactions [45,146]. Recently, advances in HA extraction technologies have led to a more stable and cost-effective form of HA [142].

HA-based hydrogels are reabsorbable and easy to scale up [98]. However, HA hydrogels have relatively weak mechanical properties [45,147], require crosslinking for stability [98], exhibit low adhesion strength in humid environments [147], and undergo rapid enzymatic degradation (through the action of hyaluronidases) in physiological environments [39,45]. However, its esterification enhances its stability, mechanical properties, and degradation rate [7,45]. Moreover, crosslinking gelatin with HA-based products allows for increasing the initial HA degradation time from one week to several months [31]. Clinical cases will be presented for two examples of medical devices that use ester derivatives of HA later on, namely, Hyalomatrix and Hyalosafe. This chemical modification reduces hydrophilic components and increases hydrophobic groups, aiming to enhance its stability while maintaining its biological properties and initial safety profile. Solubility depends on the percentage of esterification, which is a controllable process. In vivo data have demonstrated that the degradation of this material is safe and occurs through ester bond hydrolysis, resulting in gradual polymer hydration, making it increasingly resemble native HA. Furthermore, studies have confirmed its biocompatibility. Hyalomatrix and Hyalosafe are examples of two medical devices that use ester derivatives of HA. When used in burns, these are alternatives that promote functional recovery and effective repair of the burned area [7].

HA accelerates wound healing through various mechanisms [137]. It provides 3D support to the extracellular space, is involved in the proliferation and migration of cells such as fibroblasts and keratinocytes, stimulates angiogenesis, enables important complex interactions for the healing process, and contributes to the organized and structured construction of newly formed tissue. It is a valuable option, especially for application in deep burns, due to its properties that facilitate nerve regeneration while simultaneously limiting scar formation [7,19].

High concentrations of HA have been reported in processes of regeneration, remodeling, and morphogenesis. Interestingly, this molecule is involved in the early stages of wound healing and tissue repair, assisting in the organization of endothelial cells and fibroblasts at the site of injury [7].

In addition to its role in organizing the ECM, depending on the molecular weight, HA can perform different biological functions [7]. High-molecular-weight HA allows for the formation of structures with increased stability, viscosity, and viscoelasticity. It is considered low-molecular-weight if values are below $1–25 \times 10^4$ Da, and high-molecular-weight if values are above 1×10^6 Da [45].

After injury, platelets release high-molecular-weight HA, which accumulates at the wound site and leads to fibrinogen deposition and clot formation. HA recruits neutrophils involved in the removal of dead tissue and phagocytosis of debris and induces the release of interleukin 1β, interleukin 8, and tumor necrosis factor-α. It modulates the inflammatory phase of the wound healing process, exerting an anti-inflammatory effect and regulating early inflammation [45,137]. Experimental models have demonstrated that high-molecular-weight HA inhibits angiogenesis, preventing the supply of oxygen and nutrients, and consequently, tissue regeneration [7,148]. The secretion of inflammatory cytokines contributes to the fragmentation of high-molecular-weight HA into low-molecular-weight HA, which, in turn, recruits monocytes and leukocytes. It modulates inflammation by its activity on free radicals, its antioxidant effect, and the exclusion of lytic enzymes from the cell. Low-molecular-weight HA is pro-angiogenic, stimulates the production of pro-inflammatory cytokines and growth factors responsible for ECM remodeling, and regulates the migration and proliferation of fibroblasts and keratinocytes, aiding in the proliferative and remodeling phases of the wound healing process. Fibronectin and low-molecular-weight HA play an important role in wound contraction, as they induce their differentiation into myofibroblasts and guide fibroblast proliferation, essential for collagen deposition, which will form the new granulation tissue matrix, rich in HA itself. In the re-epithelialization phase, low-molecular-weight HA interacts with keratinocytes, regulating the re-epithelialization process [7,45,137]. Furthermore, previous studies have demonstrated that enzymes involved in the degradation of HA promote cell proliferation, providing additional evidence that HA must be broken down to enhance cell growth [148].

Some studies suggest that HA is not suitable for cell adhesion and proliferation [45,145]. While some authors attribute the inability to support cell attachment to insufficient strength [145], others presume that it is due to the thermodynamic polyanionic and hydrophilic characteristics of HA materials, hindering the adhesion of cells to anionic surfaces. Biomaterials based on high-molecular-weight HA have shown lower adhesion compared to lower-molecular-weight counterparts and are, therefore, used in situations where preventing adhesions is desired, such as post-surgery [31]. Cross-linking it with biopolymers containing free amine groups, such as gelatin, can enhance its cell adhesion and proliferation properties due to the much more stable amide bond compared to ester bonds [31]. The main agents for cross-linking HA chains include carbodiimides, polyfunctional epoxides, hydrazides, sulfides, and aldehydes. Carbodiimides (water-soluble) are the most commonly used due to their non-cytotoxicity, biocompatibility, and the fact that they are not incorporated into the structure after cross-linking. Additionally, their cross-linking reaction is gentle and easily controlled [31].

The rheological properties of HA depend on the pH, temperature, and ionic strength of the solution. HA undergoes hydrolytic degradation when the pH of the solution is below 4 or above 11, resulting in a decrease in viscosity and the integrity of the polymer network. This aspect is significant because the pH of the wound varies during the healing stages. After an injury, the pH increases at the wound site, reaching approximately eight and gradually decreasing as the healing process progresses until it reaches five when healing is complete [45].

Based on current clinical experience, medical devices incorporating HA are considered a safe and effective therapeutic alternative in burn treatment, demonstrating superior outcomes in wound healing compared to standard care [7].

In the context of wound healing, exogenous HA has been the subject of investigation, with promising results emerging. Preliminary in vivo studies have shown that the topical application of HA promotes skin regeneration in hamsters and rats. However, this polymer possesses limited properties regarding its residence time and solubility, necessitating chemical modifications to enhance resistance to degradation, prolong the in vivo residence period, achieve other physicochemical characteristics, and explore different production methods [7].

A retrospective study involving 11 burn treatment centers was conducted, encompassing 57 patients, with 31 presenting deep partial-thickness burns and 22 full-thickness burns. Hyalomatrix was applied, and medical follow-up was extended for 37 days. Notably, after 7 days, enhanced re-epithelialization was observed in cases of deep partial-thickness burns. By the 37th day, complete wound closure had been achieved in 85.7% of patients. Hyalomatrix, an advanced, flexible, and conformable dressing, consists of two layers: a thin, transparent silicone layer on top and a 3D fibrous matrix layer in contact with the wound, composed of ester-derived HA. Upon contact with the wound, the biodegradable fibrous matrix spontaneously integrates and undergoes hydrolysis, releasing HA [7]. In a comparable study involving 300 individuals with deep partial-thickness burns treated using Hyalomatrix, burns healed within 21 days for 83% of cases. Hypertrophic scars, evident in almost all individuals, disappeared within one year post-healing in 90% of patients and after two years in 96% of cases. The incidence of infections decreased from 29.5% to

10% [7]. The study's findings underscore the effectiveness of Hyalomatrix as a regenerative matrix, providing support for cell migration and deposition from the wound margins while promoting the organization of constituents within the matrix, including fibroblasts and endothelium in the injured area. The top layer serves as a physical barrier against microorganisms, prevents excessive fluid loss, and allows for the monitoring of the healing progress without necessitating dressing removal due to its transparent nature. Importantly, this layer was designed not to cause pain or damage to newly formed tissue upon removal, demonstrating a favorable safety profile, even in pediatric populations. For burn victims undergoing surgical wound preparation involving necrotic tissue removal, this advanced device safeguards the residual dermal layer and stimulates tissue regeneration from the wound margins and cutaneous appendages [7].

Hyalosafe, another HA ester, serves as a dressing specifically designed for effective coverage in treating first and superficial second-degree burns. This transparent HA film is directly applied to the wound, creating and maintaining an optimal level of moisture in the wound area. This, in turn, establishes favorable conditions for rapid epithelial renewal without the risk of tissue maceration. The degradation of ester bonds releases HA, actively promoting re-epithelialization. Importantly, this membrane is non-adherent, ensuring a painless removal process [7].

Facial burns pose significant challenges, often requiring hospitalization and accounting for 25% of pediatric burns. Due to the unique characteristics of facial tissue, there is a notable risk of fluid loss. A study involving 40 children demonstrated that Hyalosafe exhibits excellent wound-healing properties in second-degree facial burns, leading to significant aesthetic outcomes. Remarkably, there were no reports of wound infection [7].

4.1.6. Cellulose

Cellulose stands as the most abundant biopolymer obtained from natural sources [47,51,52,149,150]. It is considered the safest material on earth and an endless reservoir of resources to develop environmentally friendly materials. It is biocompatible, biodegradable, and possesses good mechanical strength and flexibility [149–151]. Additionally, it is eco-friendly, renewable, and low-cost [130,131,149–154]. Consisting of a linear and unbranched homopolysaccharide composed of β-D-glucopyranose units linked by β-1,4 glycosidic bonds [17,47], cellulose is readily available, with wood being the primary natural source, but it is also found in plants such as cotton and flax, as well as in vegetables, fruits, and biowaste [47,149,150]. In the majority of biowaste materials, cellulose forms fibril structures enveloped within a matrix composed of lignin and hemicellulose. Cellulose exhibits some limitations, notably low solubility in both water and most organic solvents due to the presence of strong inter- and intramolecular hydrogen bonds and van der Waals forces [149,150]. Moreover, cellulose is challenging to hydrolyze, and both acidic and enzymatic hydrolysis (cellulase) can result in the decomposition of the cellulose molecules [150].

Cellulose can also be produced by seaweed, fungi, and bacteria [36,47]. Bacterial cellulose (BC), as the name suggests, originates from bacteria [16,46].

BC is chemically equivalent to vegetal cellulose. However, it does not contain by-products such as hemicellulose, pectin, and lignin. BC is obtained through fermentation, and any secondary metabolites, nutrients, and other substances it may contain are easily removed [47,150]. Unlike vegetal cellulose, BC does not require purification because it is already obtained in a pure form (with a high degree of purity), allowing for its nearly direct use [46,47].

Other advantages of BC over vegetal cellulose include high porosity, high water absorption capacity (due to a high number of available hydroxyl groups), a higher degree of crystallinity, mechanical robustness, and permeability to gases and liquids. Despite having a molecular formula similar to that of vegetal cellulose, BC exhibits significant differences in its physical and mechanical properties. BC forms cross-linked networks of fibrils that are 100 times finer than those derived from vegetal cellulose, imparting higher elasticity, flexibility, resistance, crystallinity, and surface area to the 3D network [47,150].

Therefore, BC is a natural hydrophilic polysaccharide that is highly biocompatible, biodegradable, permeable, flexible, non-toxic, non-carcinogenic, and hypoallergenic. It possesses a high degree of native purity, high porosity, crystallinity, and the ability to retain large quantities of water. BC also has the potential for chemical modification [8,16,46,47,155]. The two most commonly used techniques for modifying and optimizing the morphological, mechanical, physical, and chemical characteristics of BC and incorporating components like antibiotics are ex situ solution impregnation, which involves the physical absorption or impregnation of molecules into the pure BC chains (without any modification), through the formation of hydrogen bonds due to the presence of hydroxyl groups in BC. Another technique is in situ modification, which involves the incorporation of molecules (included from the beginning in the culture medium) during the synthesis process of the BC fibril network itself [46].

The properties of BC make it suitable for use as a wound dressing. Its dense nanofibrillar network can retain an appropriate amount of moisture in the wound area while absorbing excess exudate [36,46]. The application of BC in hydrogels has gained prominence due to the 3D arrangement of the BC nanofibrillar network. In addition to its similarity to the ECM, its high surface area and porosity provide support for cell proliferation [8,16,46,47]. It is flexible, soft, and easily removable without interfering with wound healing [7,8,13,16,46,53]. Furthermore, it accelerates granulation [16,47], promotes the re-epithelialization process, and serves as a matrix for tissue regeneration, reducing pain and healing time [16,156,157]. Additionally, it exhibits excellent mechanical behavior, similar to synthetic polymers [8,16]. The high mechanical strength, combined with great flexibility, allows cellulose-based dressings to conform to the wound area [47]. Studies have demonstrated that BC stimulates wound re-epithelialization, efficiently improving wound healing. However, it does not possess antibacterial properties [46].

In vitro studies have demonstrated that culturing BC near cells, such as human umbilical vein endothelial cells, adipose stem cells, and fibroblasts, does not alter their morphology or proliferative capacity [36].

Kwak et al. [155] prepared bacterial cellulose membranes (BCM) for the treatment of second-degree burns in Sprague-Dawley (SD) rats. The rats were divided into two groups: one was treated with BCM, and the other with gauze. The results showed that compared to the gauze-treated group, rats treated with BCM exhibited a thicker epidermis and dermis, more blood vessels, reduced mast cell infiltration, decreased expression of VEGF and angiopoietin-1, and increased collagen expression. Additionally, metabolic enzymes indicative of kidney and liver toxicity remained at normal levels. These findings suggest that BCM can enhance the burn healing process by regulating angiogenesis, re-epithelialization, and stimulating connective tissue formation. Furthermore, they did not induce specific kidney or liver toxicity.

4.1.7. Dextran

Dextran is a linear, neutral homopolysaccharide composed of repeated D-glucopyranose units, primarily linked by α-1,6 glycosidic bonds. Additionally, it may include branching α-1,2, α-1,3, and α-1,4 linkages [54,158–167].

Dextran can be produced by various lactic acid bacteria and results from glucose condensation through the activity of a secreted dextransucrase enzyme, which transfers glucose components from sucrose and synthesizes dextrans with different structures, molecular weights, linkages, and branching patterns, depending on the bacterial genus [54,160]. Dextran produced by *Leuconostoc mesenteroides* contains about 5% of α-1,3-glycopyranosidic linkages, while that extracted from *Weissella* strains has a highly linear backbone with only 3–4% α-1,3 branching [49].

This polymer is hydrophilic, biodegradable, highly biocompatible, non-toxic, and non-immunogenic [158–162,168–170]. In addition to being sustainable, it is safe [171] and eco-friendly [133,172,173], and is well known for its low cost and wide natural abundance [130,160–162,167,173].

Lately, dextran-based hydrogels have attracted significant attention as wound-healing dressings due to their biocompatibility and flexible and soft characteristics [169]. In addition, this naturally derived material is promising for soft tissue engineering due to its similarity to the native ECM [158]. It can provide nutrients and essential elements for tissue regeneration and support cell proliferation [168].

It possesses excellent water solubility [163,167,170]. However, the high level of branching (α-1,2, α-1,3, and α-1,4 linkages) negatively impacts its solubility. It has been reported that dextrans with more than 43% of α-1,3 linkages cannot be dissolved in water [48].

This glucose-containing polysaccharide possesses three hydroxyl groups on each glucopyranose unit. These not only provide it with high hydrophilicity and good water binding capacity but also make it available for chemical functionalization and cross-linking [158,160,162,164,165]. In addition to effectively absorbing exudate and maintaining the wound bed with appropriate moisture, dextran hydrogels are transparent, facilitating the monitoring of the healing process [174].

Dextran-based hydrogels can be obtained by incorporating polymerizable groups into dextran to facilitate cross-linking [165]. Moreover, the D-glucopyranose residues can be easily chemically oxidized, resulting in oxidized dextran with aldehyde groups, which, in turn, can react with polymers such as gelatin and chitosan or their derivatives with amino side groups, forming in situ hydrogels with novel tissue adhesive characteristics and hemostatic and antibacterial properties [159].

It has been reported that dextran hydrogel scaffolds can enhance angiogenic properties. Additionally, they can serve as platforms to incorporate growth factors/drugs for skin regeneration [169]. Dextran binds to glucan-specific receptors present in human fibroblasts, which, in turn, induce the expression of growth factors that promote cell proliferation, accelerating wound healing [158,170,175].

It possesses immunomodulatory functions, and it can activate neutrophils and macrophages, stimulate cytokine secretion, and strengthen the immune system [165]. Moreover, dextran can function as a mild reactive oxygen radicals (ROSs) scavenger and potentially reduce excessive platelet activation [163].

However, pure dextran hydrogels present some limitations, particularly concerning low mechanical strength, weak tissue adhesion, poor drug loading ability, and undesirable degradation, which chemical and physical modifications help to overcome [167,168].

The biodegradation of dextran in the human body is controlled by dextranases [160,162,163]. The modification of the dextran hydrogel provided gradual material degradation, allowing for proper host tissue integration and ensuring complete skin regeneration [158].

Shen et al. [54] prepared a dextran hydrogel for the treatment of third-degree burn wounds in pigs, and the hydrogel stimulated rapid wound closure, improved re-epithelialization and ECM remodeling, and promoted angiogenesis in a third-degree porcine burn model. In another study on third-degree burns in mice, the dextran-based hydrogel exhibited good bioactivity, particularly in enhancing angiogenic responses and facilitating complete skin regeneration in vivo throughout wound healing. In addition, it recruited endothelial progenitors and cells, promoted epithelial cell migration to the wound area, and supported epithelial differentiation [54,158,166,169,170].

4.1.8. Chitosan

After cellulose, chitin is the second most abundant polysaccharide on the planet, containing amino and hydroxyl groups in its composition [51,52]. Chitosan results from the partial deacetylation of chitin [9,16,50–53]. This process can occur under harsh alkaline conditions, through treatment with sodium hydroxide (chemical hydrolysis), or in the presence of specific enzymes like chitin deacetylase (enzymatic hydrolysis) [50,52].

There are different degrees of deacetylation, which is an important chemical characteristic of this polymer. During the deacetylation process, the acetyl group ($-C_2H_3O$) is replaced by the amino group ($-NH_2$) in the polymer chain. Chitosan is a copolymer composed of units of D-glucosamine ((1,4)-2-amino-2-deoxy-β-D-glucan) and N-acetyl

D-glucosamine ((1,4)-2-acetamido-2-deoxy-β-D-glucan). To be called "chitosan," the degree of deacetylation must be at least 60%, which means it has 60% D-glucosamine units and 40% N-acetyl D-glucosamine [50–52].

Its solubility depends on the degree of deacetylation and the pH of the solution. Chitosan is soluble in slightly acidic solutions because the pKa value of its amino group is approximately 6.3. When the pH is lower, protonation of the $-NH_2$ groups occurs, converting them into $-NH_3^+$, which increases electrical repulsion, resulting in a soluble cationic polymer [50–52].

Chitosan is a linear cationic hydrophilic polysaccharide that is highly biocompatible and biodegradable [11,38,46,137]. It is low-cost [98,130,132,134,176,177] and possesses good swelling properties [78,178], providing a moist environment and promoting effective hydration of the wound area [179,180]. Furthermore, it is renewable [130,131] and eco-friendly [67,131–134,154,178].

It is an oxygen-permeable biopolymer, non-toxic, with excellent bioadhesive characteristics, tissue adhesiveness, and hemostatic properties [11,16,46,50,53,137,181]. In other words, blood coagulation occurs because chitosan, positively charged, binds to red blood cells due to the negative charge on the cell membrane residues, resulting in hemagglutination [46,50]. The higher the degree of deacetylation, the greater the number of positive charges, and thus, the better the tissue adhesiveness and hemostatic properties [52,182]. Hemostatic activity assists in blood clotting and reduces pain by blocking nerve endings [46]. On the other hand, chitosan adsorbs plasma proteins and fibrinogen, resulting in platelet aggregation [183]. Chitosan has anti-inflammatory activity and wound-healing properties, accelerating wound closure [16,46]. Chitosan speeds up tissue regeneration through wound contraction, stimulating cells such as macrophages and fibroblasts [46]. Furthermore, it exerts an immunomodulatory effect, which also contributes to wound healing [52]. Additionally, its structural similarity to the native ECM promotes cell adhesion and proliferation [137].

Moreover, it exhibits intrinsic antibacterial activity [9,11,16,46,52,53,137], which depends on its molecular weight, degree of deacetylation, concentration, and type of bacteria [184]. Several mechanisms have been proposed to elucidate this activity, with the most widely accepted one being that chitosan chains contain amino groups that can be protonated, giving them a positive charge. Electrostatic interactions occur between the protonated amino groups ($-NH_3^+$) of chitosan and the negative charges of the microbial cell membrane, leading to alteration of cell permeability, resulting in cell membrane lysis and, consequently, cell death [46,50,51,129]. Chitosan demonstrated a bacterial inhibition rate of 95.6% for *Escherichia coli* and 99.2% for *Staphylococcus aureus* [184].

Chitosan degradation produces non-toxic residues, and the degradation rate varies with the degree of deacetylation (lower degrees result in higher biodegradation) and the polymer's molecular weight (shorter chains degrade more rapidly) [50].

Chitosan-based hydrogels can be formed through physical, chemical, or metal ion coordination cross-linking. Chitosan does not require any toxic additives for gelation. Physical bonds, such as hydrogen bonds, electrostatic interactions, or hydrophobic interactions, are reversible and non-covalent. They depend on factors like temperature, pH, and others, making them relatively unstable. The polycationic nature of chitosan, under acidic conditions, allows for the formation of hydrogels through electrostatic interactions with natural or synthetic polyelectrolytes or polyanions. Among natural polyanions, proteins (such as collagen and gelatin) and polysaccharides (like alginate and HA) have been documented. Examples of synthetic polyanions include polyacrylic acid. Hydrogels formed through physical cross-linking are soft, flexible, and non-toxic. However, their main disadvantages include low mechanical strength, the possibility of uncontrolled dissolution (instability), and the difficulty of controlling pore size. Therefore, chemically cross-linked hydrogels have been investigated. Chemically cross-linked hydrogels are much more stable because covalent bonds are irreversible. In this approach, the primary structure (amino groups) undergoes chemical modifications, which can alter some chitosan properties. There may be contamination with catalysts or residues of toxic reagents. Some commonly used cross-linking agents are glutaraldehyde, glyoxal, ethylene glycol, and diglycidyl ether. However, all of these can confer toxicity to the resulting hydrogel. Genipin is a naturally derived cross-linker that has been used as an alternative, but its toxicity is not well-established. Although chemically cross-linked hydrogels may be toxic, this strategy allows for the production of soft, flexible, and stable hydrogels with better control of pore size. In addition to these two strategies, chitosan hydrogels can be formed through coordination complex cross-linking using metal ions such as Pd (II), Pt (II), and Mo (VI). However, they are less suitable for medical use [50].

Carboxymethyl chitosan is derived from chitosan. It is a biocompatible, water-soluble polymer with wound healing, hemostatic, and antibacterial properties [1,129]. In comparison to chitosan, carboxymethyl chitosan has more positive charges, leading to increased electrostatic interactions and, consequently, higher antimicrobial activity [185]. Previous studies suggest that carboxymethyl chitosan exhibits healing properties in vitro and in vivo by activating macrophages, secreting various cytokines, and stimulating fibroblast proliferation [16].

4.2. Synthetic Polymers

Synthetic polymers are also widely used in biomedical applications, particularly as wound dressings in the context of healing. Among the most commonly used synthetic polymers are polyurethane, polyethylene glycol, polyvinyl alcohol, poly-ε-caprolactone, poly-N-vinylpyrrolidone, and poly(lactic-co-glycolic) acid. Figure 2 schematically represents these synthetic polymers.

4.2.1. Polyurethane

Polyurethane (PU) contains repetitive units of the urethane group [54,55]. It is a polymeric resin [186] and has a microstructure comprising soft and hard segments, which imparts antithrombotic properties and excellent mechanical properties [55,187,188]. The soft segment (polyol part) provides better flexibility, while the hard segment (isocyanate part) imparts strength and hardness [189]. It is possible to adjust the physical properties based on the type and content of the segments used in the PU chain, according to the desired final product [190]. For example, the stiffness of the polymer depends on the extent of crosslinking [56].

This synthetic polymer is highly biocompatible [9,39,54,55,60,187,188,191] and possesses versatile features such as high strength, durability, excellent elasticity, and adjustable degradation rates according to the application [39,54,60,190]. In addition to excellent flexibility [9,55,186], PU-based hydrogels also exhibit great transparency, good adhesion, high mechanical and tensile strength, fatigue resistance, and toughness, similar to biological soft tissues [186,192,193]. PU hydrogels are non-toxic, easily modifiable [193], provide good gaseous permeability, and are more economical compared to natural polymers [9].

PU and its derivatives are generally considered biodegradable, with few exceptions [60]. The selection of synthesis compounds for PUs enables control over their degradation rate [191,194]. It forms highly porous and permeable hydrogels [188], with ease of processing and the ability to modify surface functional characteristics [188,190]. Porous tissue scaffolds possess a structure similar to that of the epidermis and are more suitable for skin regeneration [194]. The ability to control pore size and polymer chemical properties allows for the regulation of drug release rates. Therefore, PU-based dressings can be utilized as drug delivery systems [195].

It is a widely used polymer in the biomedical field, especially in wound dressings [187,188,190]. It promotes cell proliferation, and its matrix allows for the regeneration of deeper skin layers such as the dermis [11]. However, it has some limitations, such as the absence of antibacterial properties [188,194,195], and the potential for side effects from its degradation products [39].

The primary application of hydrophilic PUs is in wound dressings. The majority of PUs are hydrophobic. However, it is possible to obtain hydrophilic PUs by introducing hydrophilic groups into the soft-segment PU structure, with polyethylene glycol being the most common hydrophilic functional group [196]. It has been reported that PU cannot completely prevent water permeation due to the presence of various polar functional urethane groups in its chain, which could be critical when incorporating batteries from biological moisture or electronic circuits. One solution involves using a highly hydrophobic PU to prevent water permeability and moisture diffusion [190]. The introduction of polyethylene glycol gradients into the PU backbone enables it to absorb water effectively [193].

It is described in the literature that hydrophilic PU foams are highly absorbent and excellent candidates for use as dressings, as they provide a consistently hydrated environment in the wound area [195]. PU dressings are suitable for the management of highly exudative wounds, such as burns, because PUs effectively absorb exudate and maintain wound moisture due to their small and uniform pores. There is a balance between water absorption, moisture retention, and water loss through evaporation that makes PU effective as a wound dressing [195]. Moreover, they are cost-effective and have a simple manufacturing process [197]. In vitro studies conducted on keratinocyte HaCaT cells indicate that PU foams are non-cytotoxic [195]. Similarly, hydrophilic porous PU sponges exhibit high water absorption and moisture retention capacity, as well as good gas permeability. They are non-cytotoxic and cost-effective [197,198].

Little is known about the tissue adhesion properties of PU-based hydrogels. Wen et al. addressed this by adding tannic acid to the PU-polyethylene glycol hydrogel. Tannic acid, with its chemical structure containing multiple catechol groups that bind to urethane groups and form hydrogen bonds, imparts good adhesion to the PU-polyethylene glycol hydrogel [199].

Although PU is not considered eco-friendly, research has been conducted to develop more environmentally friendly alternatives [200,201]. The isocyanates commonly used in the production of most PUs typically derive from the reaction of amines with highly toxic phosgene compounds [55]. Moreover, the isocyanates themselves are toxic and have been associated with occupational asthma [202] due to their high volatility and the tendency to convert in vitro into aromatic diamines, compounds that are toxic, carcinogenic, and/or mutagenic to humans. Even in minimal amounts, there is a possibility of these toxic residual compounds leaching from PU-based products [55]. The trend is to replace isocyanates with alternatives that are less toxic and greener [202].

Efforts have been made to minimize toxicity, notably through the synthesis of PU using polyisocyanates derived from amino acids that undergo controlled biodegradation, resulting in non-cytotoxic byproducts. However, there is still the possibility of unreacted polyisocyanates remaining in the final product [55]. In an attempt to develop alternatives that do not involve the toxic chemistry of isocyanates, cationic, anionic, and neutral PU hydrogels were prepared through a non-isocyanate route under mild conditions, without the need for catalysts. Hydroxyurethane functions were obtained through the aminolysis of five-membered cyclic carbonates. This approach is safer and more environmentally friendly [202]. Chen et al. [203] prepared a water-borne PU nanofibrous membrane using water in its preparation instead of organic solvents, eliminating environmental and toxicity concerns. The membrane was biocompatible, non-toxic, and biodegraded within 4 to 6 weeks. Waterborne PU is easily functionalized due to its numerous functional groups ($–COONH^-$ and $–COO^-$), allowing it to interact with other chemicals effectively [204]. It has been widely used as an adhesive. It offers several advantages, including good toughness, strong mechanical properties, wear resistance, and excellent adhesion [205]. Additionally, waterborne PU is eco-friendly [189,205–208].

4.2.2. Polyethylene Glycol

Polyethylene glycol (PEG) is a synthetic polyether [54,209] that can have linear or branched structures, including multi-arm or star configurations. The basic structure of PEG comprises a PEG diol with two terminal hydroxyl groups. These groups can be converted into various other functional groups, such as amine, carboxyl, vinyl sulfone, thiol, acetylene, acrylate, and aldehyde groups, enabling their conjugation with other materials [209,210].

PEG is considered harmless to the human body [58], and is a recognized eco-friendly polymer [211]. It is highly biocompatible, non-toxic, non-immunogenic [17,41,54,60,209,212–214], a low-cost polymer [215,216], with good water solubility [17,41,60,212,213], flexibility [60], great permeability [217], and adjustable mechanical properties [209,218]. In addition to being soluble in water, it is also soluble in various organic solvents, such as ethanol and acetone [214]. Moreover, PEG demonstrates good large-scale production capabilities, and hemocompatibility [219].

PEG is often referred to as a "stealth material" [220], due to its resistance to protein adsorption [17,39,40], and it generally resists cell attachment because of its higher affinity for water binding [221]. Regarding its tissue adhesion capability, little is known about PEG on its own. However, Krishnadoss et al. [222] incorporated choline molecules into synthetic polymers such as PEG to enhance their adhesive strength. Another group of researchers reported that PEG alone cannot adhere to tissue wounds; however, PEG derivatives, specifically polyethylene glycol diacrylate, have been utilized as tissue glue, owing to their capacity to adhere to both skin and injured tissues [223].

Studies have shown that PEG can improve the mechanical strength, stability, and degradation rate of hydrogels [54,57,209,210]. However, it has also been reported that PEG hydrogels are brittle and weak [223]. Furthermore, the ease of controlling its chemical composition and scaffold architecture makes this polymer attractive for tissue engineering applications [40]. PEG-based hydrogels exhibit very small variations between batches, providing robust and reproducible substrates [218]. Some authors consider PEG as biodegradable [54,58,214,223], while others assert that it is non-biodegradable [39,40].

This polymer has been studied in the context of wound healing due to its high biocompatibility, transparency, and ability to provide support for healing [41,54,224]. It is a hydrophilic material capable of maintaining a moist environment while absorbing wound exudate [19,30,212]. PEG hydrogels exhibit a high swelling rate [223], allow for gaseous permeation [30], eliminate bacteria and prevent their proliferation [2]. Furthermore, PEG hydrogels can also be utilized as drug delivery systems and for cell encapsulation [218,225].

PEG has some limitations, notably its tendency to swell and undergo oxidative degradation in aqueous environments, compromising its mechanical durability and limiting its long-term effectiveness. Additionally, PEG lacks antibacterial properties [226]. Furthermore, due to their bioinert nature, PEG hydrogels alone are unable to provide an optimal environment to support cell adhesion and tissue formation [98].

4.2.3. Polyvinyl Alcohol

Polyvinyl alcohol (PVA), an eco-friendly polymer [86,227–229], is commonly used as a wound dressing in the treatment of cutaneous wounds [11,40,54]. Structurally, it is a vinyl polymer interconnected through carbon–carbon bonds. The presence of hydroxyl groups in the side chains allows PVA to self-crosslink [40,230].

PVA is a synthetic polymer that is highly biocompatible, biodegradable, and hydrophilic, with high water affinity [17,39,60,119,230–235]. It is easily processed [231], possesses good water solubility [17,40,60,230,233,234,236], high crystallinity, and pH sensitivity [39,60]. Additionally, it has a low tendency for protein adhesion [59,119], good chemical stability, and is toxicologically harmless [40,236], non-carcinogenic, non-toxic, and cost-effective [17,59,215,233,237]. PVA possesses desirable physical and mechanical properties, including good gas permeability [17], and other favorable physical properties such as a rubbery nature, high expansion index, high water swelling degree, fluid absorption ability, and good transparency [59,233].

PVA is a promising candidate for the preparation of biomedical hydrogels, typically prepared by physical crosslinking [236]. PVA-based hydrogels can be prepared without the addition of a chemical crosslinking agent, through freezing–thawing methods, and are simple, safe, and non-toxic [238]. However, it was experimentally observed that pure PVA hydrogels prepared through the cyclic freezing–thawing process exhibited relatively weak mechanical performance, attributed to the limited number of crosslinking agents in the formed polymeric network. Consequently, the combination of PVA with another polymer is necessary to enhance its mechanical performance [214].

PVA hydrogels can also be formed using chemical methods and irradiation, among others [234]. For instance, Oliveira et al. [3,239–241] prepared a PVA hydrogel through gamma irradiation for burn healing and loaded it with silver nanoparticles, resulting in antimicrobial hydrogels effective against *S. aureus*, *E. coli*, and *Candida albicans*. PVA hydrogels are permeable to small molecules [242]. Moreover, characteristics such as transparency, mechanical resistance, biocompatibility, biodegradability, and the ability to maintain a hydrated environment and ensure structural stability when hydrated make PVA-based hydrogels favorable for treating wounds such as burns [54].

There is some controversy regarding certain properties of PVA, particularly its mechanical properties and tissue adhesion. Some authors argue that the polymer possesses good mechanical strength, considering it to be similar to cutaneous tissue [232,236], and it has even been described in the literature that PVA is often used to enhance the mechanical performance of other polymers in wound dressings [59], while others claim that its mechanical properties are weak [234,243,244]. Regarding adhesive properties, some authors state that PVA has good adhesive properties [235], while others assert that it exhibits weak tissue adhesion [234].

PVA has some limitations that restrict its use as a wound dressing, such as a lack of antibacterial properties [234], low elasticity, and poor adhesion to cells and tissues [11,231,234]. Some authors not only argue that PVA's elasticity is inadequate but also consider its hydrophilic characteristics to be incomplete, and its membrane to be rigid. Therefore, it needs to be combined with other polymers to overcome these limitations [11,59,234].

There are studies in which PVA was combined with chitosan [245], chitosan and honey [230], chitosan and gelatin [232,233], or BC [234] that have shown promise. Shamloo et al. [232] prepared a PVA-based hydrogel with chitosan and gelatin and found that increasing the PVA concentration improved the encapsulation efficiency, slowed the degradation rate, and reduced the drug release rate. Chopra et al. [230] developed a hydrogel based on PVA and chitosan that was physically cross-linked without the use of organic solvents or harmful chemicals.

4.2.4. Poly-ε-Caprolactone

Poly-ε-caprolactone (PCL) is a semi-crystalline linear petroleum-based polyester [39,54,60,237,246]. This aliphatic polymer [40,56,246] is eco-friendly [247,248], non-toxic, non-immunogenic, and has good biocompatibility and biodegradability. However, its degradation rate is slower than other polyesters [39,237,246,249], making it less appealing in biomedical applications. However, it is more attractive for long-term implant applications or controlled release systems [59,237]. Currently, there are strategies to accelerate the biodegradability of this polyester. This includes incorporating a more reactive hydrolytic group into the structure, using more hydrophilic and acidic end groups, or employing PCL with a lower molecular weight [250].

Additionally, it has advantageous properties such as excellent mechanical properties, high toughness, elongation, good elasticity, and flexibility [40,54,60,237,251]. Its hydrophobic nature makes it water-resistant and less prone to swelling, and it is also minimally permeable to water vapor [252]. It is cost-effective [40,253,254]. It possesses a low melting point (55–65 °C) [40,54,246,249,254], enabling the easy processing of PCL through a variety of techniques [254], and a glass transition temperature at −60 °C [40,54,246]. Moreover, PCL exhibits high physical stability, simple preparation methods [54,253], remarkable blend compatibility, good solubility in many solvents, and high ductility and plasticity, which facilitate its handling and conformation. Furthermore, PCL dressings are semi-permeable and occlusive, and are easy to remove, because they are non-adherent [246]. Moreover, it is possible to control the mechanical properties, geometry, fiber, and pore size, among other properties, of PCL scaffolds [250]. PCL is suitable for low-temperature 3D printing [255].

It is widely employed in biomedical applications [249], especially in the context of wound healing and tissue regeneration, owing to its properties of reducing inflammatory infiltration and promoting rapid wound healing [59], and as a drug delivery system for controlled release [54]. Features such as low toxicity, slow biodegradation [256–258], and great permeability to many drugs allow PCL to be used as microcarriers for the administration of active compounds over extended periods [217]. Studies indicate that PCL can remain intact in the body for long periods before being fully metabolized and excreted. The low degradation rate is attributed to the presence of five CH_2 groups in the polymer's composition, making it hydrophobic and less susceptible to hydrolysis [258]. Its degradation by-products are non-toxic [255]. Moreover, short- and long-term biocompatibility studies were conducted in animal models, and no adverse effects were observed with PCL scaffolds over 2 years [250].

However, its application in tissue engineering is primarily limited by its high hydrophobicity [39,40,60,249], as well as low bioactivity. This limitation arises from the absence of functional groups and proper cell recognition sites on the scaffold's surface [40,54,249,255]. One possible solution to address these issues is to combine PCL with other polymers [40,54,249,253]. On the one hand, the high hydrophobicity significantly limits PCL's ability to absorb wound exudates [253]. Partial hydrolysis of ester linkages does not compromise the mechanical properties and increases the concentration of free hydroxyl and carboxyl groups, enhancing the hydrophilicity and water binding capacity of PCL [253]. Polymers such as PEG create more hydrophilic structures, exhibiting improved mechanical properties and degradability [40,54,259]. Therefore, by moderately increasing hydrophilicity, the exudate absorption capacity of PCL-based dressings would be enhanced, which is highly desirable in wound healing processes [253]. On the other hand, the lack of functional groups that facilitate cellular adhesion reduces its cellular affinity, compromising tissue regeneration [249,253]. Gelatin/PCL scaffold allows for leveraging the suitable mechanical strength of PCL and capitalizing on the positive aspects of gelatin, including good biodegradation, cell adhesion, and proliferation [249]. Another drawback of PCL is the lack of antibacterial activity, which can be addressed by loading the PCL dressing with antibacterial drugs [59].

The ester linkage of PCL is prone to hydrolytic degradation by lipase enzyme, an ester-like enzyme. Bacteria-producing lipase could be employed for specific and localized delivery of antibacterial drugs at the infection site [260].

Gupta et al. [261] prepared a PEG-PVA-based hydrogel for burn treatment. The hydrogel exhibited high elasticity, good long-term stability, robust mechanical strength, and a favorable water vapor transmission rate. Additionally, it demonstrated a significant capacity for absorbing bodily fluids (in vivo), resulting in enhanced healing efficiency.

4.2.5. Poly-N-Vinylpyrrolidone

Poly-N-vinylpyrrolidone (PVP) contains a hydrophobic polymeric chain and a hydrophilic amide carbonyl group in its structure [262].

This synthetic polymer is widely used as a skin alternative product since it does not cause skin irritation [40]. It offers several beneficial features, including good biocompatibility, biodegradability, low cytotoxicity, excellent water vapor transmission, environmental stability, and high thermal and chemical resistance [17,39,40]. Furthermore, PVP is eco-friendly [263] and demonstrates good molecular control, large-scale production capabilities, and hemocompatibility [219].

Due to its molecular structure, PVP is water-soluble and exhibits very good solubility in most organic solvents, as well as good affinity to complex hydrophobic and hydrophilic substances [17,39,40,262].

The adhesive properties of PVP surpass those of PEG. While PEG relies on hydrogen bonds provided by repeated ether groups for adhesion, PVP offers tertiary amides that provide multiple sites for hydrogen bond formation, significantly enhancing adhesive properties [264]. Moreover, it is low-cost [263] and possesses hemocompatibility, temperature resistance, and pH stability [17].

This hydrophilic polymer has been employed in drug delivery systems and tissue engineering [262], and in the biomedical field for wound treatment, due to its biocompatibility, biodegradability, and low toxicity [265,266]. Moreover, it has been reported that PVP can be used in the treatment of first- and second-degree burns, as well as severe sunburns, in the form of PVP-based hydrogels [54,262].

PVP hydrogels, formed through the crosslinking of PVP under ionizing radiation, resulting in crystal-clear hydrogels with outstanding transparency [54,262]. However, there is no consensus regarding their swelling capacity. Some authors claim that the properties are good [262], while others assert that the swelling capacity of these hydrogels is limited [54]. In vitro and in vivo studies suggest that PVP biomaterials obtained through ionizing radiation did not exhibit toxic effects and are considered safe. Therefore, they can be used in contact with the skin, such as wound dressings [262].

The mechanical properties also generate some controversy. On the one hand, some authors argue that PVP has good mechanical properties [266,267]. On the other hand, some authors contend that its mechanical properties are weak [54], and PVP needs to be mixed with polysaccharides [54] or hydrophilic plasticizers such as PEG to overcome these limitations [268]. Additionally, the combination of PVP with other polymers allows it to acquire useful characteristics for wound healing, such as the ability to prevent microbial penetration [40,54].

4.2.6. Poly(lactic-co-glycolic) Acid

Polylactic acid (PLA) is an aliphatic polyester that exhibits excellent biocompatibility, biodegradability, and good integrity, along with tunable mechanical properties, making it suitable for various biomedical applications [9,39,40,60,237]. Moreover, PLA is hydrophobic, non-toxic, chemically inert, structurally highly stable, and inexpensive [9,39,40,60]. Unlike some natural polymers, PLA can be used on contaminated wounds as it effectively inhibits the propagation of bacteria [237]. PLA and its copolymers are considered eco-friendly [237,269]. The ester backbone is naturally biodegraded through enzymatic action or hydrolysis under physiological conditions. This process leads to non-harmful and non-toxic products that are easily absorbed by the body through natural metabolic pathways [39,40,56]. Its degradation can be properly controlled, and it is relatively moderate [237]. PLA can be employed as drug delivery systems, with low-molecular-weight PLAs degrading more slowly and proving more efficient in drug release [237].

Polyglycolic acid (PGA) is harmless, non-toxic [270], and eco-friendly [270,271]. PGA is another linear aliphatic polyester. This polymer is biodegradable and biocompatible, with high tensile strength and excellent mechanical properties. While more hydrophilic than PLA, PGA is rapidly biodegraded through hydrolysis, yielding glycolic acid. Excessive production of glycolic acid leads to a decrease in pH in its microenvironment due to carbon dioxide production, resulting in local necrosis of cells and tissues. Additionally, it can trigger an inflammatory response [39,40,251]. In addition to tissue inflammation, PGA can induce a severe foreign body reaction [237]. Moreover, the glycolic acid released from PGA may exhibit bacteriostatic properties [237].

Poly(lactic-co-glycolic) acid (PLGA) is a linear aliphatic polyester copolymer that possesses good biocompatibility, biodegradability, easy handling, and mechanical strength [40,54,57,60]. It is also eco-friendly [272] and exhibits weak tissue adhesion [237]. PLGA is relatively hydrophobic [56,237], making it soluble in organic solvents such as acetone, ethyl acetate, tetrahydrofuran, and chlorinated solvents [56]. If the polymer contains less than 70% PLA, it is considered unstructured [56]. Its glass transition temperature is between 40 and 60 °C, and it undergoes hydrolysis in the body [60], and its hydrolysis products can be uptaken in the cellular metabolic pathway [57]. The hydrolysis of PLGA results in the formation of natural metabolic monomers (salt form of lactic acid and glycolic acid), which are relatively harmless, minimizing the risk of systemic toxicity [273]. The degradation rate can be adjusted by varying monomer ratios [9,39]. The healing process occurs within a specific time frame and depends on the synchronization of the epithelialization rate with the degradation rate of PLGA [54]. In vitro studies have shown that PLGA exhibits excellent cytocompatibility in fibroblasts, with minimal toxicity [237]. It is widely used in tissue engineering and as a drug delivery system, primarily due to its beneficial characteristics such as adjustable and controllable mechanical properties [40,54,237,273,274]. It can transport both hydrophobic or hydrophilic drugs, small or macromolecules [56,273], and protect them from degradation and control their release [273]. PLGA films are hydrophobic, stiff, and semi-permeable. They lack the ability to absorb exudates or provide a humid microenvironment. On the other hand, PLGA nanofibers are highly dense, allowing them to prevent bacterial invasion. PLGA possesses characteristics that make it suitable for use as an outer layer, capable of preserving moisture content and isolating a hydrogel matrix from the external environment [274]. Wang et al. [274] prepared a bilayer membrane scaffold using 3D printing, consisting of an outer layer made of PLGA and a lower layer of alginate hydrogel, mimicking the epidermis and dermis, respectively. The multiporous alginate hydrogel was employed to enhance in vitro cell adhesion and proliferation, while the PLGA membrane served to prevent bacterial invasion, reduce evaporation, and maintain the moisture content of the hydrogel.

5. Overview of Critical Attributes of Polymers

Table 3 compiles key information on critical attributes of polymers as candidates for the ideal wound dressing. The terms used in the table, though relative, aim to facilitate the interpretation of individual properties of each polymer and enable comparison. To the best of our knowledge, the literature lacks comprehensive information on all the features we consider critical for all polymers, both individually and in the form of hydrogels. The table presents a relative critical analysis on a scale from "-" to "+++" where "-" indicates the polymer lacks a specific property, and "+++" signifies excellent performance.

It should be noted that regarding moisture, the literature does not clearly distinguish between the different abilities of hydrogels to provide a hydrated environment in the wound bed. Similarly, concerning transparency and cost-effectiveness, there is also no clear distinction among polymers in the literature. Regarding protection, although it is generally known that hydrogels, to some extent, prevent the entry of microorganisms by forming a physical barrier between the wound and the external environment, here, we aim to convey the capacity for protection in terms of having intrinsic antibacterial properties. Regarding safety, we considered the concept of "non-toxic". A more in-depth discussion of this issue may be necessary. Finally, concerning the availability of synthetic polymers, since they are chemically synthesized, it was not considered. In addition to the commonly considered critical quality attributes for an ideal wound dressing, we also considered the environmental aspect and added a column on the environmental sustainability of the polymers.

Table 3. Polymer critical analysis per critical attribute for wound healing.

	Polymers	Critical Attributes												Eco-Friendly	References
		Moisture	Absorption	Permeability	Protection	Transparency	Mechanical Robustness	Flexibility and Elasticity	Adhesiveness	Biocompatibility	Safety	Cost-Effectiveness	Availability		
Natural polymers	Collagen	++	++	NA	+	+++	+	++	+	+++	++	+	++	√	[4,16,33,37,38,42,64,67–70,73,75,76,78,79]
	Gelatin	++	+++	NA	+	+++	+	++	+	+++	+++	++	++	√	[17,27,31,67,83–97,99]
	Silk Fibroin	++	++	++	+	CV	+++	+++	++	+++	+++	++	++	√	[16,17,54,57,60,102–110,112–117,119–127]
	Alginate	++	+++	++	+	+++	+	++	+	+++	+++	++	++	√	[1,16,21,34,38,39,44,53,128–134]
	Hyaluronic acid	++	+++	++	+	+++	+	++	+	+++	+++	+	++	√	[7,9,19,45,137,138,140–147]
	Bacterial cellulose	++	+++	++	+	NA	+++	++	NA	+++	+++	++	+++	√	[8,16,36,46,47,51,52,130,131,149–155]
	Dextran	++	+++	NA	+	+++	+	++	+	+++	+++	++	++	√	[130,133,158–162,167–174]
	Chitosan	++	++	++	++	NA	+	++	+++	+++	+++	++	+++	√	[9,11,16,38,46,50–53,67,78,98,130–134,137,154,176–181]
Synthetic polymers	Polyurethane	NA	NA *	++	+	+++	+++	+++	+	+++	+	+++	NA	X	[9,39,54,55,60,186–188,190–195,199,202]
	Polyethylene glycol	++	+++	++	+	+++	CV	++	+	+++	+++	++	NA	√	[2,17,19,30,41,54,57,58,60,209–216,223,224,226]
	Polyvinyl alcohol	++	+++	++	+	++	CV	+	CV	+++	+++	++	NA	√	[11,17,39,40,59,60,86,119,215,227–237,243,244]
	Poly-ε-caprolactone	NA	+	+	+	+	+++	++	+	++	+++	++	NA	√	[39,40,54,59,60,237,247–249,251–254]
	Poly-N-vinylpyrrolidone	NA	CV	+++	+	+++	CV	NA	++	++	+++	++	NA	√	[17,39,40,54,262–264]
	Poly(lactic-co-glycolic) acid	-	-	+	NA **	NA	++	NA	+	++	+++	NA	NA	√	[40,54,57,60,237,272–274]

Key: "-" means "does not have"; "+" means "weak"; "++" means "good"; "+++" means "excellent"; "√" means "applicable"; "X" means "not applicable"; CV—controversial; NA—not available. * There is only information about foams/sponges. ** There is only information about nanofibers.

6. Applications of Hydrogels as Wound Dressings

Hydrogels are promising candidates for the treatment of cutaneous wounds and can be approached in different ways given the potential of their characteristics. Some of the applications of hydrogels as dressings include promoting wound healing through their inherent properties, delivering substances such as drugs and growth factors, enabling cell growth within their 3D structure, which mimics the native structures of the skin, or, more recently, incorporating biosensors for real-time monitoring of wound characteristics. Figure 3 provides a schematic representation of these applications.

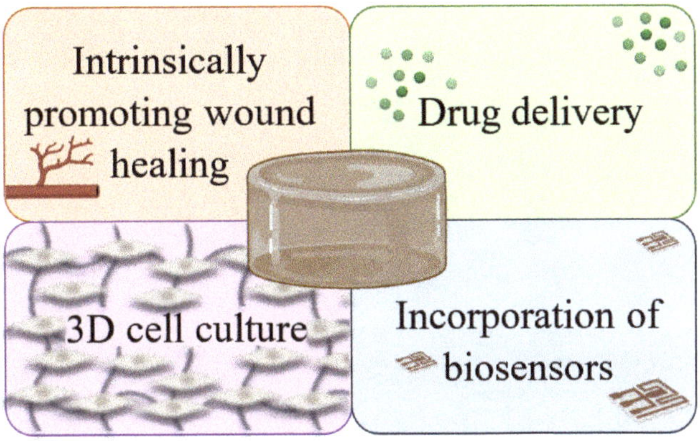

Figure 3. Representation of some of the applications of hydrogels as wound dressings: intrinsic capacity to stimulate healing; drug delivery systems or transporters of other substances; support for cell growth; real-time monitoring of the state of wounds through the incorporation of biosensors.

6.1. Hydrogels with the Intrinsic Ability to Promote Wound Healing

Hydrogels can promote wound healing through their intrinsic characteristics, notably by providing an appropriate cell-friendly microenvironment, stimulating angiogenesis, a key factor for tissue regeneration, or, for example, recruiting cells involved in the healing process [2,3,11]. For instance, Sun et al. [275] prepared a hydrogel based on dextran and PEG for the treatment of third-degree burns in mice. The results show that the hydrogel's structure allows for the infiltration of neutrophils, which, in turn, facilitates the degradation of the hydrogel during the repair phase. This results in the recruitment of endothelial progenitors and angiogenic cells, stimulating rapid in vivo neovascularization after one week of treatment. Epithelial repair was observed within 14 days, with complete skin and appendage regeneration (sebaceous glands and hair follicles) noted by the end of 21 days. Another example is the hydrogel prepared by Yang et al. [109], composed of SF, HA, and alginate. SF serves as the primary matrix for tissue formation, providing mechanical support and promoting cell adhesion and proliferation. HA enhances biocompatibility, angiogenesis, and tissue regeneration, while alginate improves biocompatibility and hydrophilic performance (Figure 4A). The hydrogel mimics the structure of the ECM in native tissue. This 3D porous microstructure possesses soft and elastic characteristics, along with good physical stability in environments simulating bodily fluids, ensuring adequate mechanical performance. These features enable the adhesion, growth, and proliferation of NIH-3T3 fibroblasts in vitro. In vivo (Figure 4B), the hydrogel creates a favorable environment for healing, facilitating enhanced re-epithelialization, increased collagen deposition, ECM remodeling, and improved angiogenesis, thereby accelerating the burn wound healing process.

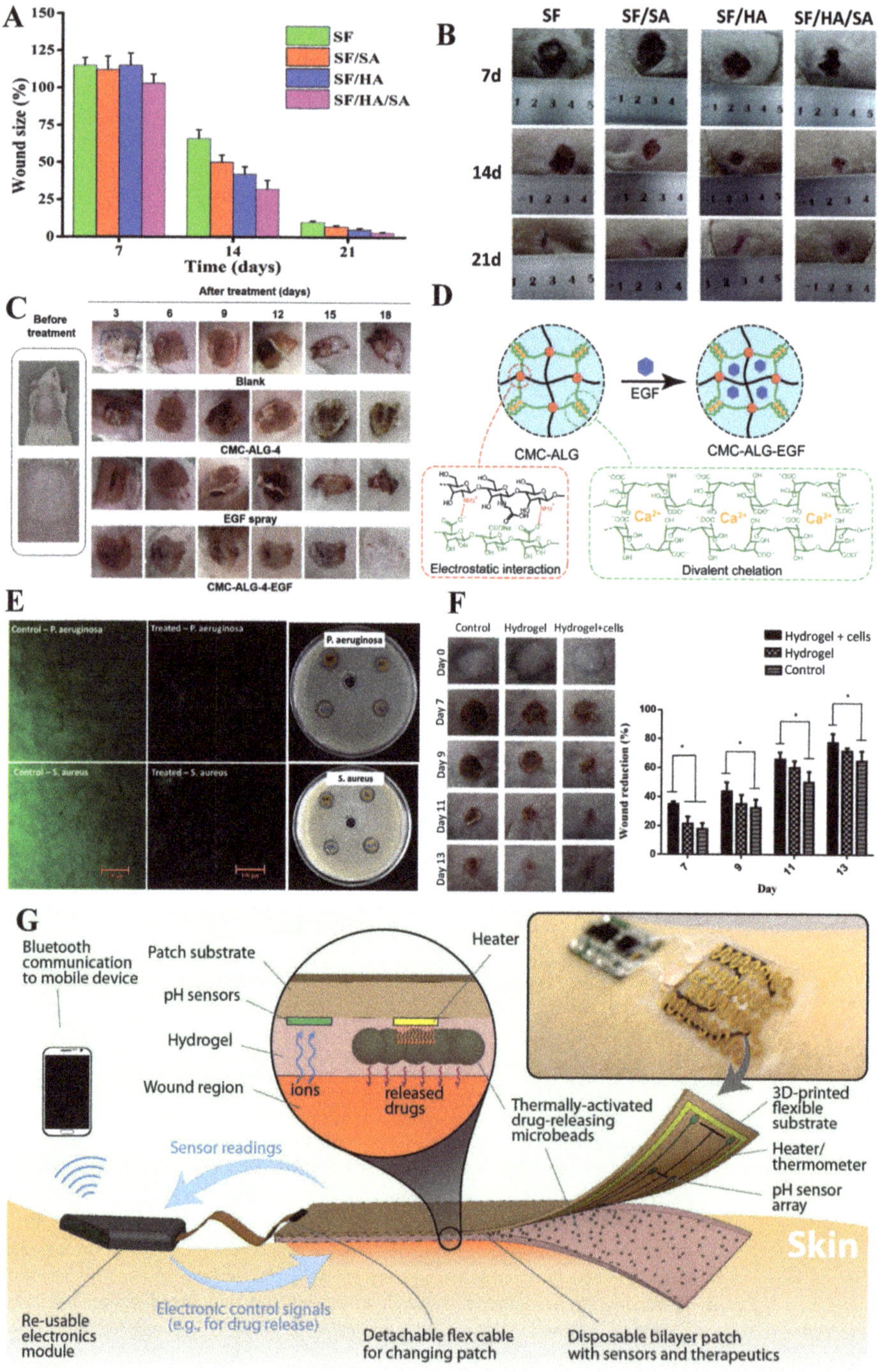

Figure 4. (**A**) Comparison of wound size over 21 days after treatment with hydrogels. (**B**) In vivo observation of burn wound healing over 21 days. (**C**) In vivo study on EGF delivery and wound healing

promotion. (**D**) Schematic representation of EGF loaded in carboxymethyl chitosan and alginate hydrogel. (**E**) In vitro inhibition of *S. aureus* and *P. aeruginosa* growth after loading ciprofloxacin in SF hydrogel. (**F**) In vivo study on partial-thickness burns treatment with hydrogels loaded with keratinocytes and fibroblasts (* means significant difference ($p < 0.05$) between the groups) (**G**) Schematic representation of the operation of a hydrogel incorporated with poly (N-isopropyl acrylamide) stimuli-responsive particles. Adapted from: [8,109,276–278].

6.2. Hydrogels as Drug Delivery Systems and Other Substance Carriers

Furthermore, the matrix of hydrogels allows for the incorporation of substances, such as growth factors, cytokines, or drugs, namely, antibacterials and anti-inflammatories, that assist in the healing process [3,11]. Numerous studies focus on incorporating growth factors into hydrogels to expedite skin healing, such as β-FGF, primary promoters of cell proliferation with chemotactic activity, and VEGF, which enhances cell proliferation and tissue remodeling in wounds [4]. Similarly, in the study conducted by Hu et al. [177] (Figure 4C), they loaded EGF into a carboxymethyl chitosan and alginate hydrogel to protect the EGF from proteolytic degradation (ensuring its bioactivity) (Figure 4D) and to allow for its gradual release, improving cell proliferation. Dong et al. [278] prepared an SF hydrogel loaded with ciprofloxacin for the healing of deep partial-thickness burns in rats. The results showed that the hydrogel effectively delivered the antibiotic, inhibiting in vitro bacterial growth and biofilm formation of *S. aureus* and *P. aeruginosa* (Figure 4E). In addition, the hydrogel promoted autolysis, a reduction in inflammation, fibroblast proliferation (attributed to the silk protein environment), increased collagen deposition, accelerated re-epithelialization, formation of granulation tissue, stimulation of angiogenesis, and showed complete reconstitution of the epidermal layer in rats after 18 days. In the study conducted by Yin et al. [114], they also prepared an SF-based hydrogel, but this time loaded it with rhein to simultaneously prevent bacterial colonization/infection and reduce inflammation. Rhein is a bioactive anthraquinone isolated from the traditional Chinese medicine rhubarb that possesses good anti-inflammatory and antibacterial properties. By incorporating it into SF hydrogels, the stability and structural integrity of rhein might be improved, thus enhancing therapeutic efficacy and minimizing negative effects. The SF/Rhein composite hydrogels combined the excellent biocompatibility and physicochemical properties of SF, along with the antibacterial and anti-inflammatory efficiency of rhein, accelerating the bacterially infected burn wound healing rate by reducing inflammation, expediting angiogenesis, and promoting skin appendages formation.

6.3. Hydrogels as 3D Scaffolds for Cell Adhesion and Proliferation

Hydrogels can serve as a platform for loading cells [3,11]. The 3D matrix of hydrogels allows for the deposition and organization of cells [8], mimicking the environment of the natural biological ECM better than two-dimensional substrates [279–281]. The architecture of the hydrogel, characterized by suitable biocompatibility, morphology, and mechanical behavior, enables it to function as a temporary support, facilitating cellular processes such as adhesion, proliferation, and differentiation for the formation of new tissue [282]. The hydrogel can be loaded with cells such as keratinocytes and fibroblasts, as demonstrated in the study by Mohamad et al. [8], where they loaded keratinocytes and fibroblasts into a hydrogel based on BC and acrylic acid. The results were very promising, achieving complete re-epithelialization of a partial-thickness burn within 13 days (Figure 4F). Furthermore, there was a more organized deposition of type I collagen fibers, attributed to the synergistic effect of the hydrogel with the incorporated cells in accelerating skin regeneration and strengthening the dermis. Additionally, the incorporated cells can be stem cells. It is considered that stem cells promote healing through differentiation into specific cell types or through paracrine effects to stimulate the host tissue regeneration [19]. Stem cell therapies have shown promising outcomes in the context of wound healing [283,284]. Dong et al. [19] demonstrated that the HA and poly (ethylene glycol) hydrogel loaded with adipose tissue-derived stem cells provided an optimized 3D microenvironment, enhancing the therapeutic efficiency of stem cell-based therapies. The hydrogel promoted cell paracrine secretion and increased the expression of pro-angiogenic growth factors and cytokines, such as angiopoietin, VEGF, platelet-derived growth factor, stromal cell-derived factor, contributing to the wound healing treatment in a burn animal model.

6.4. *Hydrogels with Integrated Biosensors*

Moreover, the incorporation of biosensors into hydrogels can solve various challenges associated with wounds, such as the early detection of infections and the acquisition of relevant information about the wound microenvironment in real-time. This information can be used to provide timely and accurate reports on the evolution of the healing process [9,11]. Villanueva et al. [285] developed a smart antimicrobial wound dressing based on keratin hydrogels with zinc oxide nanoparticles (nZnO), taking advantage of the pH-responsive behavior of keratin and the antimicrobial activity of nZnO. Infected wounds exhibit alkaline pH due to the by-products of bacterial metabolism. As the wound undergoes healing, the pH becomes acidic. In a clean wound, the dressing acts as a barrier, isolating the injury from the external environment and protecting it from microbial contamination. In a bacterially contaminated wound, the increased pH leads to hydrogel swelling, increasing its pore size, and facilitating the release of the antimicrobial agent into the medium, thereby controlling the infection. Mostafalu et al. [276] developed an alginate hydrogel sheet incorporated with poly (N-isopropyl acrylamide) stimuli-responsive particles (a drug-releasing system), loaded with cefazolin for real-time monitoring of the wound environment for individualized treatment of chronic wounds (Figure 4G). This automated, smart, flexible wound dressing comprises pH sensors and a microheater to trigger thermo-responsive drug carriers containing antibiotics. The pH sensors are connected to a microcontroller through an electronic module that processes the data measured by the sensors. Once the pH exceeds an acceptable range, it communicates wirelessly in a closed-loop manner to smartphones or computers to remotely activate the heater and program the on-demand release of the antibacterial drug.

7. Future Prospects

As we envision the future of biomedical research, particularly in the field of hydrogel polymers, numerous promising opportunities emerge, prepared to revolutionize wound healing and tissue engineering. We will witness an intensified focus on the research and development of innovative biomaterials, with an emphasis on sustainability and environmental friendliness. Currently, the concept of sustainability is a foundational pillar across all scientific domains. The increasing demand for environmentally friendly materials has driven research towards the exploration and development of biopolymers derived from natural raw materials that are eco-friendly and sustainable. Thus, biomass-based materials have gained prominence as candidates for applications that leverage renewable resources and address environmental concerns [149,150]. Cellulose, in particular, is a biopolymer that is not only biocompatible, biodegradable, and non-toxic but also renewable, as it is present in the cell walls of a large number of plants [150]. However, further studies should be carried out to explore more alternatives to polymers. Additionally, more research is needed for the development of eco-friendly polymer synthesis and acquisition techniques. It is crucial to clarify the environmental impacts resulting from the methods used to obtain both natural and synthetic polymers. Moreover, the possibility of recycling industrial waste should be considered. As a purely illustrative example, collagen can be extracted from fish scales. With hundreds of millions of tons of fish annually, a significant amount of waste is generated in fish shops and processing factories, where scales constitute a major solid waste. Improper disposal techniques lead to unpleasant odors and environmental pollution. Therefore, it would be worthwhile to explore solutions to optimize the potential of such waste for the production of valuable end products. On the other hand, further in-depth studies on the degradation pathways of polymers, both natural and synthetic, are necessary, as well as the assessment of the potential formation of toxic compounds and the possible risks of degradation products. Due to certain limitations of polymers, most studies involve polymer blends, and the individual potentials and risks of each polymer are not always clear. Therefore, additional research is required to elucidate the effective contribution of polymers to skin wound healing. However, the current well-established knowledge about the effectiveness of hydrogels and their properties implies new challenges. Future research should focus on evaluating their biocompatibility and biodegradability to ensure they are safe for use in patients and degrade naturally without causing environmental harm. The adjustment of the physicochemical properties of hydrogels to achieve the desired outcomes, specific needs, and the characteristics of the patient remains a challenge. In the future, this may be controlled through modifications in the chemical structure of molecules or by selecting molecular weights and types of cross-linking, thus expanding the currently available repertoire of natural sources [34]. The incorporation of bioactive agents promoting wound healing will enable dressings to assume a more active role. Hydrogels capable of releasing drugs in response to specific stimuli could be employed to design active reservoirs for therapeutic cells. The dynamic control of drug release using sensors to determine specific parameters and trigger the release of cells stored in the hydrogel has the potential to enhance both effectiveness and safety, providing a platform for innovative therapeutics [34]. In light of the current understanding of wound dressings, it is anticipated that the future diagnosis and treatment of wounds will be based on a multifunctional and systematic approach, considering the stage and characteristics of the wound [5]. The incorporation of sensors into hydrogels may open doors to real-time wound monitoring without the need for expensive equipment [21]. It is important to highlight a key and pressing aspect for the advancement of the real-world application of these hydrogels: the conduct of robust clinical studies to assess the effectiveness of the hydrogels in different types of wounds, comparing them with conventional treatments. Only this approach will allow for validation of the efficacy of hydrogels and expand their clinical application.

So, the aspects that will guide the next steps in advancing the development of hydrogels for wound healing will necessarily involve the following approaches: exploration of sustainable biopolymers, recycling of waste, development of multifunctional dressings, controlled drug release, modifications of hydrogel properties, incorporations of therapeutic agents, evaluation of biocompatibility and biodegradability, and performance of robust clinical studies. The focus will be on improving their effectiveness, safety, and environmental impact.

The future of polymers is brimming with possibilities, driven by innovation, sustainability, and improved outcomes for patients. By embracing interdisciplinary collaboration and harnessing cutting-edge technologies, researchers are poised to uncover new frontiers in wound healing and tissue engineering, ushering in an era of unprecedented therapeutic efficacy and environmental responsibility.

8. Conclusions

Hydrogels are commonly regarded as one of the most promising wound dressings due to their inherent properties. The literature extensively describes features such as water absorption capacity and wound exudate retention, as well as the mechanical characteristics of each polymer. However, despite the understanding that intrinsic limitations hinder the characterization of polymers in isolation, notably the lack of appropriate mechanical properties, it is paramount to clarify, for each polymer (individually and in hydrogel form), the specific attributes corresponding to the requirements of an ideal dressing. Further studies addressing these specific topics are necessary to elucidate controversies surrounding certain properties and facilitate a more accurate comparison and approach. Despite the existing gaps and the need for additional research, this manuscript can contribute to this process and serve as a foundation for future investigations. These investigations aim to expand the potential of hydrogels as promising wound dressings in biomedical applications, while also considering environmental concerns.

Author Contributions: Conceptualization, F.M.-M., C.V. and M.S.; methodology, F.M.-M., C.V., M.S. and M.R.; validation, F.M.-M., C.V. and M.S.; formal analysis, F.M.-M., C.V., M.S. and M.R.; investigation, M.R.; resources, M.R.; writing—original draft preparation, F.M.-M., C.V. and M.S. and M.R.; writing—review and editing, F.M.-M., C.V., M.S. and M.R.; supervision, F.M.-M., C.V. and M.S. All authors have read and agreed to the published version of the manuscript.

Funding: This research received no external funding.

Institutional Review Board Statement: Not applicable.

Informed Consent Statement: Not applicable.

Data Availability Statement: No new data were created or analyzed in this study. Data sharing is not applicable to this article.

Conflicts of Interest: The authors declare no conflicts of interest.

Abbreviations

3D	Three-dimensional
BC	Bacterial cellulose
BCM	Bacterial cellulose membrane
ECM	Extracellular matrix
EGF	Epidermal growth factor
FGF	Fibroblast growth factor
GA	Glycolic acid
HA	Hyaluronic acid
NF-kB	Nuclear Factor Kappa B
PCL	Poly-ε-caprolactone
PEG	Polyethylene glycol
PGA	Polyglycolic acid
PLA	Polylactic acid
PLGA	Poly(lactic-co-glycolic) acid
PU	Polyurethane
PVA	Polyvinyl alcohol
PVP	Poly-N-vinylpyrrolidone
RGD	Arginine–glycine–aspartic acid peptide sequences
SF	Silk fibroin
TGF-β	Transforming growth factor
VEGF	Vascular endothelial growth factor

References

1. Varaprasad, K.; Jayaramudu, T.; Kanikireddy, V.; Toro, C.; Sadiku, E.R. Alginate-based composite materials for wound dressing application:A mini review. *Carbohydr. Polym.* **2020**, *236*, 116025. [CrossRef] [PubMed]
2. Zhang, L.; Yin, H.; Lei, X.; Lau, J.N.Y.; Yuan, M.; Wang, X.; Zhang, F.; Zhou, F.; Qi, S.; Shu, B.; et al. A Systematic Review and Meta-Analysis of Clinical Effectiveness and Safety of Hydrogel Dressings in the Management of Skin Wounds. *Front. Bioeng. Biotechnol.* **2019**, *7*, 342. [CrossRef]
3. Surowiecka, A.; Strużyna, J.; Winiarska, A.; Korzeniowski, T. Hydrogels in Burn Wound Management—A Review. *Gels* **2022**, *8*, 122. [CrossRef]
4. Sharma, S.; Rai, V.K.; Narang, R.K.; Markandeywar, T.S. Collagen-based formulations for wound healing: A literature review. *Life Sci.* **2022**, *290*, 120096. [CrossRef]
5. Shu, W.; Wang, Y.; Zhang, X.; Li, C.; Le, H.; Chang, F. Functional Hydrogel Dressings for Treatment of Burn Wounds. *Front. Bioeng. Biotechnol.* **2021**, *9*, 788461. [CrossRef] [PubMed]
6. Fayyazbakhsh, F.; Khayat, M.J.; Leu, M.C. 3D-Printed Gelatin-Alginate Hydrogel Dressings for Burn Wound Healing: A Comprehensive Study. *Int. J. Bioprint.* **2022**, *8*, 274–291. [CrossRef] [PubMed]
7. Longinotti, C. The use of hyaluronic acid based dressings to treat burns: A review. *Burn. Trauma* **2014**, *2*, 162–168. [CrossRef]
8. Mohamad, N.; Loh, E.Y.X.; Fauzi, M.B.; Ng, M.H.; Mohd Amin, M.C.I. In vivo evaluation of bacterial cellulose/acrylic acid wound dressing hydrogel containing keratinocytes and fibroblasts for burn wounds. *Drug Deliv. Transl. Res.* **2019**, *9*, 444–452. [CrossRef]
9. Naseri, E.; Ahmadi, A. A review on wound dressings: Antimicrobial agents, biomaterials, fabrication techniques, and stimuli-responsive drug release. *Eur. Polym. J.* **2022**, *173*, 111293. [CrossRef]

10. Stubbe, B.; Mignon, A.; Declercq, H.; Van Vlierberghe, S.; Dubruel, P. Development of Gelatin-Alginate Hydrogels for Burn Wound Treatment. *Macromol. Biosci.* **2019**, *19*, 1900123. [CrossRef]
11. Tavakoli, S.; Klar, A.S. Advanced Hydrogels as Wound Dressings. *Biomolecules* **2020**, *10*, 1169. [CrossRef]
12. He, Y.; Cen, Y.; Tian, M. Immunomodulatory hydrogels for skin wound healing: Cellular targets and design strategy. *J. Mater. Chem. B* **2024**, *12*, 2435–2458. [CrossRef] [PubMed]
13. Kharaziha, M.; Baidya, A.; Annabi, N. Rational Design of Immunomodulatory Hydrogels for Chronic Wound Healing. *Adv. Mater.* **2021**, *33*, e2100176. [CrossRef]
14. Fan, F.; Saha, S.; Hanjaya-Putra, D. Biomimetic Hydrogels to Promote Wound Healing. *Front. Bioeng. Biotechnol.* **2021**, *9*, 718377. [CrossRef] [PubMed]
15. Firlar, I.; Altunbek, M.; McCarthy, C.; Ramalingam, M.; Camci-Unal, G. Functional Hydrogels for Treatment of Chronic Wounds. *Gels* **2022**, *8*, 127. [CrossRef]
16. Mogoşanu, G.D.; Grumezescu, A.M. Natural and synthetic polymers for wounds and burns dressing. *Int. J. Pharm.* **2014**, *463*, 127–136. [CrossRef]
17. Prasathkumar, M.; Sadhasivam, S. Chitosan/Hyaluronic acid/Alginate and an assorted polymers loaded with honey, plant, and marine compounds for progressive wound healing—Know-how. *Int. J. Biol. Macromol.* **2021**, *186*, 656–685. [CrossRef]
18. Feng, X.; Zhang, X.; Li, S.; Zheng, Y.; Shi, X.; Li, F.; Guo, S.; Yang, J. Preparation of aminated fish scale collagen and oxidized sodium alginate hybrid hydrogel for enhanced full-thickness wound healing. *Int. J. Biol. Macromol.* **2020**, *164*, 626–637. [CrossRef] [PubMed]
19. Dong, Y.; Cui, M.; Qu, J.; Wang, X.; Kwon, S.H.; Barrera, J.; Elvassore, N.; Gurtner, G.C. Conformable hyaluronic acid hydrogel delivers adipose-derived stem cells and promotes regeneration of burn injury. *Acta Biomater.* **2020**, *108*, 56–66. [CrossRef]
20. Smandri, A.; Nordin, A.; Hwei, N.M.; Chin, K.-Y.; Abd Aziz, I.; Fauzi, M.B. Natural 3D-Printed Bioinks for Skin Regeneration and Wound Healing: A Systematic Review. *Polymers* **2020**, *12*, 1782. [CrossRef]
21. Zhang, M.; Zhao, X. Alginate hydrogel dressings for advanced wound management. *Int. J. Biol. Macromol.* **2020**, *162*, 1414–1428. [CrossRef]
22. Ge, B.; Wang, H.; Li, J.; Liu, H.; Yin, Y.; Zhang, N.; Qin, S. Comprehensive Assessment of Nile Tilapia Skin (*Oreochromis niloticus*) Collagen Hydrogels for Wound Dressings. *Mar. Drugs* **2020**, *18*, 178. [CrossRef]
23. Tefft, J.B.; Chen, C.S.; Eyckmans, J. Reconstituting the dynamics of endothelial cells and fibroblasts in wound closure. *APL Bioeng.* **2021**, *5*, 016102. [CrossRef]
24. Zhong, J.; Wang, H.; Yang, K.; Wang, H.; Duan, C.; Ni, N.; An, L.; Luo, Y.; Zhao, P.; Gou, Y.; et al. Reversibly immortalized keratinocytes (iKera) facilitate re-epithelization and skin wound healing: Potential applications in cell-based skin tissue engineering. *Bioact. Mater.* **2022**, *9*, 523–540. [CrossRef]
25. Yang, F.; Jin, S.; Tang, Y. Marine Collagen Peptides Promote Cell Proliferation of NIH-3T3 Fibroblasts via NF-κB Signaling Pathway. *Molecules* **2019**, *24*, 4201. [CrossRef]
26. Li, J.; Yu, F.; Chen, G.; Liu, J.; Li, X.L.; Cheng, B.; Mo, X.M.; Chen, C.; Pan, J.F. Moist-Retaining, Self-Recoverable, Bioadhesive, and Transparent in Situ Forming Hydrogels To Accelerate Wound Healing. *ACS Appl. Mater. Interfaces* **2020**, *12*, 2023–2038. [CrossRef]
27. Garcia-Orue, I.; Santos-Vizcaino, E.; Etxabide, A.; Uranga, J.; Bayat, A.; Guerrero, P.; Igartua, M.; de la Caba, K.; Hernandez, R.M. Development of Bioinspired Gelatin and Gelatin/Chitosan Bilayer Hydrofilms for Wound Healing. *Pharmaceutics* **2019**, *11*, 314. [CrossRef] [PubMed]
28. Aleem, A.R.; Shahzadi, L.; Tehseen, S.; Alvi, F.; Chaudhry, A.A.; Rehman, I.U.; Yar, M. Amino acids loaded chitosan/collagen based new membranes stimulate angiogenesis in chorioallantoic membrane assay. *Int. J. Biol. Macromol.* **2019**, *140*, 401–406. [CrossRef] [PubMed]
29. Xiong, S.; Zhang, X.; Lu, P.; Wu, Y.; Wang, Q.; Sun, H.; Heng, B.C.; Bunpetch, V.; Zhang, S.; Ouyang, H. A Gelatin-sulfonated Silk Composite Scaffold based on 3D Printing Technology Enhances Skin Regeneration by Stimulating Epidermal Growth and Dermal Neovascularization. *Sci. Rep.* **2017**, *7*, 4288. [CrossRef] [PubMed]
30. Chen, G.; Yu, Y.; Wu, X.; Wang, G.; Ren, J.; Zhao, Y. Bioinspired Multifunctional Hybrid Hydrogel Promotes Wound Healing. *Adv. Funct. Mater.* **2018**, *28*, 1801386. [CrossRef]
31. Xu, S.; Li, J.; He, A.; Liu, W.; Jiang, X.; Zheng, J.; Han, C.C.; Hsiao, B.S.; Chu, B.; Fang, D. Chemical crosslinking and biophysical properties of electrospun hyaluronic acid based ultra-thin fibrous membranes. *Polymer* **2009**, *50*, 3762–3769. [CrossRef]
32. Kumar, V.B.; Tiwari, O.S.; Finkelstein-Zuta, G.; Rencus-Lazar, S.; Gazit, E. Design of Functional RGD Peptide-Based Biomaterials for Tissue Engineering. *Pharmaceutics* **2023**, *15*, 345. [CrossRef] [PubMed]
33. Mathew-Steiner, S.S.; Roy, S.; Sen, C.K. Collagen in Wound Healing. *Bioengineering* **2021**, *8*, 63. [CrossRef] [PubMed]
34. Lee, K.Y.; Mooney, D.J. Alginate: Properties and biomedical applications. *Prog. Polym. Sci.* **2012**, *37*, 106–126. [CrossRef] [PubMed]
35. Ayello, E.A.; Loehne, H.; Chariker, M.; DiCosmo, F. Edge Effect: The Role of Collagen in Wound Healing. *Adv. Skin Wound Care* **2007**, *22*, 12–15. [CrossRef]
36. Karlsson, M.; Olofsson, P.; Steinvall, I.; Sjöberg, F.; Thorfinn, J.; Elmasry, M. Three Years' Experience of a Novel Biosynthetic Cellulose Dressing in Burns. *Adv. Wound Care* **2019**, *8*, 71–76. [CrossRef] [PubMed]
37. Catoira, M.C.; Fusaro, L.; Di Francesco, D.; Ramella, M.; Boccafoschi, F. Overview of natural hydrogels for regenerative medicine applications. *J. Mater. Sci. Mater. Med.* **2019**, *30*, 115. [CrossRef] [PubMed]

38. Chen, G.; Zhou, Y.; Dai, J.; Yan, S.; Miao, W.; Ren, L. Calcium alginate/PNIPAAm hydrogel with body temperature response and great biocompatibility: Application as burn wound dressing. *Int. J. Biol. Macromol.* **2022**, *216*, 686–697. [CrossRef]
39. Negut, I.; Dorcioman, G.; Grumezescu, V. Scaffolds for Wound Healing Applications. *Polymers* **2020**, *12*, 2010. [CrossRef]
40. Oliveira, C.; Sousa, D.; Teixeira, J.A.; Ferreira-Santos, P.; Botelho, C.M. Polymeric biomaterials for wound healing. *Front. Bioeng. Biotechnol.* **2023**, *11*, 1136077. [CrossRef]
41. Jafari, A.; Hassanajili, S.; Azarpira, N.; Bagher Karimi, M.; Geramizadeh, B. Development of thermal-crosslinkable chitosan/maleic terminated polyethylene glycol hydrogels for full thickness wound healing: In vitro and in vivo evaluation. *Eur. Polym. J.* **2019**, *118*, 113–127. [CrossRef]
42. Almeida Cruz, M.; Araujo, T.A.; Avanzi, I.R.; Parisi, J.R.; Martins de Andrade, A.L.; Muniz Rennó, A.C. Collagen from Marine Sources and Skin Wound Healing in Animal Experimental Studies: A Systematic Review. *Mar. Biotechnol.* **2021**, *23*, 1–11. [CrossRef]
43. Arora, D.; Bhunia, B.K.; Janani, G.; Mandal, B.B. Bioactive three-dimensional silk composite in vitro tumoroid model for high throughput screening of anticancer drugs. *J. Colloid Interface Sci.* **2021**, *589*, 438–452. [CrossRef]
44. Froelich, A.; Jakubowska, E.; Wojtyłko, M.; Jadach, B.; Gackowski, M.; Gadziński, P.; Napierała, O.; Ravliv, Y.; Osmałek, T. Alginate-Based Materials Loaded with Nanoparticles in Wound Healing. *Pharmaceutics* **2023**, *15*, 1142. [CrossRef] [PubMed]
45. Graça, M.F.P.; Miguel, S.P.; Cabral, C.S.D.; Correia, I.J. Hyaluronic acid—Based wound dressings: A review. *Carbohydr. Polym.* **2020**, *241*, 116364. [CrossRef] [PubMed]
46. Pasaribu, K.M.; Ilyas, S.; Tamrin, T.; Radecka, I.; Swingler, S.; Gupta, A.; Stamboulis, A.G.; Gea, S. Bioactive bacterial cellulose wound dressings for burns with collagen in-situ and chitosan ex-situ impregnation. *Int. J. Biol. Macromol.* **2023**, *230*, 123118. [CrossRef] [PubMed]
47. Portela, R.; Leal, C.R.; Almeida, P.L.; Sobral, R.G. Bacterial cellulose: A versatile biopolymer for wound dressing applications. *Microb. Biotechnol.* **2019**, *12*, 586–610. [CrossRef]
48. Wang, Q.; Qi, P.-X.; Huang, S.-X.; Hou, D.-Z.; Xu, X.-D.; Ci, L.-Y.; Chen, S. Quantitative analysis of straight-chain/branched-chain Ratio During Enzymatic Synthesis of Dextran Based on Periodate Oxidation. *Biochem. Biophys. Res. Commun.* **2020**, *523*, 573–579. [CrossRef] [PubMed]
49. Petrovici, A.R.; Anghel, N.; Dinu, M.V.; Spiridon, I. Dextran-Chitosan Composites: Antioxidant and Anti-Inflammatory Properties. *Polymers* **2023**, *15*, 1980. [CrossRef] [PubMed]
50. Croisier, F.; Jérôme, C. Chitosan-based biomaterials for tissue engineering. *Eur. Polym. J.* **2013**, *49*, 780–792. [CrossRef]
51. Kumar, S.; Mukherjee, A.; Dutta, J. Chitosan based nanocomposite films and coatings: Emerging antimicrobial food packaging alternatives. *Trends Food Sci. Technol.* **2020**, *97*, 196–209. [CrossRef]
52. Sultankulov, B.; Berillo, D.; Sultankulova, K.; Tokay, T.; Saparov, A. Progress in the development of chitosan-based biomaterials for tissue engineering and regenerative medicine. *Biomolecules* **2019**, *9*, 470. [CrossRef]
53. Yuan, N.; Shao, K.; Huang, S.; Chen, C. Chitosan, alginate, hyaluronic acid and other novel multifunctional hydrogel dressings for wound healing: A review. *Int. J. Biol. Macromol.* **2023**, *240*, 124321. [CrossRef]
54. Noor, A.; Afzal, A.; Masood, R.; Khaliq, Z.; Ahmad, S.; Ahmad, F.; Qadir, M.-B.; Irfan, M. Dressings for burn wound: A review. *J. Mater. Sci.* **2022**, *57*, 6536–6572. [CrossRef]
55. Pyo, S.-H.; Wang, P.; Hwang, H.H.; Zhu, W.; Warner, J.; Chen, S. Continuous Optical 3D Printing of Green Aliphatic Polyurethanes. *ACS Appl. Mater. Interfaces* **2017**, *9*, 836–844. [CrossRef]
56. Sasmal, P.K.; Ganguly, S. Polymer in hemostasis and follow-up wound healing. *J. Appl. Polym. Sci.* **2023**, *140*, e53559. [CrossRef]
57. Li, Q.; Wang, D.; Jiang, Z.; Li, R.; Xue, T.; Lin, C.; Deng, Y.; Jin, Y.; Sun, B. Advances of hydrogel combined with stem cells in promoting chronic wound healing. *Front. Chem.* **2022**, *10*, 1038839. [CrossRef] [PubMed]
58. Kim, H.S.; Kim, D.; Jeong, Y.W.; Choi, M.J.; Lee, G.W.; Thangavelu, M.; Song, J.E.; Khang, G. Engineering retinal pigment epithelial cells regeneration for transplantation in regenerative medicine using PEG/Gellan gum hydrogels. *Int. J. Biol. Macromol.* **2019**, *130*, 220–228. [CrossRef] [PubMed]
59. Alven, S.; Aderibigbe, B.A. Fabrication of Hybrid Nanofibers from Biopolymers and Poly (Vinyl Alcohol)/Poly (ε-Caprolactone) for Wound Dressing Applications. *Polymers* **2021**, *13*, 2104. [CrossRef] [PubMed]
60. Das, P.; Manna, S.; Roy, S.; Nandi, S.K.; Basak, P. Polymeric biomaterials-based tissue engineering for wound healing: A systemic review. *Burn. Trauma* **2023**, *11*, tkac058. [CrossRef]
61. Heczko, D.; Hachuła, B.; Maksym, P.; Kamiński, K.; Zięba, A.; Orszulak, L.; Paluch, M.; Kamińska, E. The Effect of Various Poly (*N*-vinylpyrrolidone) (PVP) Polymers on the Crystallization of Flutamide. *Pharmaceuticals* **2022**, *15*, 971. [CrossRef]
62. Raj, T.; Chandrasekhar, K.; Naresh Kumar, A.; Kim, S.-H. Lignocellulosic biomass as renewable feedstock for biodegradable and recyclable plastics production: A sustainable approach. *Renew. Sustain. Energy Rev.* **2022**, *158*, 112130. [CrossRef]
63. Amirrah, I.N.; Lokanathan, Y.; Zulkiflee, I.; Wee, M.F.M.R.; Motta, A.; Fauzi, M.B. A Comprehensive Review on Collagen Type I Development of Biomaterials for Tissue Engineering: From Biosynthesis to Bioscaffold. *Biomedicines* **2022**, *10*, 2307. [CrossRef]
64. Cavallo, A.; Al Kayal, T.; Mero, A.; Mezzetta, A.; Pisani, A.; Foffa, I.; Vecoli, C.; Buscemi, M.; Guazzelli, L.; Soldani, G.; et al. Marine Collagen-Based Bioink for 3D Bioprinting of a Bilayered Skin Model. *Pharmaceutics* **2023**, *15*, 1331. [CrossRef] [PubMed]
65. Salvatore, L.; Gallo, N.; Natali, M.L.; Campa, L.; Lunetti, P.; Madaghiele, M.; Blasi, F.S.; Corallo, A.; Capobianco, L.; Sannino, A. Marine collagen and its derivatives: Versatile and sustainable bio-resources for healthcare. *Mater. Sci. Eng. C* **2020**, *113*, 110963. [CrossRef]

66. Fertala, A. Three Decades of Research on Recombinant Collagens: Reinventing the Wheel or Developing New Biomedical Products? *Bioengineering* **2020**, *7*, 155. [CrossRef] [PubMed]
67. Das, P.; Chakravarty, T.; Roy, A.J.; Manna, S.; Nandi, S.K.; Basak, P. Sustainable development of Draksha- Beeja extract loaded gelatin and starch-based green and biodegradable mats for potential tissue engineering applications. *Sustain. Chem. Pharm.* **2023**, *34*, 101134. [CrossRef]
68. He, S.; Li, H.; Chi, B.; Zhang, X.; Wang, Y.; Wu, J.; Huang, Q. Construction of a dual-component hydrogel matrix for 3D biomimetic skin based on photo-crosslinked chondroitin sulfate/collagen. *Int. J. Biol. Macromol.* **2024**, *254*, 127940. [CrossRef] [PubMed]
69. Lin, K.; Zhang, D.; Macedo, M.H.; Cui, W.; Sarmento, B.; Shen, G. Advanced Collagen-Based Biomaterials for Regenerative Biomedicine. *Adv. Funct. Mater.* **2019**, *29*, 1804943. [CrossRef]
70. Egorikhina, M.N.; Aleynik, D.Y.; Rubtsova, Y.P.; Levin, G.Y.; Charykova, I.N.; Semenycheva, L.L.; Bugrova, M.L.; Zakharychev, E.A. Hydrogel scaffolds based on blood plasma cryoprecipitate and collagen derived from various sources: Structural, mechanical and biological characteristics. *Bioact. Mater.* **2019**, *4*, 334–345. [CrossRef]
71. Sorushanova, A.; Delgado, L.M.; Wu, Z.; Shologu, N.; Kshirsagar, A.; Raghunath, R.; Mullen, A.M.; Bayon, Y.; Pandit, A.; Raghunath, M.; et al. The Collagen Suprafamily: From Biosynthesis to Advanced Biomaterial Development. *Adv. Mater.* **2019**, *31*, e1801651. [CrossRef]
72. Jafari, H.; Lista, A.; Siekapen, M.M.; Ghaffari-Bohlouli, P.; Nie, L.; Alimoradi, H.; Shavandi, A. Fish Collagen: Extraction, Characterization, and Applications for Biomaterials Engineering. *Polymers* **2020**, *12*, 2230. [CrossRef] [PubMed]
73. Liu, Y.; Fan, D. Novel hyaluronic acid-tyrosine/collagen-based injectable hydrogels as soft filler for tissue engineering. *Int. J. Biol. Macromol.* **2019**, *141*, 700–712. [CrossRef] [PubMed]
74. Zhang, Y.; Wang, Y.; Li, Y.; Yang, Y.; Jin, M.; Lin, X.; Zhuang, Z.; Guo, K.; Zhang, T.; Tan, W. Application of Collagen-Based Hydrogel in Skin Wound Healing. *Gels* **2023**, *9*, 185. [CrossRef] [PubMed]
75. Zhu, S.; Yu, J.; Xiong, S.; Ding, Y.; Zhou, X.; Hu, Y.; Chen, W.; Lin, Y.; Dao, L. Fabrication and insights into the mechanisms of collagen-based hydrogels with the high cell affinity and antimicrobial activity. *J. Appl. Polym. Sci.* **2022**, *139*, 51623. [CrossRef]
76. Li, J.; Zhai, Y.-N.; Xu, J.-P.; Zhu, X.-Y.; Yang, H.-R.; Che, H.-J.; Liu, C.-K.; Qu, J.-B. An injectable collagen peptide-based hydrogel with desirable antibacterial, self-healing and wound-healing properties based on multiple-dynamic crosslinking. *Int. J. Biol. Macromol.* **2024**, *259*, 129006. [CrossRef] [PubMed]
77. Tsegay, F.; Elsherif, M.; Butt, H. Smart 3D Printed Hydrogel Skin Wound Bandages: A Review. *Polymers* **2022**, *14*, 1012. [CrossRef]
78. Aghamirsalim, M.; Mobaraki, M.; Soltani, M.; Shahvandi, M.K.; Jabbarvand, M.; Afzali, E.; Raahemifar, K. 3D Printed Hydrogels for Ocular Wound Healing. *Biomedicines* **2022**, *10*, 1562. [CrossRef]
79. Osidak, E.O.; Karalkin, P.A.; Osidak, M.S.; Parfenov, V.A.; Sivogrivov, D.E.; Pereira, F.D.A.S.; Gryadunova, A.A.; Koudan, E.V.; Khesuani, Y.D.; Kasyanov, V.A.; et al. Viscoll collagen solution as a novel bioink for direct 3D bioprinting. *J. Mater. Sci. Mater. Med.* **2019**, *30*, 31. [CrossRef]
80. Rosenquist, J.; Folkesson, M.; Höglund, L.; Pupkaite, J.; Hilborn, J.; Samanta, A. An Injectable, Shape-Retaining Collagen Hydrogel Cross-linked Using Thiol-Maleimide Click Chemistry for Sealing Corneal Perforations. *ACS Appl. Mater. Interfaces* **2023**, *15*, 34407–34418. [CrossRef]
81. Zheng, J.; Tian, X.; Xu, B.; Yuan, F.; Gong, J.; Yang, Z. Collagen Peptides from Swim Bladders of Giant Croaker (*Nibea japonica*) and Their Protective Effects against H_2O_2-Induced Oxidative Damage toward Human Umbilical Vein Endothelial Cells. *Mar. Drugs* **2020**, *18*, 430. [CrossRef]
82. Helary, C.; Zarka, M.; Giraud-Guille, M.M. Fibroblasts within concentrated collagen hydrogels favour chronic skin wound healing. *J. Tissue Eng. Regen. Med.* **2012**, *6*, 225–237. [CrossRef]
83. Huang, Y.; Bai, L.; Yang, Y.; Yin, Z.; Guo, B. Biodegradable gelatin/silver nanoparticle composite cryogel with excellent antibacterial and antibiofilm activity and hemostasis for Pseudomonas aeruginosa-infected burn wound healing. *J. Colloid Interface Sci.* **2021**, *608*, 2278–2289. [CrossRef]
84. Yin, R.; Zhang, C.; Shao, J.; Chen, Y.; Yin, A.; Feng, Q.; Chen, S.; Peng, F.; Ma, X.; Xu, C.-Y.; et al. Integration of flexible, recyclable, and transient gelatin hydrogels toward multifunctional electronics. *J. Mater. Sci. Technol.* **2023**, *145*, 83–92. [CrossRef]
85. Chen, C.; Li, D.; Yano, H.; Abe, K. Insect Cuticle-Mimetic Hydrogels with High Mechanical Properties Achieved via the Combination of Chitin Nanofiber and Gelatin. *J. Agric. Food Chem.* **2019**, *67*, 5571–5578. [CrossRef] [PubMed]
86. Sun, L.; Yao, Y.; Dai, L.; Jiao, M.; Ding, B.; Yu, Q.; Tang, J.; Liu, B. Sustainable and high-performance Zn dual-ion batteries with a hydrogel-based water-in-salt electrolyte. *Energy Storage Mater.* **2022**, *47*, 187–194. [CrossRef]
87. Asadi, N.; Mehdipour, A.; Ghorbani, M.; Mesgari-Abbasi, M.; Akbarzadeh, A.; Davaran, S. A novel multifunctional bilayer scaffold based on chitosan nanofiber/alginate-gelatin methacrylate hydrogel for full-thickness wound healing. *Int. J. Biol. Macromol.* **2021**, *193*, 734–747. [CrossRef] [PubMed]
88. Han, K.; Bai, Q.; Wu, W.; Sun, N.; Cui, N.; Lu, T. Gelatin-based adhesive hydrogel with self-healing, hemostasis, and electrical conductivity. *Int. J. Biol. Macromol.* **2021**, *183*, 2142–2151. [CrossRef] [PubMed]
89. Ma, N.; Li, X.; Ding, Z.; Tao, J.; Xu, G.; Wang, Y.; Huang, Y.; Liu, J. A polyacrylic acid/polyacrylamide-based hydrogel electrolyte containing gelatin for efficient electrochromic device with outstanding cycling stability and flexible compatibility. *Eur. Polym. J.* **2023**, *190*, 112024. [CrossRef]

90. Ionescu, O.M.; Mignon, A.; Minsart, M.; Van Hoorick, J.; Gardikiotis, I.; Caruntu, I.D.; Giusca, S.E.; Van Vlierberghe, S.; Profire, L. Gelatin-Based Versus Alginate-Based Hydrogels: Providing Insight in Wound Healing Potential. *Macromol. Biosci.* **2021**, *21*, 2100230. [CrossRef]
91. Sharifi, S.; Islam, M.M.; Sharifi, H.; Islam, R.; Koza, D.; Reyes-Ortega, F.; Alba-Molina, D.; Nilsson, P.H.; Dohlman, C.H.; Mollnes, T.E.; et al. Tuning gelatin-based hydrogel towards bioadhesive ocular tissue engineering applications. *Bioact. Mater.* **2021**, *6*, 3947–3961. [CrossRef]
92. Chiaoprakobkij, N.; Seetabhawang, S.; Sanchavanakit, N.; Phisalaphong, M. Fabrication and characterization of novel bacterial cellulose/alginate/gelatin biocomposite film. *J. Biomater. Sci. Polym. Ed.* **2019**, *30*, 961–982. [CrossRef]
93. Dou, C.; Li, Z.; Luo, Y.; Gong, J.; Li, Q.; Zhang, J.; Zhang, Q.; Qiao, C. Bio-based poly (γ-glutamic acid)-gelatin double-network hydrogel with high strength for wound healing. *Int. J. Biol. Macromol.* **2022**, *202*, 438–452. [CrossRef] [PubMed]
94. Han, Y.; Zheng, L.; Wang, Y.; Fan, K.; Guo, S.; Kang, H.; Lin, J.; Xue, Y.; Liu, Z.; Li, C. Corneal stromal filler injection of gelatin-based photocurable hydrogels for maintaining the corneal thickness and reconstruction of corneal stroma. *Compos. Part B Eng.* **2023**, *266*, 111004. [CrossRef]
95. Ahmed, A.; Nath, J.; Baruah, K.; Rather, M.A.; Mandal, M.; Dolui, S.K. Development of mussel mimetic gelatin based adhesive hydrogel for wet surfaces with self-healing and reversible properties. *Int. J. Biol. Macromol.* **2023**, *228*, 68–77. [CrossRef]
96. Wang, Y.; Xiao, D.; Quan, L.; Chai, H.; Sui, X.; Wang, B.; Xu, H.; Mao, Z. Mussel-inspired adhesive gelatin–polyacrylamide hydrogel wound dressing loaded with tetracycline hydrochloride to enhance complete skin regeneration. *Soft Matter* **2022**, *18*, 662–674. [CrossRef] [PubMed]
97. Li, B.; Li, H.; Yang, H.; Shu, Y.; Li, K.; Chen, K.; Xiao, W.; Liao, X. Preparation and antibacterial properties of an AgBr@SiO_2/GelMA composite hydrogel. *Biomed. Mater.* **2022**, *17*, 025005. [CrossRef] [PubMed]
98. Ngadimin, K.D.; Stokes, A.; Gentile, P.; Ferreira, A.M. Biomimetic hydrogels designed for cartilage tissue engineering. *Biomater. Sci.* **2021**, *9*, 4246–4259. [CrossRef] [PubMed]
99. Xiao, Y.; Lu, C.; Yu, Z.; Lian, Y.; Ma, Y.; Chen, Z.; Jiang, X.; Zhang, Y. Transparent, High Stretchable, Environmental Tolerance, and Excellent Sensitivity Hydrogel for Flexible Sensors and Capacitive Pens. *ACS Appl. Mater. Interfaces* **2023**, *15*, 44280–44293. [CrossRef]
100. Ren, J.; Wang, X.; Zhao, L.; Li, M.; Yang, W. Effective Removal of Dyes from Aqueous Solutions by a Gelatin Hydrogel. *J. Polym. Environ.* **2021**, *29*, 3497–3508. [CrossRef]
101. He, X.; Liu, X.; Yang, J.; Du, H.; Chai, N.; Sha, Z.; Geng, M.; Zhou, X.; He, C. Tannic acid-reinforced methacrylated chitosan/methacrylated silk fibroin hydrogels with multifunctionality for accelerating wound healing. *Carbohydr. Polym.* **2020**, *247*, 116689. [CrossRef]
102. Indrakumar, S.; Joshi, A.; Dash, T.K.; Mishra, V.; Tandon, B.; Chatterjee, K. Photopolymerized silk fibroin gel for advanced burn wound care. *Int. J. Biol. Macromol.* **2023**, *233*, 123569. [CrossRef]
103. Li, Z.; Zheng, A.; Mao, Z.; Li, F.; Su, T.; Cao, L.; Wang, W.; Liu, Y.; Wang, C. Silk fibroin–gelatin photo-crosslinked 3D-bioprinted hydrogel with MOF-methylene blue nanoparticles for infected wound healing. *Int. J. Bioprint.* **2023**, *9*, 459–473. [CrossRef]
104. Maity, B.; Alam, S.; Samanta, S.; Prakash, R.G.; Govindaraju, T. Antioxidant Silk Fibroin Composite Hydrogel for Rapid Healing of Diabetic Wound. *Macromol. Biosci.* **2022**, *22*, e2200097. [CrossRef]
105. Pires, P.C.; Mascarenhas-Melo, F.; Pedrosa, K.; Lopes, D.; Lopes, J.; Macário-Soares, A.; Peixoto, D.; Giram, P.S.; Veiga, F.; Paiva-Santos, A.C. Polymer-based biomaterials for pharmaceutical and biomedical applications: A focus on topical drug administration. *Eur. Polym. J.* **2023**, *187*, 111868. [CrossRef]
106. Pourjabbar, B.; Biazar, E.; Heidari Keshel, S.; Baradaran-Rafii, A. Improving the properties of fish skin collagen/silk fibroin dressing by chemical treatment for corneal wound healing. *Int. Wound J.* **2023**, *20*, 484–498. [CrossRef] [PubMed]
107. Shefa, A.A.; Amirian, J.; Kang, H.J.; Bae, S.H.; Jung, H.-I.; Choi, H.-J.; Lee, S.Y.; Lee, B.-T. In vitro and in vivo evaluation of effectiveness of a novel TEMPO-oxidized cellulose nanofiber-silk fibroin scaffold in wound healing. *Carbohydr. Polym.* **2017**, *177*, 284–296. [CrossRef] [PubMed]
108. Sun, X.; Zhang, Y.; Cui, J.; Zhang, C.; Xing, C.; Bian, H.; Lv, J.; Chen, D.; Xiao, L.; Su, J.; et al. Advanced multilayer composite dressing with co-delivery of gelsevirine and silk fibroin for burn wound healing. *Compos. Part B Eng.* **2023**, *253*, 110549. [CrossRef]
109. Yang, W.; Xu, H.; Lan, Y.; Zhu, Q.; Liu, Y.; Huang, S.; Shi, S.; Hancharou, A.; Tang, B.; Guo, R. Preparation and characterisation of a novel silk fibroin/hyaluronic acid/sodium alginate scaffold for skin repair. *Int. J. Biol. Macromol.* **2019**, *130*, 58–67. [CrossRef]
110. Zahra, D.; Shokat, Z.; Ahmad, A.; Javaid, A.; Khurshid, M.; Ashfaq, U.A.; Nashwan, A.J. Exploring the recent developments of alginate silk fibroin material for hydrogel wound dressing: A review. *Int. J. Biol. Macromol.* **2023**, *248*, 125989. [CrossRef] [PubMed]
111. Zhu, Z.; Liu, Y.; Chen, J.; He, Z.; Tan, P.; He, Y.; Pei, X.; Wang, J.; Tan, L.; Wan, Q. Structural-Functional Pluralistic Modification of Silk Fibroin via MOF Bridging for Advanced Wound Care. *Adv. Sci.* **2022**, *9*, e2204553. [CrossRef] [PubMed]
112. Zhang, F.; Yin, C.; Qi, X.; Guo, C.; Wu, X. Silk Fibroin Crosslinked Glycyrrhizic Acid and Silver Hydrogels for Accelerated Bacteria-Infected Wound Healing. *Macromol. Biosci.* **2022**, *22*, 2100407. [CrossRef]
113. Özen, N.; Özbaş, Z.; İzbudak, B.; Emik, S.; Özkahraman, B.; Bal-Öztürk, A. Boric acid-impregnated silk fibroin/gelatin/hyaluronic acid-based films for improving the wound healing process. *J. Appl. Polym. Sci.* **2022**, *139*, 51715. [CrossRef]
114. Yin, C.; Han, X.; Lu, Q.; Qi, X.; Guo, C.; Wu, X. Rhein incorporated silk fibroin hydrogels with antibacterial and anti-inflammatory efficacy to promote healing of bacteria-infected burn wounds. *Int. J. Biol. Macromol.* **2022**, *201*, 14–19. [CrossRef] [PubMed]

115. Qian, Y.; Xu, C.; Xiong, W.; Jiang, N.; Zheng, Y.; He, X.; Ding, F.; Lu, X.; Shen, J. Dual cross-linked organic-inorganic hybrid hydrogels accelerate diabetic skin wound healing. *Chem. Eng. J.* **2021**, *417*, 129335. [CrossRef]
116. Yu, R.; Yang, Y.; He, J.; Li, M.; Guo, B. Novel supramolecular self-healing silk fibroin-based hydrogel via host–guest interaction as wound dressing to enhance wound healing. *Chem. Eng. J.* **2020**, *417*, 128278. [CrossRef]
117. Li, Z.; Song, J.; Zhang, J.; Hao, K.; Liu, L.; Wu, B.; Zheng, X.; Xiao, B.; Tong, X.; Dai, F. Topical application of silk fibroin-based hydrogel in preventing hypertrophic scars. *Colloids Surfaces B Biointerfaces* **2020**, *186*, 110735. [CrossRef] [PubMed]
118. Guan, Y.; Sun, F.; Zhang, X.; Peng, Z.; Jiang, B.; Liang, M.; Wang, Y. Silk fibroin hydrogel promote burn wound healing through regulating TLN1 expression and affecting cell adhesion and migration. *J. Mater. Sci. Mater. Med.* **2020**, *31*, 48. [CrossRef]
119. Alavi, M.; Nokhodchi, A. Antimicrobial and wound healing activities of electrospun nanofibers based on functionalized carbohydrates and proteins. *Cellulose* **2022**, *29*, 1331–1347. [CrossRef]
120. Zhang, X.-Y.; Liu, C.; Fan, P.-S.; Zhang, X.-H.; Hou, D.-Y.; Wang, J.-Q.; Yang, H.; Wang, H.; Qiao, Z.-Y. Skin-like wound dressings with on-demand administration based on in situ peptide self-assembly for skin regeneration. *J. Mater. Chem. B* **2022**, *10*, 3624–3636. [CrossRef]
121. Malinowski, C.; He, F.; Zhao, Y.; Chang, I.; Hatchett, D.W.; Zhai, S.; Zhao, H. Nanopatterned silk fibroin films with high transparency and high haze for optical applications. *RSC Adv.* **2019**, *9*, 40792–40799. [CrossRef]
122. Zhao, Y.; Guan, J.; Wu, S.J. Highly Stretchable and Tough Physical Silk Fibroin–Based Double Network Hydrogels. *Macromol. Rapid Commun.* **2019**, *40*, e1900389. [CrossRef]
123. Xiao, Y.; Wu, Y.; Si, P.; Zhang, D. Tough silk fibroin hydrogel via polypropylene glycol (PPG) blending for wearable sensors. *J. Appl. Polym. Sci.* **2023**, *140*, e54689. [CrossRef]
124. Guo, Z.; Yan, L.; Zhou, B.; Zhao, P.; Wang, W.; Dong, S.; Cheng, B.; Yang, J.; Li, B.; Wang, X. In situ photo-crosslinking silk fibroin based hydrogel accelerates diabetic wound healing through antibacterial and antioxidant. *Int. J. Biol. Macromol.* **2023**, *242*, 125028. [CrossRef]
125. Zhang, J.; Wang, L.; Xu, C.; Cao, Y.; Liu, S.; Reis, R.L.; Kundu, S.C.; Yang, X.; Xiao, B.; Duan, L. Transparent silk fibroin film-facilitated infected-wound healing through antibacterial, improved fibroblast adhesion and immune modulation. *J. Mater. Chem. B* **2023**, *12*, 475–488. [CrossRef] [PubMed]
126. Tang, X.; Chen, X.; Zhang, S.; Gu, X.; Wu, R.; Huang, T.; Zhou, Z.; Sun, C.; Ling, J.; Liu, M.; et al. Silk-Inspired In Situ Hydrogel with Anti-Tumor Immunity Enhanced Photodynamic Therapy for Melanoma and Infected Wound Healing. *Adv. Funct. Mater.* **2021**, *31*, 2101320. [CrossRef]
127. Mitropoulos, A.N.; Marelli, B.; Ghezzi, C.E.; Applegate, M.B.; Partlow, B.P.; Kaplan, D.L.; Omenetto, F.G. Transparent, Nanostructured Silk Fibroin Hydrogels with Tunable Mechanical Properties. *ACS Biomater. Sci. Eng.* **2015**, *1*, 964–970. [CrossRef]
128. Abbasi, A.R.; Sohail, M.; Minhas, M.U.; Khaliq, T.; Kousar, M.; Khan, S.; Hussain, Z.; Munir, A. Bioinspired sodium alginate based thermosensitive hydrogel membranes for accelerated wound healing. *Int. J. Biol. Macromol.* **2020**, *155*, 751–765. [CrossRef]
129. He, Y.; Zhao, W.; Dong, Z.; Ji, Y.; Li, M.; Hao, Y.; Zhang, D.; Yuan, C.; Deng, J.; Zhao, P.; et al. A biodegradable antibacterial alginate/carboxymethyl chitosan/Kangfuxin sponges for promoting blood coagulation and full-thickness wound healing. *Int. J. Biol. Macromol.* **2021**, *167*, 182–192. [CrossRef]
130. Yuan, Z.; Liu, H.; Wu, H.; Wang, Y.; Liu, Q.; Wang, Y.; Lincoln, S.F.; Guo, X.; Wang, J. Cyclodextrin Hydrogels: Rapid Removal of Aromatic Micropollutants and Adsorption Mechanisms. *J. Chem. Eng. Data* **2020**, *65*, 678–689. [CrossRef]
131. Benhalima, T.; Ferfera-Harrar, H. Eco-friendly porous carboxymethyl cellulose/dextran sulfate composite beads as reusable and efficient adsorbents of cationic dye methylene blue. *Int. J. Biol. Macromol.* **2019**, *132*, 126–141. [CrossRef] [PubMed]
132. Hamza, M.F.; Hamad, N.A.; Hamad, D.M.; Khalafalla, M.S.; Abdel-Rahman, A.A.-H.; Zeid, I.F.; Wei, Y.; Hessien, M.M.; Fouda, A.; Salem, W.M. Synthesis of Eco-Friendly Biopolymer, Alginate-Chitosan Composite to Adsorb the Heavy Metals, Cd(II) and Pb(II) from Contaminated Effluents. *Materials* **2021**, *14*, 2189. [CrossRef] [PubMed]
133. Shukla, A.; Mehta, K.; Parmar, J.; Pandya, J.; Saraf, M. Depicting the exemplary knowledge of microbial exopolysaccharides in a nutshell. *Eur. Polym. J.* **2019**, *119*, 298–310. [CrossRef]
134. Tang, S.; Yang, J.; Lin, L.; Peng, K.; Chen, Y.; Jin, S.; Yao, W. Construction of physically crosslinked chitosan/sodium alginate/calcium ion double-network hydrogel and its application to heavy metal ions removal. *Chem. Eng. J.* **2020**, *393*, 124728. [CrossRef]
135. Zdiri, K.; Cayla, A.; Elamri, A.; Erard, A.; Salaun, F. Alginate-Based Bio-Composites and Their Potential Applications. *J. Funct. Biomater.* **2022**, *13*, 117. [CrossRef] [PubMed]
136. Bahadoran, M.; Shamloo, A.; Nokoorani, Y.D. Development of a polyvinyl alcohol/sodium alginate hydrogel-based scaffold incorporating bFGF-encapsulated microspheres for accelerated wound healing. *Sci. Rep.* **2020**, *10*, 7342. [CrossRef] [PubMed]
137. Bazmandeh, A.Z.; Mirzaei, E.; Fadaie, M.; Shirian, S.; Ghasemi, Y. Dual spinneret electrospun nanofibrous/gel structure of chitosan-gelatin/chitosan-hyaluronic acid as a wound dressing: In-vitro and in-vivo studies. *Int. J. Biol. Macromol.* **2020**, *162*, 359–373. [CrossRef]
138. Eskandarinia, A.; Kefayat, A.; Rafienia, M.; Agheb, M.; Navid, S.; Ebrahimpour, K. Cornstarch-based wound dressing incorporated with hyaluronic acid and propolis: In vitro and in vivo studies. *Carbohydr. Polym.* **2019**, *216*, 25–35. [CrossRef]
139. Alven, S.; Aderibigbe, B.A. Hyaluronic Acid-Based Scaffolds as Potential Bioactive Wound Dressings. *Polymers* **2021**, *13*, 2102. [CrossRef]

140. Ganjoo, R.; Sharma, S.; Verma, C.; Quraishi, M.A.; Kumar, A. Heteropolysaccharides in sustainable corrosion inhibition: 4E (Energy, Economy, Ecology, and Effectivity) dimensions. *Int. J. Biol. Macromol.* **2023**, *235*, 123571. [CrossRef]
141. Meng, Q.; Zhong, S.; He, S.; Gao, Y.; Cui, X. Constructing of pH and reduction dual-responsive folic acid-modified hyaluronic acid-based microcapsules for dual-targeted drug delivery via sonochemical method. *Colloid Interface Sci. Commun.* **2021**, *44*, 100503. [CrossRef]
142. Ding, Y.-W.; Wang, Z.-Y.; Ren, Z.-W.; Zhang, X.-W.; Wei, D.-X. Advances in modified hyaluronic acid-based hydrogels for skin wound healing. *Biomater. Sci.* **2022**, *10*, 3393–3409. [CrossRef] [PubMed]
143. Fernandes-Cunha, G.M.; Jeong, S.H.; Logan, C.M.; Le, P.; Mundy, D.; Chen, F.; Chen, K.M.; Kim, M.; Lee, G.-H.; Na, K.-S.; et al. Supramolecular host-guest hyaluronic acid hydrogels enhance corneal wound healing through dynamic spatiotemporal effects. *Ocul. Surf.* **2022**, *23*, 148–161. [CrossRef] [PubMed]
144. Wang, S.; Chi, J.; Jiang, Z.; Hu, H.; Yang, C.; Liu, W.; Han, B. A self-healing and injectable hydrogel based on water-soluble chitosan and hyaluronic acid for vitreous substitute. *Carbohydr. Polym.* **2021**, *256*, 117519. [CrossRef]
145. Ying, H.; Zhou, J.; Wang, M.; Su, D.; Ma, Q.; Lv, G.; Chen, J. In situ formed collagen-hyaluronic acid hydrogel as biomimetic dressing for promoting spontaneous wound healing. *Mater. Sci. Eng. C* **2019**, *101*, 487–498. [CrossRef]
146. Chen, W.; Zhu, Y.; Zhang, Z.; Gao, Y.; Liu, W.; Borjihan, Q.; Qu, H.; Zhang, Y.; Zhang, Y.; Wang, Y.-J.; et al. Engineering a multifunctional N-halamine-based antibacterial hydrogel using a super-convenient strategy for infected skin defect therapy. *Chem. Eng. J.* **2019**, *379*, 122238. [CrossRef]
147. Gwak, M.A.; Hong, B.M.; Seok, J.M.; Park, S.A.; Park, W.H. Effect of tannic acid on the mechanical and adhesive properties of catechol-modified hyaluronic acid hydrogels. *Int. J. Biol. Macromol.* **2021**, *191*, 699–705. [CrossRef]
148. Passi, A.; Vigetti, D. Hyaluronan as tunable drug delivery system. *Adv. Drug Deliv. Rev.* **2019**, *146*, 83–96. [CrossRef] [PubMed]
149. Zainal, S.H.; Mohd, N.H.; Suhaili, N.; Anuar, F.H.; Lazim, A.M.; Othaman, R. Preparation of cellulose-based hydrogel: A review. *J. Mater. Res. Technol.* **2021**, *10*, 935–952. [CrossRef]
150. Bhaladhare, S.; Das, D. Cellulose: A fascinating biopolymer for hydrogel synthesis. *J. Mater. Chem. B* **2022**, *10*, 1923–1945. [CrossRef] [PubMed]
151. Xu, T.; Liu, K.; Sheng, N.; Zhang, M.; Liu, W.; Liu, H.; Dai, L.; Zhang, X.; Si, C.; Du, H.; et al. Biopolymer-based hydrogel electrolytes for advanced energy storage/conversion devices: Properties, applications, and perspectives. *Energy Storage Mater.* **2022**, *48*, 244–262. [CrossRef]
152. Tanan, W.; Panichpakdee, J.; Saengsuwan, S. Novel biodegradable hydrogel based on natural polymers: Synthesis, characterization, swelling/reswelling and biodegradability. *Eur. Polym. J.* **2019**, *112*, 678–687. [CrossRef]
153. Tanan, W.; Panichpakdee, J.; Suwanakood, P.; Saengsuwan, S. Biodegradable hydrogels of cassava starch-g-polyacrylic acid/natural rubber/polyvinyl alcohol as environmentally friendly and highly efficient coating material for slow-release urea fertilizers. *J. Ind. Eng. Chem.* **2021**, *101*, 237–252. [CrossRef]
154. Caruso, M.R.; D'agostino, G.; Milioto, S.; Cavallaro, G.; Lazzara, G. A review on biopolymer-based treatments for consolidation and surface protection of cultural heritage materials. *J. Mater. Sci.* **2023**, *58*, 12954–12975. [CrossRef]
155. Kwak, M.H.; Kim, J.E.; Go, J.; Koh, E.K.; Song, S.H.; Son, H.J.; Kim, H.S.; Yun, Y.H.; Jung, Y.J.; Hwang, D.Y. Bacterial cellulose membrane produced by Acetobacter sp. A10 for burn wound dressing applications. *Carbohydr. Polym.* **2015**, *122*, 387–398. [CrossRef]
156. Popa, L.; Ghica, M.V.; Tudoroiu, E.-E.; Ionescu, D.-G.; Dinu-Pîrvu, C.-E. Bacterial Cellulose—A Remarkable Polymer as a Source for Biomaterials Tailoring. *Materials* **2022**, *15*, 1054. [CrossRef]
157. Horue, M.; Silva, J.M.; Berti, I.R.; Brandão, L.R.; Barud, H.d.S.; Castro, G.R. Bacterial Cellulose-Based Materials as Dressings for Wound Healing. *Pharmaceutics* **2023**, *15*, 424. [CrossRef] [PubMed]
158. Andrabi, S.M.; Majumder, S.; Gupta, K.C.; Kumar, A. Dextran based amphiphilic nano-hybrid hydrogel system incorporated with curcumin and cerium oxide nanoparticles for wound healing. *Colloids Surfaces B Biointerfaces* **2020**, *195*, 111263. [CrossRef]
159. Du, X.; Liu, Y.; Wang, X.; Yan, H.; Wang, L.; Qu, L.; Kong, D.; Qiao, M.; Wang, L. Injectable hydrogel composed of hydrophobically modified chitosan/oxidized-dextran for wound healing. *Mater. Sci. Eng. C* **2019**, *104*, 109930. [CrossRef]
160. Jung, S.A.; Malyaran, H.; Demco, D.E.; Manukanc, A.; Häser, L.S.; Kučikas, V.; van Zandvoort, M.; Neuss, S.; Pich, A. Fibrin–Dextran Hydrogels with Tunable Porosity and Mechanical Properties. *Biomacromolecules* **2023**, *24*, 3972–3984. [CrossRef]
161. Luo, H.-C.; Mai, K.-J.; Liu, E.; Chen, H.; Xie, Y.-J.; Zheng, Y.-X.; Lin, R.; Zhang, L.-M.; Zhang, Y. Efficiency and Safety of Dextran-PAMAM/siMMP-9 Complexes for Decreasing Matrix Metalloproteinase-9 Expression and Promoting Wound Healing in Diabetic Rats. *Bioconjug. Chem.* **2022**, *33*, 2398–2410. [CrossRef]
162. Nonsuwan, P.; Matsugami, A.; Hayashi, F.; Hyon, S.-H.; Matsumura, K. Controlling the degradation of an oxidized dextran-based hydrogel independent of the mechanical properties. *Carbohydr. Polym.* **2019**, *204*, 131–141. [CrossRef] [PubMed]
163. Qiu, X.; Zhang, J.; Cao, L.; Jiao, Q.; Zhou, J.; Yang, L.; Zhang, H.; Wei, Y. Antifouling Antioxidant Zwitterionic Dextran Hydrogels as Wound Dressing Materials with Excellent Healing Activities. *ACS Appl. Mater. Interfaces* **2021**, *13*, 7060–7069. [CrossRef] [PubMed]
164. Rohiwal, S.S.; Ellederova, Z.; Tiwari, A.P.; Alqarni, M.; Elazab, S.T.; El-Saber Batiha, G.; Pawar, S.H.; Thorat, N.D. Self-assembly of bovine serum albumin (BSA)–dextran bio-nanoconjugate: Structural, antioxidant and in vitro wound healing studies. *RSC Adv.* **2021**, *11*, 4308–4317. [CrossRef] [PubMed]

165. Sun, Y.; Li, D.; Yu, Y.; Zheng, Y. Insights into the Role of Natural Polysaccharide-Based Hydrogel Wound Dressings in Biomedical Applications. *Gels* **2022**, *8*, 646. [CrossRef] [PubMed]
166. Wang, P.; Huang, S.; Hu, Z.; Yang, W.; Lan, Y.; Zhu, J.; Hancharou, A.; Guo, R.; Tang, B. In situ formed anti-inflammatory hydrogel loading plasmid DNA encoding VEGF for burn wound healing. *Acta Biomater.* **2019**, *100*, 191–201. [CrossRef] [PubMed]
167. Zhang, M.; Huang, Y.; Pan, W.; Tong, X.; Zeng, Q.; Su, T.; Qi, X.; Shen, J. Polydopamine-incorporated dextran hydrogel drug carrier with tailorable structure for wound healing. *Carbohydr. Polym.* **2021**, *253*, 117213. [CrossRef]
168. Yin, Y.; Xu, Q.; Wei, X.; Ma, Q.; Li, D.; Zhao, J. Rosmarinic Acid-Grafted Dextran/Gelatin Hydrogel as a Wound Dressing with Improved Properties: Strong Tissue Adhesion, Antibacterial, Antioxidant and Anti-Inflammatory. *Molecules* **2023**, *28*, 4034. [CrossRef]
169. Alibolandi, M.; Mohammadi, M.; Taghdisi, S.M.; Abnous, K.; Ramezani, M. Synthesis and preparation of biodegradable hybrid dextran hydrogel incorporated with biodegradable curcumin nanomicelles for full thickness wound healing. *Int. J. Pharm.* **2017**, *532*, 466–477. [CrossRef]
170. Mostafavi Esfahani, M.; Koupaei, N.; Hassanzadeh-Tabrizi, S.A. Synthesis and characterization of polyvinyl alcohol/dextran/Zataria wound dressing with superior antibacterial and antioxidant properties. *J. Vinyl Addit. Technol.* **2023**, *29*, 380–394. [CrossRef]
171. Yang, J.; Zhang, X.; Lu, Q.; Wang, L.; Hu, X.; Zhang, H. Preparation, flocculation and application in sugar refining of eco-friendly dextran-polylysine complex flocculant. *Sep. Purif. Technol.* **2023**, *306*, 122673. [CrossRef]
172. Maingret, V.; Courrégelongue, C.; Schmitt, V.; Héroguez, V. Dextran-Based Nanoparticles to Formulate pH-Responsive Pickering Emulsions: A Fully Degradable Vector at a Day Scale. *Biomacromolecules* **2020**, *21*, 5358–5368. [CrossRef]
173. Trejo-Caballero, M.E.; Díaz-Patiño, L.; González-Reynac, M.; Molina, G.A.; López-Miranda, J.L.; Esparza, R.; España-Sánchez, B.L.; Arjona, N.; Estevez, M. Biopolymeric hydrogel electrolytes obtained by using natural polysaccharide-poly(itaconic acid-*co*-2-hydroxyethyl methacrylate) in deep eutectic solvents for rechargeable Zn-air batteries. *Green Chem.* **2023**, *25*, 6784–6796. [CrossRef]
174. De Cicco, F.; Reverchon, E.; Adami, R.; Auriemma, G.; Russo, P.; Calabrese, E.C.; Porta, A.; Aquino, R.P.; Del Gaudio, P. In situ forming antibacterial dextran blend hydrogel for wound dressing: SAA technology vs. spray drying. *Carbohydr. Polym.* **2014**, *101*, 1216–1224. [CrossRef]
175. Zheng, C.; Liu, C.; Chen, H.; Wang, N.; Liu, X.; Sun, G.; Qiao, W. Effective wound dressing based on Poly (vinyl alcohol)/Dextran-aldehyde composite hydrogel. *Int. J. Biol. Macromol.* **2019**, *132*, 1098–1105. [CrossRef]
176. Bochani, S.; Zarepour, A.; Kalantari-Hesari, A.; Haghi, F.; Shahbazi, M.-A.; Zarrabi, A.; Taheri, S.; Maleki, A. Injectable, antibacterial, and oxygen-releasing chitosan-based hydrogel for multimodal healing of bacteria-infected wounds. *J. Mater. Chem. B* **2023**, *11*, 8056–8068. [CrossRef]
177. You, S.; Huang, Y.; Mao, R.; Xiang, Y.; Cai, E.; Chen, Y.; Shen, J.; Dong, W.; Qi, X. Together is better: Poly(tannic acid) nanorods functionalized polysaccharide hydrogels for diabetic wound healing. *Ind. Crop. Prod.* **2022**, *186*, 115273. [CrossRef]
178. Eivazzadeh-Keihan, R.; Noruzi, E.B.; Mehrban, S.F.; Aliabadi, H.A.M.; Karimi, M.; Mohammadi, A.; Maleki, A.; Mahdavi, M.; Larijani, B.; Shalan, A.E. Review: The latest advances in biomedical applications of chitosan hydrogel as a powerful natural structure with eye-catching biological properties. *J. Mater. Sci.* **2022**, *57*, 3855–3891. [CrossRef]
179. Song, K.; Hao, Y.; Liu, Y.; Cao, R.; Zhang, X.; He, S.; Wen, J.; Zheng, W.; Wang, L.; Zhang, Y. Preparation of pectin-chitosan hydrogels based on bioadhesive-design micelle to prompt bacterial infection wound healing. *Carbohydr. Polym.* **2023**, *300*, 120272. [CrossRef] [PubMed]
180. Khorasani, M.T.; Joorabloo, A.; Adeli, H.; Mansoori-Moghadam, Z.; Moghaddam, A. Design and optimization of process parameters of polyvinyl (alcohol)/chitosan/nano zinc oxide hydrogels as wound healing materials. *Carbohydr. Polym.* **2019**, *207*, 542–554. [CrossRef] [PubMed]
181. Wang, Y.; Xie, R.; Li, Q.; Dai, F.; Lan, G.; Shang, S.; Lu, F. A self-adapting hydrogel based on chitosan/oxidized konjac glucomannan/AgNPs for repairing irregular wounds. *Biomater. Sci.* **2020**, *8*, 1910–1922. [CrossRef]
182. Guo, S.; Ren, Y.; Chang, R.; He, Y.; Zhang, D.; Guan, F.; Yao, M. Injectable Self-Healing Adhesive Chitosan Hydrogel with Antioxidative, Antibacterial, and Hemostatic Activities for Rapid Hemostasis and Skin Wound Healing. *ACS Appl. Mater. Interfaces* **2022**, *14*, 34455–34469. [CrossRef]
183. Fan, P.; Zeng, Y.; Zaldivar-Silva, D.; Agüero, L.; Wang, S. Chitosan-Based Hemostatic Hydrogels: The Concept, Mechanism, Application, and Prospects. *Molecules* **2023**, *28*, 1473. [CrossRef] [PubMed]
184. Li, J.; Fang, T.; Yan, W.; Zhang, F.; Xu, Y.; Du, Z. Structure and Properties of Oxidized Chitosan Grafted Cashmere Fiber by Amide Covalent Modification. *Molecules* **2020**, *25*, 3812. [CrossRef] [PubMed]
185. Yan, D.; Li, Y.; Liu, Y.; Li, N.; Zhang, X.; Yan, C. Antimicrobial Properties of Chitosan and Chitosan Derivatives in the Treatment of Enteric Infections. *Molecules* **2021**, *26*, 7136. [CrossRef]
186. Liu, F.; Han, R.; Naficy, S.; Casillas, G.; Sun, X.; Huang, Z. Few-Layered Boron Nitride Nanosheets for Strengthening Polyurethane Hydrogels. *ACS Appl. Nano Mater.* **2021**, *4*, 7988–7994. [CrossRef]
187. He, M.; Hou, Y.; Zhu, C.; He, M.; Jiang, Y.; Feng, G.; Liu, L.; Li, Y.; Chen, C.; Zhang, L. 3D-Printing Biodegradable PU/PAAM/Gel Hydrogel Scaffold with High Flexibility and Self-Adaptibility to Irregular Defects for Nonload-Bearing Bone Regeneration. *Bioconjug. Chem.* **2021**, *32*, 1915–1925. [CrossRef]

188. Lin, Y.-J.; Lee, G.-H.; Chou, C.-W.; Chen, Y.-P.; Wu, T.-H.; Lin, H.-R. Stimulation of wound healing by PU/hydrogel composites containing fibroblast growth factor-2. *J. Mater. Chem. B* **2015**, *3*, 1931–1941. [CrossRef]
189. Zhao, B.; Yan, J.; Long, F.; Qiu, W.; Meng, G.; Zeng, Z.; Huang, H.; Wang, H.; Lin, N.; Liu, X.Y. Bioinspired Conductive Enhanced Polyurethane Ionic Skin as Reliable Multifunctional Sensors. *Adv. Sci.* **2023**, *10*, e2300857. [CrossRef] [PubMed]
190. Jeong, H.I.; An, D.H.; Lim, J.W.; Oh, T.; Lee, H.; Park, S.-M.; Jeong, J.H.; Chung, J.W. Hydrogel Surface-Modified Polyurethane Copolymer Film with Water Permeation Resistance and Biocompatibility for Implantable Biomedical Devices. *Micromachines* **2021**, *12*, 447. [CrossRef]
191. Iga, C.; Pawel, S.; Marcin, L.; Justyna, K.L. Polyurethane Composite Scaffolds Modified with the Mixture of Gelatin and Hydroxyapatite Characterized by Improved Calcium Deposition. *Polymers* **2020**, *12*, 410. [CrossRef]
192. Xiang, S.L.; Su, Y.X.; Yin, H.; Li, C.; Zhu, M.Q. Visible-light-driven isotropic hydrogels as anisotropic underwater actuators. *Nano Energy* **2021**, *85*, 105965. [CrossRef]
193. Fang, Y.; Xu, J.; Gao, F.; Du, X.; Du, Z.; Cheng, X.; Wang, H. Self-healable and recyclable polyurethane-polyaniline hydrogel toward flexible strain sensor. *Compos. Part B Eng.* **2021**, *219*, 108965. [CrossRef]
194. Kucinska-Lipka, J.; Gubanska, I.; Lewandowska, A.; Terebieniec, A.; Przybytek, A.; Cieśliński, H. Antibacterial polyurethanes, modified with cinnamaldehyde, as potential materials for fabrication of wound dressings. *Polym. Bull.* **2019**, *76*, 2725–2742. [CrossRef]
195. Xiao, L.; Ni, W.; Zhao, X.; Guo, Y.; Li, X.; Wang, F.; Luo, G.; Zhan, R.; Xu, X. A moisture balanced antibacterial dressing loaded with lysozyme possesses antibacterial activity and promotes wound healing. *Soft Matter* **2021**, *17*, 3162–3173. [CrossRef]
196. Oprea, S.; Potolinca, V.O. Synthesis and properties of water-dispersible polyurethanes based on various diisocyanates and PEG as the hard segment. *J. Appl. Polym. Sci.* **2023**, *140*, e53948. [CrossRef]
197. Song, J.; Li, L.; Niu, Y.H.; Ke, R.Y.; Zhao, X. Preparation of humic acid water-retaining agent-modified polyurethane sponge as a soilless culture material. *J. Appl. Polym. Sci.* **2022**, *139*, 52182. [CrossRef]
198. Chen, S.; Li, S.; Ye, Z.; Zhang, Y.; Gao, S.; Rong, H.; Zhang, J.; Deng, L.; Dong, A. Superhydrophobic and superhydrophilic polyurethane sponge for wound healing. *Chem. Eng. J.* **2022**, *446*, 136985. [CrossRef]
199. Wen, J.; Zhang, X.; Pan, M.; Yuan, J.; Jia, Z.; Zhu, L. A Robust, Tough and Multifunctional Polyurethane/Tannic Acid Hydrogel Fabricated by Physical-Chemical Dual Crosslinking. *Polymers* **2020**, *12*, 239. [CrossRef] [PubMed]
200. Giroto, A.S.; do Valle, S.F.; Ribeiro, T.; Ribeiro, C.; Mattoso, L.H.C. Towards urea and glycerol utilization as "building blocks" for polyurethane production: A detailed study about reactivity and structure for environmentally friendly polymer synthesis. *React. Funct. Polym.* **2020**, *153*, 104629. [CrossRef]
201. Zanini, N.C.; de Souza, A.G.; Barbosa, R.F.C.; Rosa, D.S.; Mulinari, D.R. Eco-friendly composites of polyurethane and sheath palm residues. *J. Cell. Plast.* **2021**, *58*, 139–158. [CrossRef]
202. Bourguignon, M.; Thomassin, J.M.; Grignard, B.; Vertruyen, B.; Detrembleur, C. Water-Borne Isocyanate-Free Polyurethane Hydrogels with Adaptable Functionality and Behavior. *Macromol. Rapid Commun.* **2021**, *42*, e2000482. [CrossRef]
203. Chen, S.-H.; Chou, P.-Y.; Chen, Z.-Y.; Lin, F.-H. Electrospun Water-Borne Polyurethane Nanofibrous Membrane as a Barrier for Preventing Postoperative Peritendinous Adhesion. *Int. J. Mol. Sci.* **2019**, *20*, 1625. [CrossRef]
204. Hou, Y.; Jiang, N.; Sun, D.; Wang, Y.; Chen, X.; Zhu, S.; Zhang, L. A fast UV-curable PU-PAAm hydrogel with mechanical flexibility and self-adhesion for wound healing. *RSC Adv.* **2020**, *10*, 4907–4915. [CrossRef] [PubMed]
205. Wang, F.; Zhang, H.; Sun, Y.; Wang, S.; Zhang, L.; Wu, A.; Zhang, Y. Superhydrophilic quaternized calcium alginate based aerogel membrane for oil-water separation and removal of bacteria and dyes. *Int. J. Biol. Macromol.* **2023**, *227*, 1141–1150. [CrossRef] [PubMed]
206. Zhang, Y.; Liu, B.; Huang, K.; Wang, S.; Quirino, R.L.; Zhang, Z.-X.; Zhang, C. Eco-Friendly Castor Oil-Based Delivery System with Sustained Pesticide Release and Enhanced Retention. *ACS Appl. Mater. Interfaces* **2020**, *12*, 37607–37618. [CrossRef] [PubMed]
207. Lei, W.; Zhou, X.; Fang, C.; Song, Y.; Li, Y. Eco-friendly waterborne polyurethane reinforced with cellulose nanocrystal from office waste paper by two different methods. *Carbohydr. Polym.* **2019**, *209*, 299–309. [CrossRef] [PubMed]
208. Man, L.; Feng, Y.; Hu, Y.; Yuan, T.; Yang, Z. A renewable and multifunctional eco-friendly coating from novel tung oil-based cationic waterborne polyurethane dispersions. *J. Clean. Prod.* **2019**, *241*, 118341. [CrossRef]
209. Li, H.; Zhou, X.; Luo, L.; Ding, Q.; Tang, S. Bio-orthogonally crosslinked catechol–chitosan hydrogel for effective hemostasis and wound healing. *Carbohydr. Polym.* **2022**, *281*, 119039. [CrossRef] [PubMed]
210. Zhou, Q.; Zhou, X.; Mo, Z.; Zeng, Z.; Wang, Z.; Cai, Z.; Luo, L.; Ding, Q.; Li, H.; Tang, S. A PEG-CMC-THB-PRTM hydrogel with antibacterial and hemostatic properties for promoting wound healing. *Int. J. Biol. Macromol.* **2023**, *224*, 370–379. [CrossRef]
211. Peng, L.; Chang, L.; Si, M.; Lin, J.; Wei, Y.; Wang, S.; Liu, H.; Han, B.; Jiang, L. Hydrogel-Coated Dental Device with Adhesion-Inhibiting and Colony-Suppressing Properties. *ACS Appl. Mater. Interfaces* **2020**, *12*, 9718–9725. [CrossRef] [PubMed]
212. Li, C.; Jiang, T.; Zhou, C.; Jiang, A.; Lu, C.; Yang, G.; Nie, J.; Wang, F.; Yang, X.; Chen, Z. Injectable self-healing chitosan-based POSS-PEG hybrid hydrogel as wound dressing to promote diabetic wound healing. *Carbohydr. Polym.* **2023**, *299*, 120198. [CrossRef] [PubMed]
213. Masood, N.; Ahmed, R.; Tariq, M.; Ahmed, Z.; Masoud, M.S.; Ali, I.; Asghar, R.; Andleeb, A.; Hasan, A. Silver nanoparticle impregnated chitosan-PEG hydrogel enhances wound healing in diabetes induced rabbits. *Int. J. Pharm.* **2019**, *559*, 23–36. [CrossRef] [PubMed]

214. Xiao, X.; Wu, G.; Zhou, H.; Qian, K.; Hu, J. Preparation and Property Evaluation of Conductive Hydrogel Using Poly (Vinyl Alcohol)/Polyethylene Glycol/Graphene Oxide for Human Electrocardiogram Acquisition. *Polymers* **2017**, *9*, 259. [CrossRef] [PubMed]
215. Falqi, F.H.; Bin-Dahman, O.A.; Khair, A.; Al-Harthi, M.A. PVA/PEG/graphene shape memory composites responsive to multi-stimuli. *Appl. Phys. A* **2022**, *128*, 427. [CrossRef]
216. Liu, L.; Fan, X.; Zhang, Y.; Zhang, S.; Wang, W.; Jin, X.; Tang, B. Novel bio-based phase change materials with high enthalpy for thermal energy storage. *Appl. Energy* **2020**, *268*, 114979. [CrossRef]
217. Liu, M.; Chen, W.; Zhang, X.; Su, P.; Yue, F.; Zeng, S.; Du, S. Improved surface adhesion and wound healing effect of madecassoside liposomes modified by temperature-responsive PEG-PCL-PEG copolymers. *Eur. J. Pharm. Sci.* **2020**, *151*, 105373. [CrossRef]
218. Qayyum, A.S.; Jain, E.; Kolar, G.; Kim, Y.; Sell, S.A.; Zustiak, S.P. Design of electrohydrodynamic sprayed polyethylene glycol hydrogel microspheres for cell encapsulation. *Biofabrication* **2017**, *9*, 025019. [CrossRef]
219. Chen, Q.; Passos, A.; Balabani, S.; Chivu, A.; Zhao, S.; Azevedo, H.S.; Butler, P.; Song, W. Semi-interpenetrating network hyaluronic acid microgel delivery systems in micro-flow. *J. Colloid Interface Sci.* **2018**, *519*, 174–185. [CrossRef] [PubMed]
220. Chaturvedi, R.; Kang, Y.; Eom, Y.; Torati, S.R.; Kim, C. Functionalization of Biotinylated Polyethylene Glycol on Live Magnetotactic Bacteria Carriers for Improved Stealth Properties. *Biology* **2021**, *10*, 993. [CrossRef] [PubMed]
221. Nishida, K.; Anada, T.; Kobayashi, S.; Ueda, T.; Tanaka, M. Effect of bound water content on cell adhesion strength to water-insoluble polymers. *Acta Biomater.* **2021**, *134*, 313–324. [CrossRef]
222. Krishnadoss, V.; Melillo, A.; Kanjilal, B.; Hannah, T.; Ellis, E.; Kapetanakis, A.; Hazelton, J.; San Roman, J.; Masoumi, A.; Leijten, J.; et al. Bioionic Liquid Conjugation as Universal Approach To Engineer Hemostatic Bioadhesives. *ACS Appl. Mater. Interfaces* **2019**, *11*, 38373–38384. [CrossRef]
223. Shen, C.; Li, Y.; Meng, Q. Adhesive polyethylene glycol-based hydrogel patch for tissue repair. *Colloids Surfaces B Biointerfaces* **2022**, *218*, 112751. [CrossRef] [PubMed]
224. Teng, Y.Y.; Zou, M.L.; Liu, S.Y.; Jia, Y.; Zhang, K.W.; Yuan, Z.D.; Wu, J.J.; Ye, J.X.; Yu, S.; Li, X.; et al. Dual-Action Icariin-Containing Thermosensitive Hydrogel for Wound Macrophage Polarization and Hair-Follicle Neogenesis. *Front. Bioeng. Biotechnol.* **2022**, *10*, 902894. [CrossRef]
225. Sun, S.; Cui, Y.; Yuan, B.; Dou, M.; Wang, G.; Xu, H.; Wang, J.; Yin, W.; Wu, D.; Peng, C. Drug delivery systems based on polyethylene glycol hydrogels for enhanced bone regeneration. *Front. Bioeng. Biotechnol.* **2023**, *11*, 1117647. [CrossRef] [PubMed]
226. Lee, S.H.; Kang, M.; Jang, H.; Kondaveeti, S.; Sun, K.; Kim, S.; Park, H.H.; Jeong, H.E. Bifunctional Amphiphilic Nanospikes with Antifogging and Antibiofouling Properties. *ACS Appl. Mater. Interfaces* **2022**, *14*, 39478–39488. [CrossRef] [PubMed]
227. Si, R.; Wu, C.; Yu, D.; Ding, Q.; Li, R. Novel TEMPO-oxidized cellulose nanofiber/polyvinyl alcohol/polyethyleneimine nanoparticles for Cu^{2+} removal in water. *Cellulose* **2021**, *28*, 10999–11011. [CrossRef]
228. Rahmani, S.; Olad, A.; Rahmani, Z. Preparation of self-healable nanocomposite hydrogel based on Gum Arabic/gelatin and graphene oxide: Study of drug delivery behavior. *Polym. Bull.* **2023**, *80*, 4117–4138. [CrossRef]
229. Takács, T.; Abdelghafour, M.M.; Lamch, Ł.; Szenti, I.; Sebők, D.; Janovák, L.; Kukovecz, Á. Facile modification of hydroxyl group containing macromolecules provides autonomously self-healing polymers through the formation of dynamic Schiff base linkages. *Eur. Polym. J.* **2022**, *168*, 111086. [CrossRef]
230. Chopra, H.; Bibi, S.; Kumar, S.; Khan, M.S.; Kumar, P.; Singh, I. Preparation and Evaluation of Chitosan/PVA Based Hydrogel Films Loaded with Honey for Wound Healing Application. *Gels* **2022**, *8*, 111. [CrossRef]
231. Montaser, A.S.; Rehan, M.; El-Naggar, M.E. pH-Thermosensitive hydrogel based on polyvinyl alcohol/sodium alginate/N-isopropyl acrylamide composite for treating re-infected wounds. *Int. J. Biol. Macromol.* **2019**, *124*, 1016–1024. [CrossRef]
232. Shamloo, A.; Sarmadi, M.; Aghababaie, Z.; Vossoughi, M. Accelerated full-thickness wound healing via sustained bFGF delivery based on a PVA/chitosan/gelatin hydrogel incorporating PCL microspheres. *Int. J. Pharm.* **2018**, *537*, 278–289. [CrossRef]
233. Shamloo, A.; Aghababaie, Z.; Afjoul, H.; Jami, M.; Bidgoli, M.R.; Vossoughi, M.; Ramazani, A.; Kamyabhesari, K. Fabrication and evaluation of chitosan/gelatin/PVA hydrogel incorporating honey for wound healing applications: An in vitro, in vivo study. *Int. J. Pharm.* **2021**, *592*, 120068. [CrossRef] [PubMed]
234. Yi, X.; He, J.; Wei, X.; Li, H.; Liu, X.; Cheng, F. A polyphenol and ε-polylysine functionalized bacterial cellulose/PVA multifunctional hydrogel for wound healing. *Int. J. Biol. Macromol.* **2023**, *247*, 125663. [CrossRef] [PubMed]
235. Zhang, Q.; Zhang, M.; Wang, T.; Chen, X.; Li, Q.; Zhao, X. Preparation of aloe polysaccharide/honey/PVA composite hydrogel: Antibacterial activity and promoting wound healing. *Int. J. Biol. Macromol.* **2022**, *211*, 249–258. [CrossRef] [PubMed]
236. Wang, J.; Zhang, C.; Yang, Y.; Fan, A.; Chi, R.; Shi, J.; Zhang, X. Poly (vinyl alcohol) (PVA) hydrogel incorporated with Ag/TiO2 for rapid sterilization by photoinspired radical oxygen species and promotion of wound healing. *Appl. Surf. Sci.* **2019**, *494*, 708–720. [CrossRef]
237. Li, J.; Feng, X.; Liu, B.; Yu, Y.; Sun, L.; Liu, T.; Wang, Y.; Ding, J.; Chen, X. Polymer materials for prevention of postoperative adhesion. *Acta Biomater.* **2017**, *61*, 21–40. [CrossRef]
238. Li, Y.; Zhu, C.; Fan, D.; Fu, R.; Ma, P.; Duan, Z.; Li, X.; Lei, H.; Chi, L. Construction of porous sponge-like PVA-CMC-PEG hydrogels with pH-sensitivity via phase separation for wound dressing. *Int. J. Polym. Mater. Polym. Biomater.* **2020**, *69*, 505–515. [CrossRef]

239. Cobos, M.; De-La-Pinta, I.; Quindós, G.; Fernández, M.J.; Fernández, M.D. Synthesis, Physical, Mechanical and Antibacterial Properties of Nanocomposites Based on Poly(vinyl alcohol)/Graphene Oxide–Silver Nanoparticles. *Polymers* **2020**, *12*, 723. [CrossRef]
240. Jackson, J.; Burt, H.; Lange, D.; Whang, I.; Evans, R.; Plackett, D. The Design, Characterization and Antibacterial Activity of Heat and Silver Crosslinked Poly(Vinyl Alcohol) Hydrogel Forming Dressings Containing Silver Nanoparticles. *Nanomaterials* **2021**, *11*, 96. [CrossRef]
241. Krishnan, P.D.; Banas, D.; Durai, R.D.; Kabanov, D.; Hosnedlova, B.; Kepinska, M.; Fernandez, C.; Ruttkay-Nedecky, B.; Nguyen, H.V.; Farid, A.; et al. Silver Nanomaterials for Wound Dressing Applications. *Pharmaceutics* **2020**, *12*, 821. [CrossRef] [PubMed]
242. Bercea, M.; Gradinaru, L.-M.; Morariu, S.; Plugariu, I.-A.; Gradinaru, R.V. Tailoring the properties of PVA/HPC/BSA hydrogels for wound dressing applications. *React. Funct. Polym.* **2022**, *170*, 105094. [CrossRef]
243. Xie, J.; Qin, Y.; Zeng, Y.; Yuan, R.; Lu, X.; Yang, X.; Wei, E.; Cui, C. Phytic acid/tannic acid reinforced hydrogels with ultra-high strength for human motion monitoring and arrays. *Soft Matter* **2024**, *20*, 640–650. [CrossRef] [PubMed]
244. Guo, Z.; Wang, Z.; Pan, W.; Zhang, J.; Qi, Y.; Qin, Y.; Zhang, Y. Fiber-reinforced polyvinyl alcohol hydrogel via in situ fiber formation. *e-Polymers* **2023**, *23*, 20230056. [CrossRef]
245. Kalantari, K.; Mostafavi, E.; Saleh, B.; Soltantabar, P.; Webster, T.J. Chitosan/PVA hydrogels incorporated with green synthesized cerium oxide nanoparticles for wound healing applications. *Eur. Polym. J.* **2020**, *134*, 109853. [CrossRef]
246. Souza, S.O.L.; Cotrim, M.A.P.; Oréfice, R.L.; Carvalho, S.G.; Dutra, J.A.P.; de Paula Careta, F.; Resende, J.A.; Villanova, J.C.O. Electrospun poly(ε-caprolactone) matrices containing silver sulfadiazine complexed with β-cyclodextrin as a new pharmaceutical dosage form to wound healing: Preliminary physicochemical and biological evaluation. *J. Mater. Sci. Mater. Med.* **2018**, *29*, 67. [CrossRef] [PubMed]
247. Joseph, B.; James, J.; Grohens, Y.; Kalarikkal, N.; Thomas, S. Additive Manufacturing of Poly (ε-Caprolactone) for Tissue Engineering. *JOM* **2020**, *72*, 4127–4138. [CrossRef]
248. Thangunpai, K.; Hu, D.; Kajiyama, M.; Neves, M.A.; Enomae, T. Effects of Grafting Maleic Anhydride onto Poly-ℰ-caprolactone on Facilitative Enzymatic Hydrolysis. *Macromol. Mater. Eng.* **2023**, *308*, 2300067. [CrossRef]
249. Salehi, M.; Niyakan, M.; Ehterami, A.; Haghi-Daredeh, S.; Nazarnezhad, S.; Abbaszadeh-Goudarzi, G.; Vaez, A.; Hashemi, S.F.; Rezaei, N.; Mousavi, S.R. Porous electrospun poly(ε-caprolactone)/gelatin nanofibrous mat containing cinnamon for wound healing application: In vitro and in vivo study. *Biomed. Eng. Lett.* **2019**, *10*, 149–161. [CrossRef]
250. Van Rie, J.; Declercq, H.; Van Hoorick, J.; Dierick, M.; Van Hoorebeke, L.; Cornelissen, R.; Thienpont, H.; Dubruel, P.; Van Vlierberghe, S. Cryogel-PCL combination scaffolds for bone tissue repair. *J. Mater. Sci. Mater. Med.* **2015**, *26*, 123. [CrossRef]
251. Mbese, Z.; Alven, S.; Aderibigbe, B.A. Collagen-Based Nanofibers for Skin Regeneration and Wound Dressing Applications. *Polymers* **2021**, *13*, 4368. [CrossRef] [PubMed]
252. Gutiérrez, T.J.; Alvarez, V.A. Eco-friendly films prepared from plantain flour/PCL blends under reactive extrusion conditions using zirconium octanoate as a catalyst. *Carbohydr. Polym.* **2017**, *178*, 260–269. [CrossRef]
253. Augustine, R.; Dan, P.; Schlachet, I.; Rouxel, D.; Menu, P.; Sosnik, A. Chitosan ascorbate hydrogel improves water uptake capacity and cell adhesion of electrospun poly(epsilon-caprolactone) membranes. *Int. J. Pharm.* **2019**, *559*, 420–426. [CrossRef]
254. Sowmya, B.; Hemavathi, A.B.; Panda, P.K. Poly (ε-caprolactone)-based electrospun nano-featured substrate for tissue engineering applications: A review. *Prog. Biomater.* **2021**, *10*, 91–117. [CrossRef] [PubMed]
255. Ji, X.; Yuan, X.; Ma, L.; Bi, B.; Zhu, H.; Lei, Z.; Liu, W.; Pu, H.X.; Jiang, J.; Jiang, X.; et al. Mesenchymal stem cell-loaded thermosensitive hydroxypropyl chitin hydrogel combined with a three-dimensional-printed poly(ε-caprolactone) /nano-hydroxyapatite scaffold to repair bone defects via osteogenesis, angiogenesis and immunomodulation. *Theranostics* **2020**, *10*, 725–740. [CrossRef] [PubMed]
256. Fox, K.; Ratwatte, R.; Booth, M.A.; Tran, H.M.; Tran, P.A. High Nanodiamond Content-PCL Composite for Tissue Engineering Scaffolds. *Nanomaterials* **2020**, *10*, 948. [CrossRef] [PubMed]
257. Behl, A.; Parmar, V.S.; Malhotra, S.; Chhillar, A.K. Biodegradable diblock copolymeric PEG-PCL nanoparticles: Synthesis, characterization and applications as anticancer drug delivery agents. *Polymer* **2020**, *207*, 122901. [CrossRef]
258. Rangel, A.; Nguyen, T.N.; Egles, C.; Migonney, V. Different real-time degradation scenarios of functionalized poly(ε-caprolactone) for biomedical applications. *J. Appl. Polym. Sci.* **2021**, *138*, 50479. [CrossRef]
259. Arbade, G.K.; Srivastava, J.; Tripathi, V.; Lenka, N.; Patro, T.U. Enhancement of hydrophilicity, biocompatibility and biodegradability of poly(ε-caprolactone) electrospun nanofiber scaffolds using poly(ethylene glycol) and poly(L-lactide-co-ε-caprolactone-co-glycolide) as additives for soft tissue engineering. *J. Biomater. Sci. Polym. Ed.* **2020**, *31*, 1648–1670. [CrossRef]
260. Ullah, N.; Khan, D.; Ahmed, N.; Zafar, A.; Shah, K.U.; ur Rehman, A. Lipase-sensitive fusidic acid polymeric nanoparticles based hydrogel for on-demand delivery against MRSA-infected burn wounds. *J. Drug Deliv. Sci. Technol.* **2023**, *80*, 104110. [CrossRef]
261. Gupta, A.; Upadhyay, N.K.; Parthasarathy, S.; Rajagopal, C.; Roy, P.K. Nitrofurazone-loaded PVA–PEG semi-IPN for application as hydrogel dressing for normal and burn wounds. *J. Appl. Polym. Sci.* **2013**, *128*, 4031–4039. [CrossRef]
262. Timaeva, O.; Pashkin, I.; Mulakov, S.; Kuzmicheva, G.; Konarev, P.; Terekhova, R.; Sadovskaya, N.; Czakkel, O.; Prevost, S. Synthesis and physico-chemical properties of poly(N-vinyl pyrrolidone)-based hydrogels with titania nanoparticles. *J. Mater. Sci.* **2020**, *55*, 3005–3021. [CrossRef] [PubMed]

263. Veeramuthu, L.; Liang, F.C.; Zhang, Z.X.; Cho, C.J.; Ercan, E.; Chueh, C.C.; Chen, W.C.; Borsali, R.; Kuo, C.C. Improving the Performance and Stability of Perovskite Light-Emitting Diodes by a Polymeric Nanothick Interlayer-Assisted Grain Control Process. *ACS Omega* **2020**, *5*, 8972–8981. [CrossRef]
264. Nam, H.G.; Nam, M.G.; Yoo, P.J.; Kim, J.-H. Hydrogen bonding-based strongly adhesive coacervate hydrogels synthesized using poly(*N*-vinylpyrrolidone) and tannic acid. *Soft Matter* **2019**, *15*, 785–791. [CrossRef]
265. Kouser, S.; Prabhu, A.; Prashantha, K.; Nagaraja, G.K.; D'souza, J.N.; Meghana Navada, K.; Qurashi, A.; Manasa, D.J. Modified halloysite nanotubes with Chitosan incorporated PVA/PVP bionanocomposite films: Thermal, mechanical properties and biocompatibility for tissue engineering. *Colloids Surfaces A Physicochem. Eng. Asp.* **2022**, *634*, 127941. [CrossRef]
266. Sun, S.; Hao, M.; Ding, C.; Zhang, J.; Ding, Q.; Zhang, Y.; Zhao, Y.; Liu, W. SF/PVP nanofiber wound dressings loaded with phlorizin: Preparation, characterization, in vivo and in vitro evaluation. *Colloids Surfaces B Biointerfaces* **2022**, *217*, 112692. [CrossRef]
267. Ajaz, N.; Khan, I.U.; Khalid, I.; Khan, R.U.; Khan, H.A.; Asghar, S.; Khalid, S.H.; Shahzad, Y.; Yousaf, A.M.; Hussain, T.; et al. In vitro and toxicological assessment of dexamethasone sodium phosphate loaded pH sensitive Pectin-g-poly(AA)/PVP semi interpenetrating network. *Mater. Today Commun.* **2020**, *25*, 101325. [CrossRef]
268. Pushp, P.; Bhaskar, R.; Kelkar, S.; Sharma, N.; Pathak, D.; Gupta, M.K. Plasticized poly(vinylalcohol) and poly(vinylpyrrolidone) based patches with tunable mechanical properties for cardiac tissue engineering applications. *Biotechnol. Bioeng.* **2021**, *118*, 2312–2325. [CrossRef] [PubMed]
269. Wang, B.-W.; Liu, H.; Ying, J.; Liu, C.-T.; Shen, C.-Y.; Wang, Y.-M. Effect of Physical Aging on Heterogeneity of Poly(ε-caprolactone) Toughening Poly(lactic acid) Probed by Nanomechanical Mapping. *Chin. J. Polym. Sci.* **2023**, *41*, 143–152. [CrossRef]
270. Sun, Y.; Huang, Y.; Wang, X.Y.; Wu, Z.Y.; Weng, Y.X. Kinetic analysis of PGA/PBAT plastic films for strawberry fruit preservation quality and enzyme activity. *J. Food Compos. Anal.* **2022**, *108*, 104439. [CrossRef]
271. Chen, L.; Sun, X.; Ren, Y.; Wang, R.; Sun, M.; Liang, W. Enhancing melt strength of polyglycolic acid by reactive extrusion with chain extenders. *J. Appl. Polym. Sci.* **2022**, *139*, 51796. [CrossRef]
272. Álvarez, I.; Gutiérrez, C.; de Lucas, A.; Rodríguez, J.; García, M. Measurement, correlation and modelling of high-pressure phase equilibrium of PLGA solutions in CO_2. *J. Supercrit. Fluids* **2020**, *155*, 104637. [CrossRef]
273. Zuhour, M.; Güneş, C.; Fındık, S.; Dündar, M.A.; Gök, O.; Altuntaş, Z. Effect of methylprednisolone loaded poly lactic-co-glycolic acid (PLGA) bioabsorbable nanofibers on tendon healing and adhesion formation. *J. Drug Deliv. Sci. Technol.* **2023**, *89*, 104988. [CrossRef]
274. Wang, S.; Xiong, Y.; Chen, J.; Ghanem, A.; Wang, Y.; Yang, J.; Sun, B. Three Dimensional Printing Bilayer Membrane Scaffold Promotes Wound Healing. *Front. Bioeng. Biotechnol.* **2019**, *7*, 348. [CrossRef] [PubMed]
275. Sun, G.; Zhang, X.; Shen, Y.-I.; Sebastian, R.; Dickinson, L.E.; Fox-Talbot, K.; Reinblatt, M.; Steenbergen, C.; Harmon, J.W.; Gerecht, S. Dextran hydrogel scaffolds enhance angiogenic responses and promote complete skin regeneration during burn wound healing. *Proc. Natl. Acad. Sci. USA* **2011**, *108*, 20976–20981. [CrossRef] [PubMed]
276. Mostafalu, P.; Tamayol, A.; Rahimi, R.; Ochoa, M.; Khalilpour, A.; Kiaee, G.; Yazdi, I.K.; Bagherifard, S.; Dokmeci, M.R.; Ziaie, B.; et al. Smart Bandage for Monitoring and Treatment of Chronic Wounds. *Small* **2018**, *14*, e1703509. [CrossRef]
277. Hu, Y.; Zhang, Z.; Li, Y.; Ding, X.; Li, D.; Shen, C.; Xu, F.J. Dual-Crosslinked Amorphous Polysaccharide Hydrogels Based on Chitosan/Alginate for Wound Healing Applications. *Macromol. Rapid Commun.* **2018**, *39*, e1800069. [CrossRef]
278. Dong, M.; Mao, Y.; Zhao, Z.; Zhang, J.; Zhu, L.; Chen, L.; Cao, L. Novel fabrication of antibiotic containing multifunctional silk fibroin injectable hydrogel dressing to enhance bactericidal action and wound healing efficiency on burn wound: In vitro and in vivo evaluations. *Int. Wound J.* **2022**, *19*, 679–691. [CrossRef]
279. Hu, X.; Xia, Z.; Cai, K. Recent advances in 3D hydrogel culture systems for mesenchymal stem cell-based therapy and cell behavior regulation. *J. Mater. Chem. B* **2022**, *10*, 1486–1507. [CrossRef]
280. Morales, X.; Cortés-Domínguez, I.; Ortiz-De-Solorzano, C. Modeling the Mechanobiology of Cancer Cell Migration Using 3D Biomimetic Hydrogels. *Gels* **2021**, *7*, 17. [CrossRef]
281. Habanjar, O.; Diab-Assaf, M.; Caldefie-Chezet, F.; Delort, L. 3D Cell Culture Systems: Tumor Application, Advantages, and Disadvantages. *Int. J. Mol. Sci.* **2021**, *22*, 12200. [CrossRef]
282. Hernández-Rangel, A.; Martin-Martinez, E.S. Collagen based electrospun materials for skin wounds treatment. *J. Biomed. Mater. Res. Part A* **2021**, *109*, 1751–1764. [CrossRef] [PubMed]
283. Sharma, P.; Kumar, A.; Dey, A.D.; Behl, T.; Chadha, S. Stem cells and growth factors-based delivery approaches for chronic wound repair and regeneration: A promise to heal from within. *Life Sci.* **2021**, *268*, 118932. [CrossRef] [PubMed]
284. Wang, M.; Xu, X.; Lei, X.; Tan, J.; Xie, H. Mesenchymal stem cell-based therapy for burn wound healing. *Burn. Trauma* **2021**, *9*, tkab002. [CrossRef] [PubMed]
285. Villanueva, M.E.; Cuestas, M.L.; Pérez, C.J.; Campo Dall' Orto, V.; Copello, G.J. Smart release of antimicrobial ZnO nanoplates from a pH-responsive keratin hydrogel. *J. Colloid Interface Sci.* **2019**, *536*, 372–380. [CrossRef]

MDPI AG
Grosspeteranlage 5
4052 Basel
Switzerland
Tel.: +41 61 683 77 34

Gels Editorial Office
E-mail: gels@mdpi.com
www.mdpi.com/journal/gels

www.ingramcontent.com/pod-product-compliance
Lightning Source LLC
LaVergne TN
LVHW070412100526
838202LV00014B/1443
9783725816460